Collaborative Bioethics

Volume 6

Series Editor
Insoo Hyun, Harvard Medical School
Boston, USA

The aim of **Collaborative Bioethics** , is to draw attention to an underexplored but increasingly important area of scholarly thought and action: bioethics as a co-creative activity of ethicists working with scientists rather than as always a reaction to biomedical developments after the fact. The scope of this series is determined by each major subfield of science and medicine that raises ethical uncertainties for researchers, regulators, and the public.

Collaborative Bioethics is a series that will provide a central hub for timely publications addressing ethical issues that are emerging right alongside the science. As such, this series will be of interest to a wide swath of readers: bioengineers and scientists at all professional levels; bioethicists intrigued by bioengineering and medical advances; research regulators and funders; and the general public.

Leander Penaso Marquez
King Lau Chow
Alexander Atrio Lim Lopez
Editors

COVID-19 in the Pacific Rim and Beyond

Policies, Practices, and Perspectives from Public Health Ethics

Springer

Editors
Leander Penaso Marquez
College of Education
University of the Philippines Diliman
Quezon City, Philippines

Alexander Atrio Lim Lopez
School of Law
Ateneo de Manila University
Makati City, Philippines

King Lau Chow
Division of Public Policy, Division of Life Science, Department of Chemical and Biological Engineering
Hong Kong University of Science and Technology
Clear Water Bay, Hong Kong

ISSN 2730-5619 ISSN 2730-5627 (electronic)
Collaborative Bioethics
ISBN 978-3-032-21217-7 ISBN 978-3-032-21218-4 (eBook)
https://doi.org/10.1007/978-3-032-21218-4

This Springer imprint is published by the registered company Springer Nature Switzerland AG
The registered company address is: Gewerbestrasse 11, 6330 Cham, Switzerland

"To my late father, Lauro, who is not just another statistic of the COVID-19 pandemic. We miss you so much." —Leander Penaso Marquez.

"To all (medical personnel, volunteers, and many administrators) who have put their lives at risk to get us through this last pandemic. May their effort and dedication inspire more to join the effort to safeguard a healthy future of the community." —King Lau Chow.

"To Mama, Papa, Lyka, and Tita Be who braved through the onslaught of the pandemic." —Alexander Atrio Lim Lopez.

Preface

If you are reading this book, then chances are that you have lived through the COVID-19 pandemic and are reflecting on whether we would be able to manage the situation if another pandemic hits us. You may also be wondering why people are still writing about the pandemic when the pandemic feels like a distant memory. The fact is that COVID-19 still exists and there may still be more stories to tell. Definitely, there are still many lessons to be learned. More importantly, this is not just another book on bioethics or public health ethics. This book tells stories—of challenges, failures, struggles, realizations, and victories—triggered by the same event and taking place in this historical episode, albeit in relatively different and yet intertwining contexts.

Our book, *COVID-19 in the Pacific Rim and Beyond: Policies, Practices, and Perspectives from Public Health Ethics*, is a collection of lived experiences of territories and economies along the Pacific Rim regarding COVID-19 policies and practices analyzed from the perspective of public health ethics. The majority of the chapters were written by members of the Association of Pacific Rim Universities (APRU) Global Health Program network. The chapters featured in the following pages generally focus on public health ethics analyses of each economy's policies (for instance, travel restrictions, mask implementation and lifting, vaccination rate, etc.), practices (such as the spectrum of how much intervention was implemented, availability of vaccines, communication with the public, enforcement, vaccine rollout, how sick people were handled in the hospital, hospitalization rate, maternity rate, etc.), private sector initiatives vis-à-vis government programs, community-led interventions and views from the community, and influence of cultural aspects, among others.

As a general guideline, the contributed chapters start with a comprehensive account of how COVID-19 first entered the economy and how it spread over time. What did the government and private sector do? How did these actions affect the behavior of the people? How effective was the COVID-19 response? What were the impacts of deaths related to COVID-19? What were the issues encountered? What are the important elements of the COVID-19 policy and practices? The chapters then proceed with a normative analysis of these policies and practices, considering several elements of public health ethics, such as prevention, the common good, societal responsibility, government action, and collective effort, to name a few. Each chapter concludes by answering the question "what can be learned from these in preparing for future pandemics?"

Nonetheless, what makes this collection unique is the fact that the guideline is just that—a guide. Due to different contexts, some contributors have followed the guideline to the letter, while others only included what was applicable to their experiences. Each chapter tends to have some different emphases, reflecting the perception or sentiment of the authors towards the status of their respective economies. You might find the contrast to be even more interesting as it gives each chapter a strong cultural or national flavor. Many did their analysis close to the ground, while there are those who analyzed from a more conceptual level. This "dynamic" resulted in an interestingly diverse collection of papers, where some are more extensive and nuanced, others highlighted diverse issues, several focused on particular aspects like education, and quite a few shared success stories.

Chapter 1 provides an introduction of preexisting public health ethics challenges exposed and intensified by the COVID-19 pandemic. It sets the stage for the book by highlighting central public health ethics issues in the Pacific Rim and the importance of inclusive, resilient health systems. It also emphasizes the need to translate ethical principles into practice. Following this introductory chapter are five clusters of chapters focusing on a specific region within the Pacific Rim.

The first cluster focuses on East Asia. Chapter 2 examines the COVID-19 response of China through a People-Centered Integrated Management and Control Model. It highlights strong governance, coordinated policies, and societal sacrifice. The chapter also underscores how prioritizing public health led to reforms and strengthened national health systems. Meanwhile, Chapter 3 analyzes the multi-wave pandemic experience of Hong Kong, while identifying key factors that shaped policy decisions. It argues that effective governance requires evidence-based action, resource coordination, and balancing public health with economic and social concerns. At the same time, Chapter 4 explores the pandemic strategy of Japan, which relied on voluntary compliance rather than strict enforcement. The chapter highlights strengths in social cooperation but also reveals weaknesses in testing capacity, data systems, and legal protections against discrimination. As the final chapter within this cluster, Chapter 5 discusses the K-Quarantine model of South Korea. It notes the effectiveness of the model in controlling the virus through technology and rapid response. However, the chapter raises ethical concerns about privacy, equity, and political influences on public health decisions.

The second cluster of chapters zeroes in on Southeast Asia. Chapter 6 looks into the pandemic response of Malaysia, emphasizing the tension between disease control and social equity. It highlights the disproportionate impact on vulnerable groups and calls for more inclusive, equity-centered policies in future crises. Similarly, Chapter 7 evaluates the COVID-19 response of the Philippines. Focusing on effectiveness, proportionality, and the protection of vulnerable populations, the chapter offers ethical recommendations for improving future pandemic policies and governance. Forward looking, Chapter 8 reflects on the transition to a "new normal" of Thailand by assessing its policies, multi-sectoral efforts, and reopening strategies. It identifies areas for improvement and aims to provide lessons for future pandemic preparedness. The last chapter of this cluster, Chapter 9, examines the use of the precautionary principle in COVID-19 strategies of Vietnam. The chapter reviews the early successes but later shortcomings due to rigid policies and governance issues. It leaves a call for stronger ethical frameworks and more open dialogue.

The third cluster is made up of a single chapter – a success story – and is the lone entry from Oceania. Chapter 10 describes the successful early containment strategy through strict border controls and high vaccination rates of Cook Islands. It provides a narrative account of leadership and community trust in managing the pandemic. The high vaccination rate of the population and their confidence in their leaders prove to be significant factors in holding the frontline.

The fourth cluster of chapters centers on North America. Chapter 11 discusses the COVID-19 story of Costa Rica by highlighting how international guidelines were applied without sufficient local adaptation. It also stresses the need for context-sensitive policies, better communication, and stronger support for vulnerable populations. In addition, Chapter 12 reflects on the experiences of Mexico during the pandemic in using core biomedical ethics principles. It assesses policies such as vaccination and public communication while offering recommendations for more equitable and ethically grounded strategies. Lastly, Chapter 13 provides a critique of the pandemic policies and practices of the United States through a relational justice lens. The chapter focuses on systemic inequalities, leadership failures, and individualism. Furthermore, it proposes reforms to promote equity, solidarity, and collective responsibility.

The fifth and final cluster highlights South America. Chapter 14 scrutinizes the impact of COVID-19 on education through the experiences of teachers in Chile. It emphasizes disruptions to student behavior, socioemotional development, and school dynamics following prolonged closures. Meanwhile, Chapter 15 analyzes the top-down approach of Colombia to fight the pandemic and its ethical implications. It identifies four key themes, namely, workforce effects, decision-making approaches, defining the "public" in health interventions, and inter-organizational activities. These novel themes offer lessons for more inclusive and ethical crisis management. Finally, Chapter 16 details the pandemic experience of Ecuador. Marked by overwhelmed health systems, high mortality, and governance challenges, the chapter illustrates the complex interaction of health, political, and social factors in crisis response.

The concluding chapter, Chapter 17, informed by the stories and analyses of the preceding chapters, provides a global overview of the COVID-19 pandemic's phases and impacts. It features a special focus on inequities in vaccine access and distribution. Ultimately, the chapter leaves a lingering question about global preparedness for future pandemics, especially in the Asia-Pacific region.

We understand that there is no perfect policy or flawless practice. This work aimed to learn from the experiences of these neighboring economies during the COVID-19 pandemic, with the goal of gathering information that may be useful for the next pandemic, particularly in terms of public health ethics. However, this book may just be the single largest collection of public health ethics analyses on COVID-19 policies, practices, and perspectives on critical issues, conducted by individuals who evaluated these areas from their own unique positionalities, and informed by specific local contexts—in the Pacific Rim, at least.

We hope that you enjoy and learn from these stories as much as (if not more than) we did.

Quezon City, Philippines Leander Penaso Marquez
Clear Water Bay, Hong Kong King Lau Chow
Makati City, Philippines Alexander Atrio Lim Lopez
2026

Acknowledgments

Alexander and Leander would like to acknowledge the APRU Global Health Program Working Group on Bioethics and their most important bioethics mentor, Dr. Leonardo D. de Castro.

Leander would also like to thank his wife, Rola, for all the support, patience, and understanding throughout the course of this book project.

King would like to acknowledge the sacrifices of all medical personnel, volunteers, and administrators during the COVID-19 pandemic.

The editors would like to thank Dr. Mellissa Withers, Director of the APRU Global Health Program, for her leadership, and especially for her support for this book project and the activities of the Working Group on Bioethics.

Contents

About the Editors

Leander Penaso Marquez is an Assistant Professor of Philosophy of Education and Values Education at the College of Education, University of the Philippines Diliman. He currently holds graduate degrees in Philosophy, Education, and Health Research Ethics. He has published several articles in local and international journals as well as presented some of his research papers in various conferences in Asia and Europe. He sits as co-chair of the Working Group on Bioethics of the Association of Pacific Rim Universities (APRU) Global Health Program. He is also serving as Junior Ambassador to the Philippines for the Universal Scientific Education and Research Network (USERN). He is also a member of international organizations such as the International Association for Education in Ethics (IAEE) and the Southeast Asia Bioethics Network (SEABioN) as well as national organizations such as the Philosophical Inquiry in Schools Initiative Philippines, Inc. (ISIP Pinas) and the Philosophy and Values Education Society, Inc. (PAVESOC, Inc.). His research interests include Ethics, Epistemology, Bioethics, Philosophy for Children, Philosophy of Education, Philosophy and Popular Culture, Education Policy, and Values Education.

King Lau Chow is a Professor of Life Science, Chemical & Biological Engineering and Public Policy. He served as the Academic Director of Common Core Program, the Director of Interdisciplinary Programs Office overseeing Environment and Sustainability, Public Policy, Risk Management, Technology Management, and various innovation programs at The Hong Kong University of Science and Technology. He heads the Dean of Students Office supporting various student-oriented schemes beyond academic issues. While his research focuses on genetics, development, evolution, and synthetic biology, he engages in various teaching pedagogy initiatives and curricular development projects, spearheads liberal arts and interdisciplinary education at the juncture cutting across science, engineering, social science, and humanity on all levels. He was the inaugural School of Science Teaching Award winner and was awarded the Michael G. Gale Medal of distinguished teaching at HKUST. He serves as a member of the Advisory Group of the Association of Pacific Rim Universities (APRU) Global Health Program and in the Global Future Council of Generative Biology, the World Economic Forum.

Alexander Atrio Lim Lopez is currently pursuing his Juris Doctor degree at the Ateneo de Manila University School of Law. Prior to entering law school, he taught philosophy at the University of the Philippines Diliman. He has published in bioethics, philosophy and video games, and philosophy for children. He earned his MA Philosophy degree from UP Diliman where he also graduated with a BA Philosophy degree as summa cum laude and valedictory speaker. His first degree is BS Basic Medical Sciences from UP Manila. He currently serves as Co-Chair of the Working Group on Bioethics of the Association of Pacific Rim Universities (APRU) Global Health Program. He is also a board member of the Philosophical Inquiry in Schools Initiative Philippines, Inc. (ISIP Pinas).

Contributors

Bangyuan An Division of Public Policy, The Hong Kong University of Science and Technology, Kowloon, Hong Kong, China

School of Public Administration and Policy, Renmin University of China, Beijing, China

Metua Bates Public Health Te Marae Ora Ministry of Health Cook Islands, Rarotonga, Cook Islands

Carlos F. Cáceres-Palacios Cayetano Heredia Peruvian University, Lima, Peru

Qiuru Chen School of Public Health, Peking University, Beijing, China

King L. Chow Division of Public Policy, Division of Life Science, Department of Chemical and Biological Engineering, Hong Kong University of Science and Technology, Clear Water Bay, Hong Kong

Hai Thanh Doan University of Otago, Dunedin, New Zealand

Independent Researcher, Ho Chi Minh City, Vietnam

Isidora Espinoza-Carrillo Universidad de Chile, Santiago, Chile

Ricardo J. Estrada-Mendizabal Division of Allergy, Asthma, and Clinical Immunology, Mayo Clinic, Scottsdale, AZ, USA

Teariki Faireka Primary Healthcare Te Marae Ora Ministry of Health Cook Islands, Rarotonga, Cook Islands

Ralema Makena Geno ICT Te Marae Ora Ministry of Health Cook Islands, Rarotonga, Cook Islands

Varaalakshmy Gokilavanan Hospital Kuala Lumpur, Kuala Lumpur, Malaysia

Ingrid Gómez-Duarte Center for Research in Nursing and Healthcare, University of Costa Rica, San Pedro, Costa Rica

Carlos A. Gonzalez-Assad Tecnológico de Monterrey, Escuela de Medicina y Ciencias de la Salud, Monterrey, NL, México

Jonathan Guillemot Escuela de Medicina, Colegio de Ciencias de la Salud, Universidad San Francisco de Quito USFQ, Quito, Ecuador

Instituto de Medicina Social & Desafíos Globales, Universidad San Francisco de Quito USFQ, Quito, Ecuador

Garth Henderson Ministry of Finance and Economic Management, Rarotonga, Cook Islands

Francisco E. Herrera-Carrillo Department of Pediatrics, Boston Children's Hospital, Harvard Medical School, Boston, MA, USA

Carl Hunter The Office of the Public Service Commission, Rarotonga, Cook Islands

Cheol Kang Graduate School of Public Health, Yonsei University, Seoul, Korea

Arantza Kessi-Gutiérrez Universidad de Chile, Santiago, Chile

Junko Kiriya Department of Community and Global Health, Graduate School of Medicine, The University of Tokyo, Tokyo, Japan

Ilhak Lee Department of Medical Humanities and Social Sciences, College of Medicine, Yonsei University, Seoul, Korea

Jeffrey Soon-Yit Lee Clinical Research Centre, Hospital Sibu, Sarawak, Malaysia

Alexander Atrio Lim Lopez School of Law, Ateneo de Manila University, Makati City, Philippines

Wendy López-Vargas Center for Research in Nursing and Healthcare, University of Costa Rica, San Pedro, Costa Rica

Leander Penaso Marquez College of Education, University of the Philippines Diliman, Quezon City, Philippines

Yin Yin May Hospital Health Services Te Marae Ora Ministry of Health Cook Islands, Rarotonga, Cook Islands

María de Jesús Medina-Arellano Instituto de Investigaciones Jurídicas, Universidad Nacional Autónoma de México, Mexico City, Mexico

Pumari Mena-Gutiérrez Universidad de Chile, Santiago, Chile

Karen Ngamata Public Health Te Marae Ora Ministry of Health Cook Islands, Rarotonga, Cook Islands

Parau Ngamata Public Health Te Marae Ora Ministry of Health Cook Islands, Rarotonga, Cook Islands

Kai Hong Ooi Bioscience and Biomedical Engineering Thrust, System Hub, Hong Kong University of Science and Technology, Guangzhou, China

Esteban Ortíz-Prado One Health Group, Universidad de las Américas, Quito, Ecuador

Ana Francisca Pérez-Jara Universidad de Chile, Santiago, Chile

Constantine S. Psimopoulos Department of Epidemiology, Harvard T.H. Chan School of Public Health, Boston, MA, USA

Department of Global Health and Social Medicine, Harvard Medical School, Boston, MA, USA

Human Flourishing Program, Harvard University, Cambridge, MA, USA

Diego Ramonfaur Department of Internal Medicine, Cleveland Clinic, Cleveland, OH, USA

Laura Rueda-Castro Facultad de Medicina, Universidad de Chile, Santiago, Chile

Rocío Sáenz-Madrigal Center for Research in Nursing and Healthcare, University of Costa Rica, San Pedro, Costa Rica

Jean-Paul Sarrazin Department of Sociology, University of Antioquia, Medellín, Colombia

Diego Sepúlveda-González Tecnológico de Monterrey, Escuela de Medicina y Ciencias de la Salud, Monterrey, NL, México

Nuhisifa Seve Auckland, New Zealand

Akira Shibanuma Department of Community and Global Health, Graduate School of Medicine, The University of Tokyo, Tokyo, Japan

Luis Fernando Solís-Calvo Center for Research in Nursing and Healthcare, University of Costa Rica, San Pedro, Costa Rica

Mark Kiak Min Tan Jeffrey Cheah School of Medicine and Health Sciences, Monash University Malaysia, Bandar Sunway, Malaysia

Enrique Teran Escuela de Medicina, Colegio de Ciencias de la Salud, Universidad San Francisco de Quito USFQ, Quito, Ecuador

Mauricio Torres-Martínez Department of International Health, Johns Hopkins Bloomberg School of Public Health, Baltimore, MD, USA

Simon Turner School of Management, University of Los Andes, Bogotá, Colombia

Eduardo Vasconez One Health Group, Universidad de las Américas, Quito, Ecuador

Pramon Viwattanakulvanid College of Public Health Sciences, Chulalongkorn University, Bangkok, Thailand

Miguel A. Walwyn-Hernández Tecnológico de Monterrey, Escuela de Medicina y Ciencias de la Salud, Monterrey, NL, México

Xinyi Wang Division of Public Policy, The Hong Kong University of Science and Technology, Kowloon, Hong Kong, China

Yuru Wang College of Public Health Sciences, Chulalongkorn University, Bangkok, Thailand

Bob Williams Te Marae Ora Ministry of Health Cook Islands, Rarotonga, Cook Islands

Mellissa Withers Keck School of Medicine, University of Southern California, Los Angeles, CA, USA

Hui Yin School of Public Health, Peking University, Beijing, China

Marielle Antoinette H. Zosa Department of Philosophy, University of the Philippines Diliman, Quezon City, Philippines

Part I

Introduction

1 COVID-19 and Public Health Ethics Challenges in the Pacific Rim

Leander Penaso Marquez, King L. Chow, and Alexander Atrio Lim Lopez

Abstract

The COVID-19 pandemic not only brought to the fore important bioethical issues and concerns but also tested the moral commitments of leaders, and the rest of the world, in addressing public health ethics challenges. This effort requires operationalizing ethics in public health governance, which means making sure that no one is left behind not just in terms of creating policies, but most importantly, in the implementation and delivery of these policies. In doing so, public health systems can evolve from reactive crisis management toward resilient structures guided by ethical responsibility. This introductory chapter highlights preexisting public health ethics challenges in the Pacific Rim that have been exacerbated by the pandemic. Its purpose is to lay the groundwork for the succeeding chapters, which are more focused and informed by the lived experiences of authors who wrote about their respective economies.

Keywords

Public health ethics · Bioethics · Pacific Rim · COVID-19 · Pandemic

---

L. P. Marquez (✉)
College of Education, University of the Philippines Diliman, Quezon City, Philippines
e-mail: lpmarquez@up.edu.ph

K. L. Chow
Division of Public Policy; Division of Life Science, Department of Chemical and Biological Engineering, Hong Kong University of Science and Technology, Clear Water Bay, Hong Kong, China

A. A. L. Lopez
School of Law, Ateneo de Manila University, Makati City, Philippines

L. P. Marquez et al. (eds.), *COVID-19 in the Pacific Rim and Beyond*, Collaborative Bioethics 6, https://doi.org/10.1007/978-3-032-21218-4_1

Introduction

Public health ethics challenges have always existed, but the COVID-19 pandemic, being a global experience, has highlighted them even more, especially in terms of how health systems, policies, and societies respond to inequities, vulnerabilities, and moral dilemmas. For instance, according to several bioethics experts from the Association of Pacific Rim Universities (APRU) Global Health Program, among those that have been severely affected by the pandemic are marginalized groups such as migrants, low-income workers, and those in fragile health systems [1]. The experiences of these marginalized groups prove how the pandemic deepened existing inequalities and disproportionately affected those who do not necessarily occupy the best positions in society. Ethical principles of justice and equity are underscored as public health responses that take precedence over containment or even social, economic, and political costs to address inequalities that lead to avoidable suffering and harm [2].

Meanwhile, in a study that focused on the interplay between public health and primary care in six economies in the Asia-Pacific [3], solidarity and preparedness, together with robust health infrastructures, proved to be essential to success in containing the virus. On the contrary, the weak integration of structures with public health values and priorities highlights the vulnerabilities in protecting not just the public, but also the health workers. This puts at the center the moral duty to invest in resilient, equitable systems that ensure continuity of care even under crisis conditions.

Other factors that may have played a role in the worsening of structural disparities during the pandemic are resource allocation, privacy in contact tracing, and vaccine distribution, which point to key public health ethics principles. Transparency, accountability, and community participation should always be at the forefront of any ethical decision-making. For instance, lack of transparency and accountability, unfair distribution, or isolated efforts run the risk of undermining established public health measures and result in further disparities. The principles of trustworthiness and stewardship are also vital as governments, elected to serve the people, should act as responsible caretakers of both scarce resources and public trust. Thus, ethical lapses in public health policies and judgments will most likely occur without the said core principles [4]. These lapses further underscore the categorical imperative that ethical public health responses must balance equity, solidarity, stewardship, and trust.

Serving as the introductory chapter of this book, the discussions that follow highlight existing literature on public health ethics challenges during the COVID-19 pandemic that are largely situated within economies and territories in the Pacific Rim. The discussions herein, therefore, are more general and approached in a thematic manner. The purpose of this chapter is to lay the groundwork for the succeeding chapters, which are more focused and informed by the lived experiences of authors who wrote about their respective economies.

Public Health Challenges in the Pacific Rim

Frontliners

The COVID-19 pandemic made everyone realize how healthcare providers, particularly frontline workers, are essential in maintaining healthcare systems, especially during a crisis and when these systems are under immense pressure. In the Pacific region, healthcare providers demonstrated that being well-trained, well-resourced, resilient, and adaptable were critical skills for managing not only health crises but also, equally importantly, everyday health services [5].

Beyond the formal healthcare settings, there are also the silent health workers inside the homes. They are those who became part of the expanded concept of "frontline" necessitated by the pandemic. In more specific instances, these silent health workers took on informal care burdens, which included eldercare and childrearing inside the homes, as these grew significantly during the pandemic; this situation highlights the frequently overlooked efforts of the unpaid workers whose contributions are equally important but undervalued [6].

A study on the experiences of nurses in quarantine facilities in the Philippines, for instance, emphasizes issues related to self-sacrifice, psychological struggle, and self-fulfillment [7]. Due to the current state of the Philippine healthcare system, Filipino nurses are forced to prioritize the calls of their duty even if it means putting aside personal needs. With an environment that lacks support, resources, and general concern for healthcare workers, they are often overworked and underpaid, with many of them feeling stressed, anxious, and morally fatigued. Despite these, there are Filipino nurses who still find purpose and pride in serving, especially during a health crisis. These findings underline the critical need for policies that offer not just material and systemic support to frontline nurses but also emotional, psychological, and moral support to help them preserve their well-being and capacity to serve [7]. This is true not only in the Philippines but in every economy and territory in the world.

Overall, across institutional and domestic frontlines, many problems were faced by the healthcare workers, raising ethical imperatives for justice, protection, and reciprocity, and calling for equitable social policies grounded in fairness and gender justice. Not only did they have to grapple with under-resourced systems, but they also had to shoulder disproportionate burdens, mostly because of their gender, as a significant number of unpaid caregivers are women. Healthcare workers struggle between altruistic service and the need for systemic support and the necessity to provide for their own families. This highlights a difficult choice between principles of beneficence and institutional accountability.

Quarantine

Quarantine policies, especially those integral to contract tracing efforts, in the Indo-Pacific region have revealed critical gaps between global guidelines and local realities. This is due to the diverse infrastructures, cultural practices, and socio-cultural landscapes of different countries [8]. Quarantine rules are often general and created as a one-size-fits-all approach. Thus, these rules failed to address nuances, especially the needs of the most vulnerable groups. This failure points to the importance of context-sensitive quarantine guidelines, strengthened and tailored training programs, effective community engagement in partnership with national government organizations (NGOs) and faith-based groups, and locally appropriate communications plans to enhance adherence and effectiveness [8].

In several countries in the region, the most common quarantine policy strictly implemented was the lockdown, characterized by stringent local community quarantines. Many of those who used to work from offices shifted to work-from-home settings to comply with the quarantine policies being implemented in their country. Hence, cities with a larger share of work-from-home-capable jobs recorded greater mobility reduction and compliance under the strictest lockdown phases [9]. This scenario sheds light on how workforce structure and economic composition affect the effectiveness of quarantine policies and general social compliance.

According to experts, complete shutting of borders proved to be one of the most effective strategies for limiting international transmission [10]. However, there are inevitable severe socio-economic consequences of these strategies if implemented for a prolonged period, necessitating a phased, risk-based kind of reopening. This type of reopening, however, must be supported by diagnostic testing, quarantine-on-arrival protocols, and regional cooperation. Ultimately, effective quarantine strikes a balance between strictness and flexibility, which balances public health priorities with economic recovery [10].

In view of this, quarantine guidelines and protocols should be able to address context-specific concerns, ensuring that there is enough room for flexibility to avoid compromising inclusivity. In doing so, local values and capacities are appropriately considered and reflected. On the ground, these considerations translate to accounting for the differential impacts of the pandemic on populations; while some individuals continue to work despite the changes in the work environment, with work-from-home setups being widely popular, others are unable to work remotely and require alternative forms of support. While maintaining context-specific considerations, quarantine guidelines should not put in the backseat evidence-based support that is sensitive to economic and social ramifications. The formulation of such guidelines necessitates enhanced collaboration to ensure the effectiveness and ethical legitimacy of quarantine strategies.

Vulnerable Populations

A study on the effects of healthcare systems and policy responses on marginalized groups across four Asia-Pacific countries (South Korea, Indonesia, the Philippines, and Singapore) shows that health disparities, especially on vulnerable populations, were amplified by the pandemic [11]. Informal workers, the elderly, migrants, and low-income communities were exposed to heightened risk of infection, with limited to no access to healthcare, and socio-economic precarity. Disproportionate socio-cultural supports, health system unresponsiveness, and bad governance worsened the severity of these impacts. As such, there is a need for coordinated international cooperation and policies that ensure health justice across diverse vulnerabilities.

Meanwhile, healthcare access is being denied to approximately 1.4 million stateless individuals in Southeast Asia because of the centrality of principles such as non-interference and state sovereignty [12]. This denial happens despite the ratification of international human rights treaties. Stateless individuals, for example, find it extremely difficult to avail of social services such as testing, vaccination, and treatment due to lack of identity documents because of the current form of governance and the framework of the Association of Southeast Asian Nations (ASEAN) that guides it. This framework inadvertently perpetuates health inequities. Such predicament calls for stronger policy commitments to safeguard the health rights of stateless populations.

Geographically isolated populations also suffer from far greater systemic health disparities, including limited to no access to culturally relevant social and medical services, inadequate healthcare infrastructure and workers, and greater prevalence of chronic illnesses, among others [13]. Since most, if not all, pandemic guidelines originated from the center or capital, many failed to account for rural-specific contexts. Hence, many of the guidelines being widely implemented do not necessarily take into consideration communication barriers and unique social structures, which play critical roles in the success of the strategies being implemented in isolated areas.

The experiences of vulnerable populations during the pandemic underscore the need for health justice to be a primary focus of public health, as these populations are often overlooked and unsupported. The needed response requires upholding human dignity and the right to health as well as ensuring fairness, inclusivity, and community engagement. In this light, public health ethics considerations must include fostering transparent and inclusive dialogue, adapting urban models to local realities, emphasizing tailored healthcare strategies, and recognizing the needs and experiences of marginalized communities.

Mental Health

The COVID-19 pandemic led to elevated mental health symptoms among the general population in Southeast Asia, specifically among younger adults, women, and those with poor social support or exposure to misinformation [14]. The results of a study in one Southeast Asian nation show that there were particularly high levels of stress, anxiety, and depression among very specific populations, particularly educators, healthcare workers, and students [15]. A study in Australia also indicates a significant increase in the number of people availing mental health services and crisis support lines, clearly indicating an escalating need across the population [16]. Thus, apart from the debilitating physical and social effects of the COVID-19 pandemic, mental health challenges also were brought to the fore, revealing deeply entrenched ethical concerns surrounding equity, access, and the evolving delivery of care.

Moreover, as previously mentioned, vulnerable populations faced the same barriers to access when it comes to conventional or digital mental health care. This issue emphasizes the ethical principle of accessibility and the need to uphold equity in all modalities of care delivery [17]. Thus, what is highlighted is the critical role played by the principle of vulnerability-sensitive justice. Policies must recognize complex patterns of vulnerability (e.g., gender, age, isolation, and misinformation) and prioritize support accordingly. Such a policy particularly targets social and demographic factors affecting mental health outcomes and ensures that public health strategies are ethically attuned to the needs of these concerned populations. Mental health interventions must ensure equitable access across digital divides, offering alternative modalities where needed. Meaning, it is necessary to know which modalities best suit the needs of vulnerable populations.

The principle of reciprocity demands that these vulnerable and highly affected populations, those who bear high psychological burdens, should be given adequate support and protection. Equitable mental health interventions should be provided to these populations as a priority. By doing so, the aggravation of inequities for those unable or unwilling to seek help is prevented. Ultimately, an ethically sound and valid mental health strategy after the COVID-19 pandemic should be considered. It assures that the most impacted are the ones supported and prioritized, without compromising equitable access for the rest of society.

Leadership

During the global pandemic, emergency care clinicians and frontline stakeholders took on the role of leaders amidst fragmented governance systems; their leadership was most seen when clinicians more actively engaged collaboratively with political and administrative leaders to prioritize safety, quality, and service integration despite systemic constraints [18]. Their experience and expertise were proven to be the catalysts of an effective leadership founded on being trustworthy, adaptive, and inclusive, and characterized by unmistakable acts of integrity, courage, and responsiveness.

Public health ethics challenges during the pandemic drew attention to the intersection of leadership and governance ethics. For instance, one of the highly talked about issues was public procurement. Given the urgency of the situation, the rapid procurement of COVID-19 supplies increased the risk of corruption since not all guidelines were necessarily met and followed to address the problem at hand quickly. Thus, situations like this demand ethical leadership through transparency, accountability, and integrity, as well as the key responsibility of leaders to safeguard public trust and maintain institutional integrity, making sure that essential services are delivered and the rights of citizens are upheld [19].

Ethical and effective leadership puts people's general well-being first and foremost. It ensures that responsible decision-making is grounded on evidence-based studies that take into account fairness and accountability. New Zealand's ethics and equity framework is an example of a useful tool in offering guidance to inform resource allocation decisions, particularly during health emergencies [20]. At the heart of the guidelines is the demand for ethical leadership from decision-makers, applying principles such as equity, stewardship, fairness, accountability, and transparency. The framework requires leaders to embed these principles in policymaking with the intent to always protect the public's welfare during times of scarcity [20]. Leadership fails when one's own interests take precedence over duty and morality, whether through neglect of duty or corruption. Conversely, ethical leadership fosters resilience, legitimacy, and solidarity, allowing for a more just crisis management.

Preparedness and Response

In countries like South Korea, Thailand, and Vietnam, effective early response coupled with strong health governance, universal health coverage, and community-backed approaches were key to their success in containing the COVID-19 virus, especially when infrastructure, collaboration, and communication are firmly in place [21]. However, the same is not necessarily true in many Western Pacific countries. Failure to effectively plan and prepare made it difficult to contain the virus and protect the people. Often, such plans lacked transparency, equity, and public involvement. In setting the priorities and planning for the implementation, the voices of the marginalized are not heard, undermining both fairness and responsiveness [22]. The situation is even worse in South America, resulting in the problems being more pronounced and widespread. Political agendas that are fragmented, primary care systems that are not sufficient, infrastructures that are not crisis-ready, and misplaced investments are some of the reasons that hinder both preparedness and response [23].

These experiences suggest that effective and sound preparedness and response demand more than technical capacity. They require leadership grounded in equity, accountability, and participation. Economies that have successfully contained the virus did so by combining effective emergency response with fairness and trust-building. On the other hand, those that have failed did so due to inefficiencies in planning, preparation, and exclusion. The key to managing current crises and building resilient, just, and trusted health systems for the future is incorporating ethical principles into preparedness and response. In the end, a renewed focus on primary healthcare systems and inclusive dialogue with communities to build resilient frameworks for future crises remain paramount [23].

Medicine and Health Systems

One of the most profound inequalities in the healthcare systems, even before the pandemic, is access to medicine, which was significantly underscored by the onslaught of COVID-19. This, in fact, tested the resilience of health systems worldwide and revealed how deeply rooted systemic inequities shaped global vaccine distribution. While wealthier nations secured the bulk of early supplies, enabling them to vaccinate and distribute earlier, low- to middle-income nations relied on delayed or limited allocations, leaving them unvaccinated and vulnerable. This situation exhibits a strong structural imbalance in global pharmaceutical markets where the driving forces to access are intellectual property, production monopolies, and geopolitical influence [24].

The same problems were more evident in the Pacific. Pacific Island Countries and Territories encountered significant delays in the delivery of essential services such as maternal and child health, immunization, and treatment for non-communicable diseases due to supply chain interruptions and workforce shortages [25]. Yet again, these disruptions are reflections of deeply rooted systemic fragilities with a particular focus on high dependence on external medical supplies due to low domestic production capacities.

A study that focuses on a local government unit shows the critical roles played by decentralized governance, community engagement, and adaptive planning in enhancing the ability to respond to COVID-19 while still sustaining other essential health services [26]. Several innovations that proved to be useful include expanded digital health platforms, strengthened local procurement mechanisms, and integration of primary care networks. These innovations not only made it easier for the constituents of the local government to access medicines and other healthcare services but also eased the workload of healthcare providers.

What can be learned here is that in order to uphold the right to health of all, it is important to establish resilient, just, and responsive health systems and doing so requires global solidarity, regional cooperation, and local governance innovations to ensure fairer access to essential medicines in future crises. Local and national resilience ought to be augmented by structural reforms in governance, financing, and international cooperation to strengthen medicine supply chains, ensure continuity of care, and support community-driven responses.

Recovery and Strengthening

The COVID-19 pandemic challenged nations to not only respond to the crises at hand but also to chart pathways toward sustainable recovery and resilience. In this regard, economic revitalization, digital transformation, and health system resilience are necessary [27]. Effective management of pandemics and long-term resilience depend on the critical role played by robust and adaptable health systems. Therefore, what must occur is the identification of gaps in financing, governance, and integration of public health functions into broader health systems [28]. Meanwhile, critical factors that support containment and recovery include early interventions, risk communication, and community mobilization. It is also important to note that there are factors that may affect the effectiveness of these measures, such as differences in governance capacity and trust in institutions [29].

An ASEAN roadmap that provides guidance for collective recovery and long-term resilience states that cross-sectoral cooperation, health system strengthening, digital innovation, and inclusivity are important factors that shape recovery [30]. This roadmap also spotlights the importance of focusing on marginalized communities and stresses the need to address existing inequalities that were worsened by the pandemic. In doing so, the welfare of the most vulnerable populations is ensured and prioritized in recovery initiatives.

The pandemic experience has put the concept of recovery under a new light as it does not merely signify going back or returning to pre-pandemic conditions but rather points toward building stronger and more inclusive systems. Universal health coverage, workforce strengthening, and transparent governance to protect populations from future pandemics must be prioritized to ensure proper recovery. These efforts require responsible, effective, and equitable utilization of public resources. To achieve this, promotion of shared resources and effective capacity-building are necessary. In more specific terms, this includes investing in accessible digital health, social protection, and regional cooperation to build stronger systems capable of managing future pandemics. Furthermore, primary healthcare and indigenous knowledge systems must also be recognized as integral components of resilience and preparedness.

Success Stories

While it is without a doubt that the COVID-19 pandemic brought about numerous challenges and struggles to the world, it cannot be denied that it has also produced notable success stories that not only provide valuable lessons but also avenues for cooperation and rebuilding. For instance, lessons from historical experiences on epidemics, geographic isolation, and community solidarity helped Pacific Island Countries and Territories to contain the virus. Effective local responses rooted in medical preparedness and external support are also worth mentioning [31].

Other effective and successful measures undertaken in the Asia-Pacific region include decisive lockdowns, coordinated risk communication, and regional knowledge-sharing platforms [32]. Through these, several governments in the region were able to limit transmission and protect the vulnerable groups (e.g., Vietnam and New Zealand). Compliance has played a key role in containment and recovery [33], highlighting the fact that mutual trust between the government and communities was a decisive factor in the successful fight against COVID-19. Indigenous knowledge and values-driven practices such as *bayanihan* (communitarian spirit or altruism) and individual-led and initiated resilience efforts such as community pantries deserve to be promoted and celebrated [34]. Lastly, these stories of success reveal that rebuilding should focus on inclusiveness and sustainability, addressing the inequalities already exposed by the pandemic.

Conclusion

This chapter highlights the preexisting public health ethics challenges in the Pacific Rim that have been exacerbated by the COVID-19 pandemic. In doing so, it also recognizes the devotion and efforts of everyone involved in coping with the pandemic, particularly healthcare workers, government leaders, community members, and the authors who contributed to this book. Despite the tragedy, the pandemic has shown, time and again, that economies and territories in the Pacific Rim (and the rest of the world) can work together in crisis contexts.

As has been shown in the preceding discussions, the practical aspect of public health ethics requires a perfect balance between immediate and effective health crisis response and systemic reforms. This goes beyond technical processes and into ethical imperatives like upholding equity, solidarity, accountability, and stewardship, among others. The goal is not just preparedness and resilience, but more importantly, the transformation of systems that either no longer work or worsen situations during public health crises. Hence, public health initiatives require the application of the ethical values of trust, participation, and accountability in public health responses to ensure that measures are not only technically sound but also socially legitimate and inclusive.

References

1. Withers M, Dawson A, Caicedo A, Ho CW, Guillemot J, de Jesus Medina-Arellano M, de Castro L (2020) Inequities, vulnerabilities and ethics in the time of COVID-19: diverse perspectives from the APRU Global Health Program. J Glob Health Sci. https://doi.org/10.35500/jghs.2020.2.e25
2. Chi C (2024) Introduction: balancing saving lives and livelihood—six Pacific rim high-income nation studies of the COVID-19 pandemic control policies' impacts on society and the economy. Asian Pac J Health Econ Policy 7(1) https://www.ihep.jp/wp-content/uploads/APJVol7No1-02.pdf
3. Noknoy S, Kassai R, Sharma N, Nicodemus L, Canhota C, Goodyear-Smith, F. (2021). Integrating public health and primary care: the response of six Asia–Pacific countries to the COVID-19 pandemic. Br J Gen Pract, 71(708). https://bjgp.org/content/bjgp/71/708/326.full.pdf
4. Kooli C (2021) COVID-19: public health issues and ethical dilemmas. Ethics Med Pub Health 17:100635. https://doi.org/10.1016/j.jemep.2021.100635
5. Brolan CE, Körver S, Phillips G, Sharma D, Herron LM, O'Reilly G et al (2022) Lessons from the frontline: the COVID-19 pandemic emergency care experience from a human resource perspective in the Pacific region. The Lancet Regional Health–Western Pacific 25. https://doi.org/10.1016/j.lanwpc.2022.100514
6. Economic and Social Commission for Asia and the Pacific (ESCAP). (2021). COVID-19 and the unpaid care economy in Asia and the Pacific. United Nations Economic and Social Commission for Asia and the Pacific. https://www.unescap.org/sites/default/d8files/knowledge-products/2021_Regional_Report_Covid19-Care-Economy.pdf
7. Sadang JM (2020) The lived experience of Filipino nurses' work in COVID-19 quarantine facilities: a descriptive phenomenological study. Pac Rim Int J Nurs Res 25(1) https://he02.tci-thaijo.org/index.php/PRIJNR/article/view/246371

8. Islam MS, Vogt F, King C, Sheel M (2024) COVID-19 contact tracing and quarantine policies in the Indo-Pacific Region: a mixed-methods study of experiences of public health professionals. PLoS Glob Public Health 4(5):e0003121. https://doi.org/10.1371/journal.pgph.000312
9. Jiang Y, Laranjo JR, Thomas M (2022) COVID-19 lockdown policy and heterogeneous responses of urban mobility: evidence from The Philippines. PLoS One 17(6):e0270555. https://doi.org/10.1371/journal.pone.0270555
10. Leadership Institute for Global Health Transformation (LIGHT). (2020). COVID-19 in Asia Pacific: border control and path to reopening. Saw Swee Hock School of Public Health, National University of Singapore. https://sph.nus.edu.sg/wp-content/uploads/2020/11/COVID-19-in-Asia-Pacific-Border-Control-and-Path-to-Reopening_Event-Report.pdf
11. Lee J, Park S, Jang S, Reyes KA, Garcia F, Canila C, Oraño J, Ballesteros AJ, Muhartini T, Frans S, Marthias T, Putri LP, Mahendradhata Y, De Foo C (2024) Differential impacts of health systems and sociocultural environment on vulnerable populations during the COVID-19 pandemic: lessons from four Asia-Pacific countries. BMC Public Health 24(1). https://doi.org/10.1186/s12889-024-18949-1
12. Suwalak P, Noichim C, Kanthawee P (2025) Scoping the ASEAN framework and the 'ASEAN Way' on healthcare policies for stateless in ASEAN member states during COVID-19. Discov Health Systems 4(28). https://doi.org/10.1007/s44250-025-00193-y
13. Erwin C, Aultman J, Harter T, Illes J, Kogan RCJ (2020) Rural and remote communities: unique ethical issues in the COVID-19 pandemic. Am J Bioeth 20(7). https://doi.org/10.1080/15265161.2020.1764139
14. Balakrishnan V, Ng KS, Kaur W et al (2023) COVID-19 mental health prevalence and its risk factors in South East Asia. Curr Psychol 42. https://doi.org/10.1007/s12144-021-02556-z
15. Ocampo JMF, Santos RAN, Sevilleja JEAD, Gloria CT (2024) The COVID-19 pandemic's effects on mental and psychosocial health in The Philippines: a scoping review. Glob Ment Health (Camb). https://doi.org/10.1017/gmh.2024.14
16. Australian Institute of Health and Welfare. (2021). Mental health impact of COVID-19. https://www.aihw.gov.au/getmedia/0e4bf78e-5ec2-4c00-93b4-52e29203b53f/mental-health-impact-of-covid-19.pdf.aspx
17. Murphy JK, Khan A, Sun Q et al (2021) Needs, gaps and opportunities for standard and e-mental health care among at-risk populations in the Asia Pacific in the context of COVID-19: a rapid scoping review. Int J Equity Health 20:161. https://doi.org/10.1186/s12939-021-01484-5
18. Phillips G, Kendino M, Brolan CE, Mitchell R, Herron L, Kõrver S, Sharma D, O'Reilly G, Poloniati P, Kafoa B, Cox M (2022) Lessons from the frontline: leadership and governance experiences in the COVID-19 pandemic response across the Pacific region. The Lancet Regional Health—Western Pacific 25. https://doi.org/10.1016/j.lanwpc.2022.100518
19. United Nations Pacific Regional Anti-Corruption. (2021). Corruption Risks in Public Procurement in the Context of COVID-19 in Pacific Island Countries. United Nations Office on Drugs and Crime (UNODC) and United Nations Development Programme (UNDP). https://www.unodc.org/roseap/uploads/archive/documents/pacific/2021/UN-PRAC_Brief_-_Corruption_Risks_in_Public_Procurement_in_the_Context_of_COVID-19_in_Pacific_Island_Countries.pdf
20. National Ethics Advisory Committee (2021) Ethics and equity: resource allocation and COVID-19. Ministry of Health, Wellington. https://neac.health.govt.nz/assets/Uploads/NEAC/publications/ethics-equity-resource-allocation-covid-19-ethics-framework-support-decision-makers-feb21.pdf
21. Osewe PL (2023) Pandemic preparedness and response strategies: COVID-19 lessons from the Republic of Korea, Thailand, and Viet Nam. Asian Development Bank. https://www.adb.org/publications/pandemic-preparedness-covid-19-lessons
22. Essue BM, Kapiriri L, Mohamud H, Vélez C, Nouvet E, Aguilera B, Williams I, Kiwanuka S (2024) Priority setting in times of crises: an analysis of priority setting for the COVID-19 response in the Western Pacific region. Health Policy 142. https://doi.org/10.1016/j.healthpol.2024.105010

23. Varela AR, Touchton M, Miranda JJ, Grueso JM, Laajaj R, Carrasquilla G, Florez MV, Gaviria AMV, Hoyos AMO, Duarte EOV, Morales AV, Velasco N, Restrepo SR (2023) Assessing pandemic preparedness, response, and lessons learned from the COVID-19 pandemic in four south American countries: agenda for the future. Front Public Health 11. https://doi.org/10.3389/fpubh.2023.1274737
24. United Nations Development Programme (2023) Inequality in access to essential health and medicine: COVID-19 vaccines. United Nations. https://www.undp.org/publications/inequality-access-essential-health-and-medicine-covid19-vaccines
25. Bell L, van Gemert C, Merilles OEJ, Cash HL, Stoové M, Hellard M (2022) The impact of COVID-19 on public health systems in the Pacific Island countries and territories. The Lancet Regional Health—Western Pacific 25. https://www.thelancet.com/journals/lanwpc/article/PIIS2666-6065(22)00113-4/fulltext
26. Escano-Arias EA, Abarquez RDG, Cruz RV, Espeleta R, Ong M, Loreche A, Pepito VCF, Gomez V, Dayrit MM (2025) Strengthening local health systems and governance for Universal Health Coverage: experiences and lessons from the COVID-19 pandemic response in Quezon City, Philippines. Health Policy Plan 40(3). https://doi.org/10.1093/heapol/czaf002
27. From Recovery to Resilience. (2024). In: *The ASEAN*. Association of Southeast Asian Nations (ASEAN). https://asean.org/wp-content/uploads/2024/05/The-ASEAN-Issue-No.-37_-From-Recovery-to-Resilience-1.pdf
28. Sheel M, Rendell N (2022) Health security in the Pacific: expert perspectives to guide health system strengthening. World Health Organization (WHO). https://ses.library.usyd.edu.au/bitstream/handle/2123/28200/HsforHS%20report_edited_FINAL_20220223.pdf?sequence=1
29. De Foo C, Verma M, Tan SM, Haldane V, Reyes KA, Garcia F, Canila C, Orano J, Ballesteros AJ, Marthias T, Mahendradhata Y, Tuangratananon T, Rajatanavin N, Poungkantha W, Oanh TM, Due O, Asgari-Jirhandeh N, Tangcharoensathien V, Legido-Quigley H (2022) COVID-19 public health and social measures: a comprehensive picture of six Asian countries. BMJ Glob Health 7(11). https://doi.org/10.1136/bmjgh-2022-009863
30. The ASEAN Secretariat. (2020). ASEAN Comprehensive Recovery Framework. Association of Southeast Asian Nations (ASEAN). https://asean.org/wp-content/uploads/2020/11/2-FINAL-ACRF_adopted-37th-ASEAN-Summit_12112020.pdf
31. Weber E, Kopf A, Vaha M (2021) COVID-19 pandemics in the Pacific Island countries and territories. In: Akhtar R (ed) Coronavirus (COVID-19) outbreaks, environment and human behaviour. Springer, Cham. https://doi.org/10.1007/978-3-030-68120-3_3
32. Macroeconomic Policy and Financing for Development Division. (2020). Combating COVID-19 in Asia and the Pacific: measures, lessons and the way forward. United Nations Economic and Social Commission for Asia and the Pacific https://www.unescap.org/sites/default/files/xPB112_Combating%20COVID-19%20in%20Asia%20and%20the%20Pacific_0.pdf
33. Srijayanth P, Sivabakya TK, Srinivasan G, Seshayyan S (2020) Success stories of COVID-19 from top countries: a review. IOSR-JDMS 19(12) https://www.iosrjournals.org/iosr-jdms/papers/Vol19-issue12/Series-14/E1912142831.pdf
34. Zosa MAH, Ombao RP (2024) Bayanihan and community pantries. In: Llanera T (ed) Resilience and the Brown Babe's Burden: writings by Filipina philosophers. Routledge India. https://doi.org/10.4324/9781003546955-10

Part II

East Asia

A People-Centered Integrated Management and Control Model (PCIMC) of Pandemic: A Case of China

2

Qiuru Chen and Hui Yin

Abstract

Since its outbreak in late 2019, coronavirus disease 2019 (COVID-19) has come to shape our lives and our world. In China, globally minimal rates of incidence and mortality from COVID-19 have been observed. The chapter summarizes a People-Centered Integrated Management and Control Model (PCIMC) of pandemic, a combination of strong political leadership; rigorous, evolving, and coordinated public health measures; adopting whole-of-government and whole-of-society approaches; and research, development, and innovation. These achievements entailed major social and economic sacrifices made in the interest of preserving life and health and putting people at the center. The lessons identified from the pandemic have triggered further reforms to strengthen the public health system and promoted a greater understanding of the role of health in national development and security.

Keywords

People-centered · Global health security · Health systems · Pandemic prevention, Preparedness and response · COVID-19

Q. Chen · H. Yin (✉)
School of Public Health, Peking University, Beijing, China
e-mail: yinhui@pku.edu.cn

L. P. Marquez et al. (eds.), *COVID-19 in the Pacific Rim and Beyond*,
Collaborative Bioethics 6, https://doi.org/10.1007/978-3-032-21218-4_2

Introduction

The COVID-19 pandemic has shaken some standard assumptions that a country's wealth ensures its health. Leadership and competence are more important than cash in the pandemic response. Many of the best examples of decisive leadership come from governments and communities in more resource-constrained environments. There is a clear opportunity to build a post-pandemic future that draws on wells of wisdom from all parts of the world.

The pandemic has put the world's population at grave risk and exposed deep inequalities. Divisions and inequalities within and between countries have deepened. The impact has been severe on vulnerable and disadvantaged populations. By the end of March 2022, more than 480 million confirmed cases of COVID-19 and more than six million deaths had been reported worldwide. The pandemic was "the worst combined health and socioeconomic crisis in living memory and a disaster on all levels" [1].

Many governments are adjusting their strategies to reflect the status of the pandemic in their respective countries and for their populations. The COVID-19 pandemic is a multifaceted phenomenon with many political, economic, and social consequences. Success in managing and controlling this pandemic depends on the coordinated efforts of many organizations and institutions. Therefore, this study aimed to identify and analyze the strategies and measures related to managing and controlling this pandemic in China. This study was conducted in January–September 2023 as retrospectively.

The results, as seen in Fig. 2.1, show that the success for China came from a People-Centered Integrated Management and Control Model (PCIMC) of pandemic, a combination of strong political leadership; rigorous, evolving, and coordinated public health measures; adopting whole-of-government and whole-of-society approaches; and research, development, and innovation. These achievements meant great social and economic sacrifices to preserve life and health. Insights from the

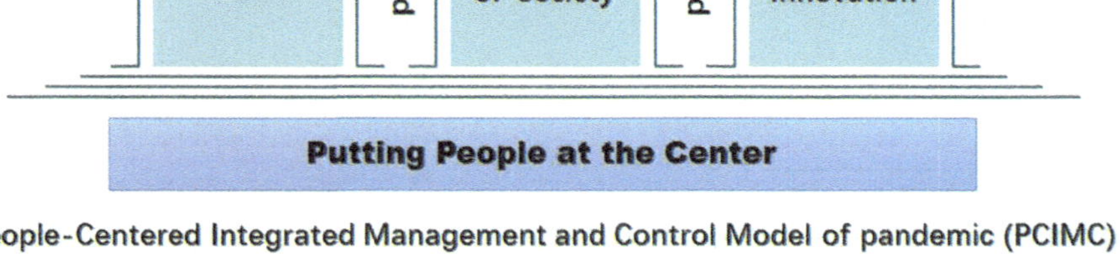

Fig. 2.1 People-Centered Integrated Management and Control Model (PCIMC) of pandemic

pandemic triggered further reforms to strengthen the public health system and fostered a better understanding of the role of health in national development and security.

Strong Political Leadership to Put People at the Center

Facing the sudden COVID-19 pandemic, China always puts people and their lives first to interpret the great love among humans. Back in May 2020, Chinese President Xi Jinping pointed to how China saves lives at all costs. "What is people-centered? So many people worked together to save a single patient. This, in essence, embodies doing whatever it takes (to save lives). We are willing to save lives at all costs. No matter how old the patients are and how serious their conditions have become, we never give up," [2] he said. Indeed, China has given every patient equal treatment irrespective of their age or wealth. As of May 22, 2020, in Hubei alone, more than 3600 COVID-19 patients over the age of 80 have been cured. In the provincial capital Wuhan, seven centenarian patients have been cured. More tellingly, China has stuck to a dynamic zero-COVID policy, calling for maximized efforts to protect people's lives and health for the past three years.

To prepare for potential risks of medical resource overloading associated with the reopening, the National Health Commission of China has taken a series of measures over the past three years to expand healthcare capacity and improve the treatment of COVID-19. Common fever clinics have been established, and the number of critical care beds has been increased to alleviate medical overloading. Medical institutions across the country have enhanced their capabilities in preventing and controlling respiratory infectious diseases, with more medical beds and equipment, as well as improved medical rescue abilities. Additionally, China's tiered diagnosis and treatment system has played an important role in meeting patient needs and effectively operating the primary medical system. Antiviral drugs for COVID-19 have been included in medical insurance through fast-approval channels, and maximum efforts are being made to provide these drugs to patients in need. The government and health departments at all levels are also providing human, financial, and material support to primary medical institutions to help establish an effective and efficient diagnosis and treatment model.

On January 28, 2020, to make sure that patients seek and receive timely medical treatment regardless of socio-economic status, as well as to ease financial pressure on hospitals, the Ministry of Finance (MOF) and National Healthcare Security Administration decided that individual expenses arising from COVID-19medical treatments would be fully covered by the government. Both confirmed and suspected patients could therefore get completely free medical care [3]. The government also encouraged manufacturers to produce more masks and other protective necessities by means of fiscal incentives and credit facilitation. Also, the Chinese government has committed to covering the cost of the vaccines, and they have been made available to the public. From 2021 to 2022, the country settled more than 150 billion yuan in COVID-19 vaccination costs, including health insurance funds and financial subsidies. These figures reflect the Chinese government's strong determination and commitment to epidemic prevention and control.

Rigorous, Evolving, and Coordinated Public Health Measures

As the pandemic evolves, each country faces different challenges. To better understand the development of the COVID-19 pandemic in China, we must know the timeline of the overall situation, infection rate, death rate, and so on since we can know the essential crises that need dealing with and trade-offs based on actual circumstances. Overall, the development of the COVID-19 pandemic in China is also closely related to strain changes. However, China's unique "Dynamic COVID-zero strategy" is an indispensable factor. Mainly, there are 5 stages in the development of the COVID-19 pandemic in mainland China.

Owing to limited medical resources, there was a serious shortage of beds for the treatment of confirmed cases. Confirmed cases couldn't be hospitalized and quarantined promptly, which further contributed to the spread, transmission, and propagation of the epidemic within families and the community. To improve patient intake rates, China decided to open a large-spaced, multi-bed "module hospital" to treat patients with confirmed mild disease.

"Module hospital" has become an important place for the designated intake and treatment of patients infected with SARS-CoV-2, and has realized good monitoring of the condition of mildly ill patients and asymptomatic infected patients, as well as effective identification of seriously ill patients and timely treatment, so as to obtain a better diagnostic and therapeutic effect [4].

However, most of "module hospitals" were built from stadiums, halls, and temporarily unused industrial buildings in a short period. Compared to normal medical institutions, "module hospitals" couldn't provide high-quality healthcare service and had a relatively high risk of hospital-acquired infections. But under specific pandemic circumstances, "module hospitals" can offer safe and scalable care for more individuals. As a result, China has achieved almost population-wide access to countermeasures.

It is worth mentioning that some of these public health measures might restrict residents' privacy and liberty to protect others in the community. But national identity was positively associated with support for and engagement with public health behaviors around the globe [5]. Chinese peoples' collectivist cultural preference and strong national identity play an important role in the implementation of public health measures. Given COVID-19's highly contagious nature, preventive actions are considered as a form of collective action, and organizations and media outlets have asked people to consider how their actions or inactions can affect others, including members of out-groups.

For people with a high collective cultural orientation, who do not focus on exerting their personalized power, adopting preventative behaviors that may restrict personalized power does not conflict with their values. Indeed, they may perceive that adhering to government policy and preventive measures is consistent with the pursuit of socialized power or with self-reliance in terms of preventing the spread of the disease and thus helping others. Thus, people with a high collective cultural orientation are likely to have a higher level of perceived control over decisions to engage in preventive measures, as adopting these measures is consistent with their desired outcomes [6].

Box 2.1 Timeline of COVID-19 Response

Stage 1: In December 2019, a series of pneumonia cases of unknown cause emerged in Wuhan, Hubei, China [7]. Unfortunately, this sudden outbreak came just in time for the Spring Festival travel rush, when the population movement was the most frequent. The epidemic rapidly spread throughout other regions in China. On January 23, 2020, the Chinese government announced a lockdown of Wuhan City. It went from the peak of a single-day increase in domestic cases of 15,152 on February 12. With very strict Public Health and Social Measures (PHSM), the outbreak was brought under control. The leading Group of the CPC Central Committee for COVID-19 Prevention and Control made it clear on March 23 that "the spread of the local epidemic in the country, with Wuhan as the primary battleground, has basically been halted" [8]. On May 22, for the first time, there were no new confirmed and fatal cases in the mainland, with 31 provinces, autonomous regions, and municipalities having zero confirmed cases for the first time.

Stage 2: China has entered a normalization stage of prevention and control since May 2020 [9]. There were outbreaks in key places and units in several provinces and cities in mainland China, and most of the outbreaks lasted for 20–30 days [10]. Most of the key places and units of the epidemic have a high concentration of people, high mobility, and are more confined, such as medical institutions, offices, public transportation, markets, and teaching institutions [11]. Overall, however, the outbreaks that occurred during the period were relatively small in scope and short in duration.

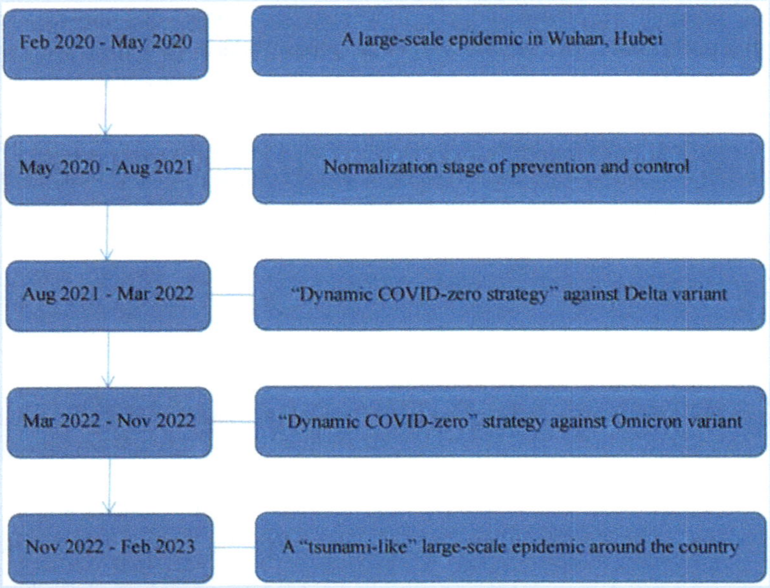

Stage 3: The Delta (B.1.617.2) variant was first reported in India in October 2020. It has "higher transmissibility and immune evasion than the early original virus strains" [12]. In order to control the epidemic at a higher level, at a lower cost, and in a shorter time, China adopted a new strategy called "Dynamic COVID-zero" from August 2021 [13]. In this stage, speed and precision were the most emphasized. The outbreak is usually contained within a maximum incubation period.

Stage 4: South Africa first reported cases of the Omicron variant to the WHO on November 24, 2021, and since then, the variant has spread rapidly to many countries around the world. The Omicron variant is characterized by a shorter incubation period for the disease, a smaller intergenerational interval, a faster spread of the virus, a higher proportion of asymptomatic infections, and a lower morbidity and mortality rate than other strains [14]. In response to the epidemic in Shanghai in the spring of 2022, Chinese authorities implemented one of the strictest lockdowns in the city. After that, China further optimized prevention and control efforts while the number of confirmed cases rose unprecedentedly.

Stage 5: On December 7, 2022, Chinese authorities abandoned "Dynamic COVID-zero" strategy. Inevitably, there was a "tsunami-like" large-scale epidemic that swept the country. A rather large mortality and morbidity happened during the transitional period. On May 15, 2023, Zhong Nanshan, a senior member of Chinese Academy of Engineering, declared that "Roughly 85% of the country's population has been infected with SARS-CoV-2 virus, with approximately 1.1 to 1.2 billion people" [15]. Oven Dyer uses models to predict there are a million or more deaths [16]. Finally, the pandemic, which lasted three years, has been fully contained.

In contrast, for people who are highly individual cultural oriented, the preventive behaviors associated with the pandemic may seem like restrictions on their personalized power. This may result in individual cultural oriented people feeling that their personalized power with regard to the prescribed preventative actions is limited. Furthermore, highly individual cultural oriented individuals may feel a lack of autonomy in choosing to comply with their respective government's suggested preventative health measures. This loss of choice is linked to perceptions of being controlled by others, which can lead to lower perceived control over decisions to engage in preventive behaviors for those who are highly individual cultural oriented.

East Asian culture, including China, is uncertainty-avoidant. This has been passed down through thousands of years of Chinese civilization and is also one of the reasons for the high savings rate of the Chinese people. For uncertainty-avoidant individuals, they tend to adopt more restrictive measures and are less proactive in taking risks. Besides, as mentioned above, East Asian culture is also a representative of collectivism. Under the collectivist culture, people will bear more moral pressure for the externalities they may bring and therefore do not want to be a source of infection. From this perspective, they are more supportive of restrictive measures.

During and after the COVID-19 pandemic, mental and psychological problems have increased significantly among different populations, including patients of COVID-19 and their families, health workers, college students, adolescents, children, and even the general population [17]. Threat of infection or re-infection of the virus, severe impact on work, study, economic income, and other aspects undoubtfully affect mental health and livelihood. Regrettably, it is hard to determine whether these impacts are caused by the COVID-19 pandemic or by public health measures.

An Integration of the Whole Government, Society, and Community

China has been enhancing its epidemic prevention and control mechanisms during and after the COVID-19 outbreak. The establishment of the "State Council Joint Prevention and Control Mechanism" has served as a multi-departmental coordination platform to effectively respond to the COVID-19 outbreak. Through this mechanism, numerous local outbreaks have been successfully contained, and guidelines have been developed by the National Health Commission in collaboration with experts in epidemiology, infectious diseases, medical treatment, and public health policy. The "Diagnosis and Treatment Plan for New Coronavirus Infection" has been updated ten times to provide the most current strategies and recommendations.

In evaluating China's performance in the fight against the COVID-19 pandemic, E. Monaco et al. state that "Actions across sectors are essential to combating communicable diseases" and "China established a paradigm to combat the virus." And that's largely thanks to China establishing Joint Prevention and Control Mechanism of the State Council.

The Chinese government took immediate action just after the emergence of the pneumonia cases of unknown cause. Wuhan Institute of Virology, Chinese Academy of Sciences confirmed on January 8, 2020, that the virus causing the outbreak was SARS-CoV-2. The National Health Commission released the first version of the Diagnosis and Treatment Scheme of COVID-19 on January 15, 2020. On January 21, 2020, it was announced that "National Health Commission has taken the lead in establishing Joint Prevention and Control Mechanism to deal with the pneumonia outbreak caused by SARS-CoV-2 infection, with a total of 32 departments as members." Guided by the Mechanism, China orderly responded to the pandemic from all sides. With the development of the epidemic and changes in the virulent strains, the Mechanism has introduced different policies accordingly. For example, the Mechanism issued Notice on Further Optimizing the Prevention and Control Measures of COVID-19 on November 11, 2022, mitigating the impact of COVID-19 policies on people's lives.

In conclusion, the Joint Prevention and Control Mechanism is the brain of China's responses to the COVID-19 pandemic. The Mechanism is well coordinated across sectors and regions and responds quickly to changes in the epidemic. There's no doubt that the Mechanism functions perfectly as the primary core component.

As China is a country with a huge population, it is hard to commit traditional governance in circumstances requiring precise prevention and control. "Grid" governance as an institution is adapted. State organs imposed strict controls, assigning community workers and medical staff to set up health and temperature check stations in all residential estates and areas. As residents were not allowed to leave their homes (and supermarkets were no longer open to individuals), government officials, property management companies, and volunteers delivered food and supplies to households. Big data and monitoring technology were deployed to track individual exposure to the virus. The community has also taken on the role of disseminating information about SARS-CoV-2 and the latest policies. Community workers usually set up chatting groups of community residents on WeChat and share essential information online.

Based on each community grid's management and control, strict segregation measures and adequate livelihoods are guaranteed throughout the whole country. However, there are problems with such governance, such as over-enforcement of control policies, which may cause residents' inconvenience.

Research, Development, and Innovation as Good Pillars

"Vaccination is the most economical and effective means to control and prevent infectious diseases" [18]. Before 2021, China had invented SARS-CoV-2-inactivated vaccine PiCoVacc, which had a protection rate of 79.34%. The world's first

SARS-CoV-2 vaccine to enter clinical trials is a human adenovirus-vectored SARS-CoV-2 vaccine from China. China's government adheres to the principles of informed consent and voluntariness, and encourages people over 3 years of age who have no contraindications to vaccination to receive vaccinations as required. Emphasis was placed on increasing the rates of full vaccination and booster vaccination for people aged 60 years and above who are at high risk of serious illnesses. As of December 23, 2022, a cumulative total of 3,469,670,000 doses of SARS-CoV-2 vaccine have been reported in mainland China.

Traditional Chinese Medicines (TCM) have developed a rich theoretical system and practical experience in fighting to infectious diseases over the past thousands of years, and have played an important role in controlling the spread owing to its unique advantages. In particular, their significant contribution to the prevention and control of COVID-19 is widely recognized. In the course of the prevention and control of this epidemic, the Chinese authorities have called for adherence to the combination of Traditional Chinese Medicines (TCMs) and Western medicines, the promotion of in-depth involvement of TCMs in the whole process of diagnosis and treatment, and the timely popularization of effective formulas and proprietary TCMs. The TCMs team issued Chinese herbal tonics or proprietary TCMs to mildly ill and suspected patients at the centralized isolation points in a timely manner, effectively easing the progression of the patients' conditions, alleviating the pressure on healthcare resources, and greatly increasing the cure rate.

In terms of surveillance, China has created a tool to make epidemiological investigation and contact tracing much more convenient and rapid: an online "health code" system. This app tracks an individual's travel history, COVID-19 contact history, and biometric data through one's smartphone. Residents entering public areas must show their health QR code. Different colored code represents different risk levels. A green one is free to travel within the city. A yellow one needs nucleic acid testing and is not allowed to enter public areas. A red one must contact medical institutions in time and should be quarantined in hospitals. Through the "health code" system, the government is able to quickly locate individuals once in close contact with the confirmed cases to prevent the occurrence and spread of cluster outbreaks.

They also shed new light on China's role in global health. In the fight against COVID-19, by October 2023, China has provided materials to support pandemic responses to more than 160 countries and international agencies and supplied more than 2 billion doses of COVID-19 vaccines to more than 120 countries and international agencies, enhancing the accessibility and affordability of vaccines in developing nations [19].

Public Health Ethics Lessons

Making progress toward the United Nations' Sustainable Development Goal 3 (Ensure health lives and promote well-being for all at all ages), including target 3.8 on universal health coverage, requires countries to move toward ensuring that all people and communities have access to health services that are high quality, safe, and acceptable. Cost-efficient, effective approaches to service delivery must be maximized for this to be attainable and sustainable. A People-Centered Integrated Management and Control Model (PCIMC) is crucial to the development of health systems that can respond to emerging and varied health challenges, especially the pandemic like COVID-19 and other healthcare crises.

There are several lessons we have learned that are meaningful for the future:

Align Health Workforce with PCIMC Services Realization of integrated people-centered management and control services will depend on health system inputs, including the availability, accessibility, and quality of health workers and the services they provide. As such, efforts have been made to form firm links between the frameworks of both pandemic prevention, preparedness, and response on integrated, people-centered health services and the health workforce strategies, including aligning human resources for health investment frameworks at national and global levels to future needs of health systems. PCIMC requires health workers with relevant skills. In addition to the benefits to communities and populations, the benefits of PCIMC also extend to health workers, including improved job satisfaction; more balanced workloads and fewer instances of burnout; and education and training opportunities to learn new skills, such as working in team-based health care environments.

Balance Pandemic Priorities with Timely Care for Other Patients Putting people at the center not only targets COVID-19 patients, but also patients with other diseases, as well as healthy individuals. COVID-19 has led to the deferral of elective procedures and created an unprecedented backlog of patients with non-COVID-19. This postponement may have potentially fatal consequences for patients. Health professionals have expressed concern that while healthcare resources are being directed toward combating the pandemic, the dual effect might result in unintended harm through delays in the treatment of other diseases. Beyond fear, these delays are partly due to changes in healthcare systems brought about by COVID-19. Some medical offices faced workforce shortages, enhanced sanitation protocols, and reserved hospital capacity protocols. These changes effectively prevented many patients from receiving care. The surge in COVID-19 cases diverted resources toward controlling the spread of the disease, leading to the postponement of elective care. Ethically, the deferral of elective procedures raises several dilemmas. There is no guarantee that short-term fixes will not cause long-term harm to patients. In fact, waiting times for care are likely to worsen, potentially affecting more patients. Unethical choices and decisions made today will add to the backlog and waiting lists. While we can immediately save the lives of COVID-19 patients, in the long run, we risk compromising the lives and well-being of millions awaiting future care. Most importantly, the potential damage caused by delayed care due to deteriorating conditions must be weighed against the risks associated with high patient and resource demands. Those who decided to delay care may not have fully considered the emotional challenges patients faced, which in some cases could be more devastating than the illness itself.

Leverage and Adapt Collectivism for Flexible Pandemic Control PCIMC also highlights culture adaptation. Chinese culture is identified as collective. Collectivism served as an institution that helped provide residents with tangible and intangible resources [20]. In the face of the COVID-19 pandemic, the community does not need the Government to issue authoritative requirements before it will take voluntary action to wear masks and receive vaccinations. Even if the government has not made it mandatory for the people to wear masks, if there are people who are not wearing masks among those who are wearing masks, they will still be required to wear masks under group pressure; if there are underreported positive cases among the masses, people will point out the anomalies for the sake of collective interests. This centralized model played a very effective role in the early stages of the epidemic when the country did not have a deep understanding of the principles of the virus and the lethality and seriousness of the virus body was high, controlling the spread of the virus and safeguarding the lives of the people. However, as the virus continues to mutate, this model has become less flexible in terms of policy implementation.

Recognize Socio-political Factors Shaping Health Policy Decisions The appropriate time window for a national policy shift is inherently shaped by socio-political and economic determinants, rather than solely dictated by scientific evidence. For example, in Brazil, while the COVID-19 pandemic disrupted health services, the reduction in testing and treatment coverage for schistosomiasis was already evident since 2017. Political and economic factors, including decentralization policies and budget cuts in public health, appear to have played a more significant role in the decline of schistosomiasis control efforts than the pandemic itself [21]. The same applies to China's response to the COVID-19 pandemic.

Sustain and Evolve Integrated Services from PCIMC Experiences From the PCIMC model, we can see that life and health are indeed placed in policy priority, receiving organizational support from other departments, and to a certain extent, integrating health into all policies. However, when the epidemic became normalized, some good practices were maintained (such as community grid workers), while others failed to be sustainable (the synergy of infectious disease prevention and treatment, the integration of public health and medical services). When dealing with future pandemics, it is necessary to guarantee the protection of personal privacy fully, freedom, and other rights, enhance public participation and communication mechanisms, and provide necessary living material support for affected individuals and families when implementing necessary isolation and other prevention and control measures, to reduce the impact on personal basic living.

Strengthen Transparency and Flexibility Under Accountability Systems In addition, China's political system has a strict hierarchy and accountability system, which in a way exacerbates the rigidity of collectivism. On February 27, 2020, at a press conference in Guangzhou, Zhong Nanshan, head of the high-level expert group of the National Health Commission (NHSC), revealed that as early as December 31, 2019, research institutes had made it clear that the outbreak was caused by a new type of coronavirus. On January 3, 2020, the virus had been successfully isolated; it was reported to the United Nations on January 7. This information was reported simultaneously to the local CDC and the national CDC, which also reported it to the Health Commission. But for more than 10 days afterward, the public knew almost nothing about the outbreak. According to Zhong Nanshan, the CDC at all levels knew about the outbreak but had no authority to release it to the public and had to report it level by level. This institutional shortcoming resulted in a 20-day delay in publicly disclosing human-to-human transmission of SARS-CoV-2. China had taken strict measures to deal with unqualified and derelict officials during the prevention and control of COVID-19. Misconduct including dereliction of duty and misappropriation of relief funds and materials will be investigated and handled in accordance with the law. Before February 5, 2020, the central government fired several high-ranking officials in Wuhan. There are 337 officials who were "penalized." Officials from the Red Cross Society in Hubei province were removed [22]. Under a strict accountability system, policy implementation inevitably shifts to greater rigor.

Engage Communities and Manage Infodemics for Inclusive Health Responses PCIMC also highlights evidence-informed risk communication, community engagement, and infodemic management strategies, developed in partnership with affected communities, which have been proven to drive more equitable, inclusive, and sustained response actions. Community engagement strategies must be complemented by infodemic management approaches, as well as improvements in the systems and structures that support agile learning, adaptation of health information and health program delivery, and priority-setting. Listening to communities, both offline and online, and obtaining regular feedback from them are essential for tailoring communication strategies.

Conclusion

Therefore, it is essential to enable all members of society to actively contribute to pandemic management by providing structured training, accurate information, and appropriate incentives. Strengthening PCIMC requires aligning health workforce capacity, cultural adaptability, governance accountability, and community engagement. Further research is needed to clarify the roles of key stakeholders and develop sustainable, evidence-based strategies for coordinated, transparent, and equitable responses in future pandemics.

References

1. Independent Panel for Pandemic Preparedness and Response (2021) COVID-19: make it the last pandemic. https://theindependentpanel.org/. Accessed 12 May 2021
2. Wang J. (2020) People-centered philosophy. https://www.chinadaily.com.cn/a/202005/23/WS5ec8c31da310a8b241157d4c.html. Accessed 30 Apr 2025
3. Liu C, Xu K, Du Q. (2020) China's flexible fiscal arrangements offer adequate medical expenses for treating COVID-19 patients. https://www.globaltimes.cn/content/1182599.shtml. Accessed 30 Apr 2025
4. Yao X, Li L et al (2022) Discussion on the characteristics and strategies of infection prevention and control in module hospitals. Chin J Nosocomiol 32(10):1441–1444. (In Chinese)
5. Van B, Jay J et al (2022) National identity predicts public health support during a global pandemic. National identity predicts public health support during a global pandemic. Nat Commun 13(1):517. https://doi.org/10.1038/s41467-021-27668-9
6. So J, Agrawal N (2023) Cultural orientation and perceived control over COVID-19 preventive behaviors. Social Influence 18(1). https://doi.org/10.1080/15534510.2023.2219462
7. Huang C et al (2020) Clinical features of patients infected with 2019 novel coronavirus in Wuhan, China. Lancet (London, England) 395(10223):497–506. https://doi.org/10.1016/S0140-6736(20)30183-5
8. Xin Hua Net. (2023) Li Keqiang presides over a meeting of the Central Leading Group for Responding to the COVID-19 pandemic. National Health Commission of the People's Republic of China. http://www.nhc.gov.cn/wjw/xwdt/202003/2d6c2b20f2bb4edf9194d1f7bcd95a48.shtml. Accessed 30 Apr 2025(In Chinese)
9. Liang WN, Yao JH, Wu J, Liu X, Liu J, Zhou L et al (2021) Experience and thinking on the normalization stage of prevention and control of COVID-19 in China. Natl Med J China 101(10):695–699. https://doi.org/10.3760/cma.j.cn112137.20210104-00008. (In Chinese)
10. Zong HY, Ning PS et al (2021) Analysis of the current status of outbreak information collection in domestic COVID-19 pandemic rebound events. Chin J Health Stat 38(06):837–847. (In Chinese)
11. Gao P, Li M et al (2022) Epidemic characteristics and key measures of prevention and control in key places and units under regular COVID-19 prevention and control—a literature analysis. Chin J Public Health 38(5):619–623. https://doi.org/10.11847/zgggws1138136
12. Tian D et al (2021) The global epidemic of the SARS-CoV-2 Delta variant, key spike mutations and immune escape. Front Immunol 12:751778. https://doi.org/10.3389/fimmu.2021.751778
13. Liu J, Liu M, Liang W (2022) The dynamic COVID-zero strategy in China. China CDC weekly 4(4):74–75. https://doi.org/10.46234/ccdcw2022.015
14. Wu Y, Liu J, Liu M, Liang W (2022) Epidemiologic features and scientific prevention and control advice of SARS-CoV-2 omicron variant. Chin J Dis Contr Prevent 26(5):497–501. https://doi.org/10.16462/j.cnki.zhjbkz.2022.05.001
15. GDToday. (2023) Zhong Nanshan: About 85% of people in China have been infected with COVID-19. https://www.newsgd.com/node_99363c4f3b/9f2c6959be.shtml. Accessed 30 Apr 2025
16. Dyer O (2023) Covid-19: China stops counting cases as models predict a million or more deaths. BMJ (Clinical research ed) 380:2. https://doi.org/10.1136/bmj.p2
17. Su S, Gong Y et al (2023) Challenges of and responses to mental health problems in the post-COVID-19 era. J Sichuan Univ (Med Sci) 54(2):217–222. https://doi.org/10.12182/20230260301. (In Chinese)
18. Jiang PP, Wang T (2022) Immunogenicity of SARS-Cov-2 and progress in the development of a COVID-19 vaccine. Chin J Pharm (Online Version) 20(01):40–46. (In Chinese)
19. Beijing: The State Council Information Office (2023) The White Paper on "The belt and road initiative: a major practice in building a community with a shared future for mankind". http://www.cidca.gov.cn/2023-10/19/c_1212291018.htm. Accessed 30 Apr 2025

20. Fowler, Jie G, James G, Rongwei C (2022) Catalytic institutions and community resilience: COVID-19, obstacles, and adaptive mechanisms in Wuhan, China. In: Community, economy and COVID-19: lessons from multi-country analyses of a global pandemic. Springer International Publishing, Cham, pp 119–135
21. de Jesus Santos CT, Graeff-Teixeira C, Di Christine Oliveira YL et al (2025) National Health Policy and the COVID-19 pandemic on human schistosomiasis mansoni in Sergipe state, Northeastern Brazil. Discov Public Health 22:179. https://doi.org/10.1186/s12982-025-00576-9
22. Xin Hua Net. (2020) China penalizes derelict officials in coronavirus fight. http://www.xinhuanet.com/english/2020-02/05/c_138755872.htm. Accessed 30 Apr 2025

Three Years of COVID in Hong Kong: Health Risk, Economy, and Trust: Ethical Balance of Group Versus Individual Interests

3

Bangyuan An, Xinyi Wang, and King L. Chow

Abstract

During the COVID-19 pandemic, five distinct waves of cases struck Hong Kong between January 2020 and May 2022, presenting significant challenges to the government. In response, the government implemented measures to mitigate these adverse effects, including immigration and border controls, mandatory viral testing, social distancing, mask mandates, information disclosure, and the adoption of vaccine passes. Against this backdrop, we analyze seven key factors contributing to the adoption of these policies, focusing on mobilization of available resources, changing economic conditions and public sentiment, expert opinions, and constraints imposed by policies in China and other jurisdictions. We will discuss their influences at specific stages and their impact on various stakeholders. Based on the collected information, we argue that effective pandemic governance requires swift and decisive action, grounded in scientific evidence,

B. An
Division of Public Policy, The Hong Kong University of Science and Technology, Kowloon, Hong Kong, China

School of Public Administration and Policy, Renmin University of China, Haidian District, Beijing, China

X. Wang
Division of Public Policy, The Hong Kong University of Science and Technology, Kowloon, Hong Kong, China

K. L. Chow (✉)
Division of Public Policy, Division of Life Science, Department of Chemical and Biological Engineering, Hong Kong University of Science and Technology, Clear Water Bay, Hong Kong
e-mail: bokchow@ust.hk

L. P. Marquez et al. (eds.), *COVID-19 in the Pacific Rim and Beyond*, Collaborative Bioethics 6, https://doi.org/10.1007/978-3-032-21218-4_3

comprehensive contingency planning, prudent resource allocation, coordinated information dissemination, and a careful balance between economic, political, and public concerns. From this analysis, we propose five targeted policy recommendations. Protecting the well-being of the majority within the community is the key principle that drives ethical practices in pandemic management for both public health practitioners and the government.

Graphical Abstract

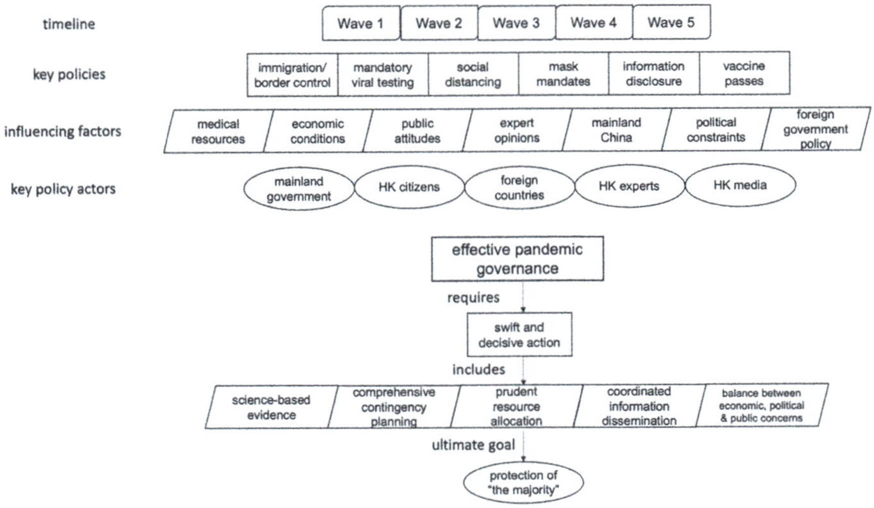

Keywords

Pandemic management · Information dissemination · Resources mobilization · Economic alignment · Prompt decision

Introduction

The COVID-19 pandemic, which emerged in late 2019, rapidly evolved into a global public health crisis, triggering widespread concern. As a densely populated metropolis, Hong Kong faced heightened vulnerability to viral transmission. The first confirmed case was reported on January 23, 2020 (First case in Wuhan, China: December 26, 2019), followed by five distinct waves of outbreaks of varying severity. By May 30, 2023, cumulative data indicated that approximately 39.4% of Hong Kong's population had contracted COVID-19 at least once, with a fatality rate of 0.18% [1]. Notably, Hong Kong recorded the world's highest mortality rate at the peak of the pandemic [2].

In response, the Hong Kong government adopted a containment strategy guided by three core principles: *rapid response, preparedness, and transparency* [3]. These measures initially proved effective in mitigating transmission. Over time, the government refined its approach, bolstering healthcare infrastructure and public health interventions. However, prolonged pandemic conditions led to diminishing public vigilance. By 2022, mounting case numbers strained medical resources, and certain policy decisions faced scrutiny from stakeholders. The quarantine order was rescinded on January 30, 2023, followed by the lifting of the mask mandate on March 1, 2023. On May 30, 2023, the official response level was downgraded from "emergency" to "alert," signaling a transition toward post-pandemic normalization.

As a global financial center, Hong Kong is the fourth most densely populated area in the world [4] with the ninth largest port globally [5]. High population mobility increases the risk of imported infections, complicating containment efforts. Consequently, Hong Kong's experience offers valuable insights for pandemic policy formulation in other densely populated trade centers.

This chapter has reviewed the important figures and policies of the Hong Kong epidemic through these five waves, highlighting (1) immigration control, (2) mandatory testing, (3) social distancing, (4) mask mandates, (5) information disclosure, and (6) vaccine passes and their implementation. Factors and concerns implicated in the decision-making process will be discussed.

Overview of Policies Implemented

The five distinct waves of COVID-19 outbreaks in Hong Kong are illustrated in Fig. 3.1, with a summary of corresponding policy measures provided in Appendix 3.1. As the pandemic progressed, the government incrementally strengthened its containment strategies, expanding policy coverage to mitigate rising transmission risks. However, authorities demonstrated a tendency to swiftly relax restrictions once outbreaks appeared to be under control, suggesting a prioritization of socioeconomic recovery over sustained caution.

Wave 1 (January–February 2020) Hong Kong's initial response to the COVID-19 outbreak was both rapid and decisive, earning commendation from both the public and national agencies. This effective early intervention can be attributed to lessons learned during the 2003 SARS epidemic, as will be elaborated in subsequent sections. The government took proactive measures by officially classifying COVID-19 as a statutory notifiable infectious disease on January 8, 2020. Following the confirmation of the first two cases on January 23 [6], the Chief Executive escalated the response level to "emergency"—the highest alert tier—within 2 days as additional cases emerged.

Within 1 week of the initial detection, the government implemented stringent border controls, including the closure of key transit points such as the Lo Wu crossing and Hong Kong-Macau Ferry Terminal [7, 8]. All incoming mainland travelers

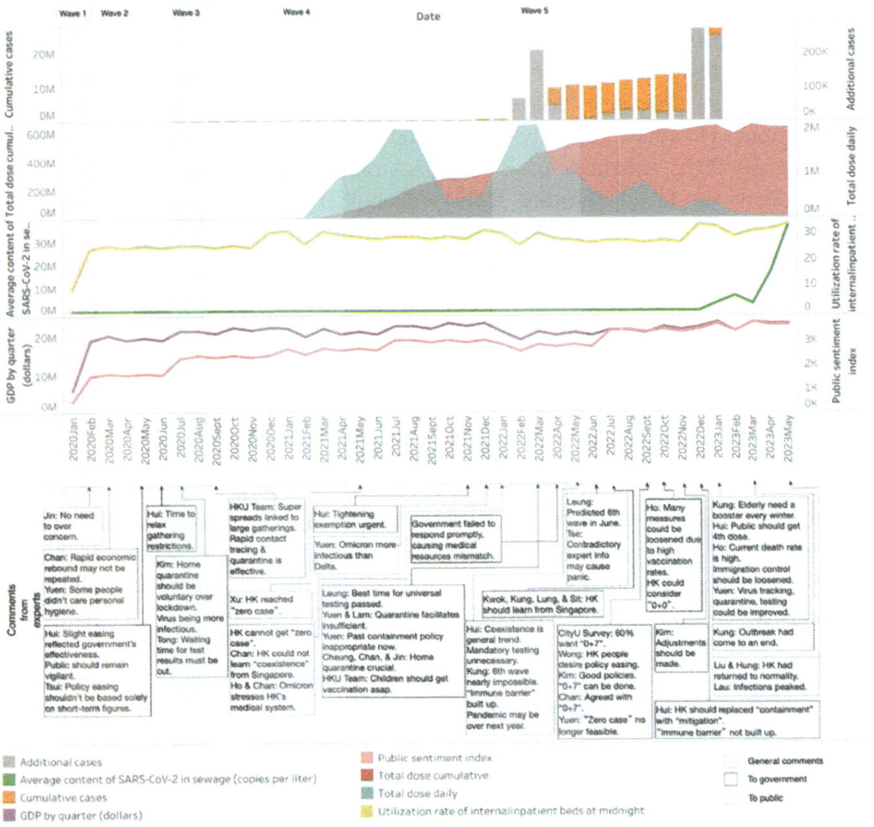

Fig. 3.1 General Statistics of Hong Kong's COVID-19 Epidemic Illustrating Six Contributing Factors to the Local Public Health Policies

were subjected to mandatory quarantine measures, though community transmission persisted [9]. The outbreak peaked on February 9 with 10 confirmed cases reported in a single day. Through coordinated efforts led by the Chief Executive's Cross-Departmental Epidemic Response Taskforce, the first wave was successfully contained within days [10].

Wave 2 (March–April 2020) The second wave emerged in Hong Kong primarily due to imported cases from returning residents and international students, coinciding with increased local transmission from social gatherings during Chinese New Year celebrations. This dual transmission pathway—combining overseas importation and community spread through activities such as wedding banquets, bar visits, and karaoke sessions—resulted in a significant surge of infections throughout March 2020. In response, authorities implemented a comprehensive package of enhanced containment measures, including extended quarantine policy for all arriving passengers [11], restrictions on gatherings, closing of public venues, and stringent operating regulations on restaurants [12]. These interventions with operational flexibility proved effective in controlling transmission. The successful containment enabled Hong Kong to implement its first temporary relaxation of restrictions by mid-May [13], marking a brief respite in pandemic control measures.

Wave 3 (July–August 2020) The third wave hit in early July, with local cases primarily concentrated in high-risk settings such as nursing homes and schools. This led to a rapid doubling of infections and a significant surge in deaths. Daily confirmed cases surpassed 100, and more than 3700 cases were reported in the following weeks. Authorities implemented strict gathering restrictions in response: closed public venues, reduced operating hours, and mandated mask-wearing [14, 15]. This wave exposed critical shortcomings in the city's virus testing capacity, hospital bed availability, and vaccination rates. As a result, the Chinese National Health Commission dispatched the "Mainland Nucleic Acid Testing Support Team" and the "Mainland Fangcang Hospital Support Team" to assist with local containment efforts [16]. These measures brought the situation under control by the end of October. Additionally, an official 14-day universal community testing program was launched on September 1.

To gradually resume cross-boundary travel between Hong Kong, Mainland China, and Macao, while balancing epidemic control with economic needs, the Hong Kong government launched the Return2hk Scheme in November [17, 18]. This scheme was limited to Hong Kong residents holding a Hong Kong Identity Card or other valid proof of residency. Eligible individuals traveling from low-risk areas and possessing a valid negative nucleic acid test result could reserve a daily quota through an online system [19]. Approximately 6000 daily entries were permitted across all border checkpoints. Upon arrival, residents confirmed their quarantine exemption via the health code system (green code) which dynamically assessed individuals' epidemic risk levels through a three-color QR code displayed via a mobile App. A green code indicated low risk, allowing free movement, and was typically assigned to individuals with no known exposure and a negative nucleic acid test result. A yellow code signified medium risk, and a red code denoted high risk, necessitating centralized quarantine, and is usually assigned to confirmed cases, or those of close contacts or staying in high-risk areas. The color classification was automatically determined based on users' travel history, test results, contact status, and official epidemic data. The scheme would be suspended if the risk level of the departure region increased.

Wave 4 (November 2020–April 2021) A burst of over 700 infected cases was reported in late November in a large dance group starting the fourth wave. Widespread locations and sectors among these cases made tracking and containment difficult. Preventive measures were tightened with an expansion of the testing scheme. With the new *delta* variant surfacing in the UK and many returnees from the UK for holidays, the strictest mandatory 14 + 7-day quarantine for these returnees was imposed.

The Hong Kong government launched the "LeaveHomeSafe" mobile App as part of its efforts to combat the COVID-19 epidemic [20]. This tracking tool was designed to facilitate contact tracing, manage vaccination and testing records, and support epidemiological investigations conducted by the health authority. The App's core functions include: (1) Venue check-in: users scan QR codes at public venues or on taxi license plates to record the time and location of visits; (2) Exposure notifications: if a user was present at a venue during the same time as a confirmed COVID-19 case, an exposure alert and health advice were issued; and (3) Vaccination and testing record management: users can upload QR codes of their vaccination certificates and test results, with privacy protected through biometric authentication or password verification. This mobile application upheld the principle of protecting personal data privacy considering public concerns.

Local lockdown of communities was implemented for selected districts [21], which had a heavy impact on consumers' sentiment. E-coupons were issued by the government to boost local consumption in April to revitalize the local economy [22]. The strict mandatory policy was relaxed only briefly and was maintained till the end of summer 2021. With the sustained zero local cases for the subsequent months, streamlined customs clearance for transit travelers between Hong Kong and mainland China was discussed in multiple rounds of meetings, leading to the adoption of the Come2hk Scheme in September when non-Hong Kong residents were allowed to enter at the border [23]. All non-residents of Hong Kong, e.g., Mainland Chinese citizens and foreign nationals residing in Guangdong or Macao, could reserve a slot while observing a daily quota (approximately 2000 per day across designated checkpoints), provided they were from low-risk regions and had a valid negative nucleic acid test result. These visitors under the Come2hk Scheme were required to undergo regular COVID-19 testing even when they were exempted from mandatory quarantine [24].

Wave 5 (January–May 2022) The Omicron variant spread rapidly throughout the community, leading to a severe outbreak toward the end of 2021. At its peak in March 2022, the city recorded over 30,000 new cases per day on March 5 [25], coupled with the highest death rate (the overall case fatality rate is 0.36%). Senior homes were particularly affected due to the low vaccination coverage (the first-dose vaccination rate among people aged 60 and above is 78.12%) in this high-risk groups, the elderly. Specific communities were placed under timely lockdowns, and city-wide tightened policies were implemented later in January 2022 [26]. However, during the gap between the fourth and fifth waves, the government was unprepared for the surge. The sharp increase in cases overwhelmed the system, making it difficult to track cases for effective isolation, management, and support. A severe shortage of virus testing capacity, hospital beds, and medical resources further exacerbated the situation.

To address these challenges, the government introduced the "StayHomeSafe" App [27] and "Recovery Record" QR Code [28] to encourage infected individuals to opt for home quarantine. A system of self-declaration for positive cases was adopted, allowing patients with low CT (Cycle Threshold) values from Rapid Antigen Test (RAT) [29] results to recover at home. To sustain economic activity, a second batch of E-coupons was issued [30], and mandatory vaccination with three doses was enforced for entry into specific public premises [31].

The StayHomeSafe Scheme was implemented to monitor individuals under mandatory home quarantine and ensure they remained at their designated locations, preventing the spread of COVID-19 [32]. Quarantined individuals wore an electronic wristband or connected to the government system via a mobile app (e.g., the "StayHomeSafe" app). In some cases, they scanned QR codes to confirm their presence at home and uploaded real-time location data. During quarantine, individuals were strictly prohibited from leaving their residences and reported their body temperature and health status daily via the App.

The Recovery Record QR Code was introduced as a digital health certificate for identity verification [28, 33]. It served as proof of completed quarantine, COVID-19 recovery, or vaccination, meeting entry requirements for public venues. The QR code was integrated with the "LeaveHomeSafe" App and Mainland China's health code system, enabling cross-boundary travel. Individuals scanned the code to enter restaurants, malls, and public transport.

For the first time since early 2020, China's top leaders publicly addressed Hong Kong's epidemic prevention efforts, issuing clear management instructions. This marked indirect criticism of local authorities. Three batches of expert teams were sent to assist in epidemic prevention in February and March [34–36]. While the Chief Executive announced a universal testing plan for late March, it was rejected by experts and the medical community due to a shortage of isolation facilities and manpower.

The fifth wave highlighted a dilemma between "zero cases" and a "coexistence with the virus" strategy. Many experts and citizens argued that preventing viral spread was no longer feasible or cost-effective, advocating for a shift to focus on high-risk groups and severe cases. However, the Chief Executive stood firm on the "zero case" policy, as explored in Section III.

By March, with declining cases, restrictive policies relaxed. Inbound traveler quotas increased [37], quarantine periods shortened, and tracking Apps were gradually discontinued. Four ports with Mainland China and Macau reopened [38], the mask mandate was lifted [39], and a third round of E-coupons was issued to boost the economy [30]. By May 30, 2022, the government lowered the emergency response level to "alert" status [40]. Epidemic coordination was transferred to the Health Bureau (HB) and relevant departments. The Central Command and Coordination Group (CCG) and COVID-19 Expert Advisory Panel were dissolved when society quickly returned to normal [41].

Analysis of Policies

Many scholars [42, 43] have categorized the local epidemic prevention measures on six aspects: (a) immigration/border control, (b) mandatory viral testing, (c) social distancing, (d) mask mandates, (e) information disclosure, and (f) vaccine passes. Appendix 3.1 shows chronologically the key steps taken, and the analysis goes as follows.

Immigration Control

Closing the six border customs services between Hong Kong, Mainland China, and Macau, along with mandatory quarantine for foreign travelers, was the first epidemic prevention policy issued by the government during the initial outbreak, even before the World Health Organization declared COVID-19 a "public health emergency of international concern." The Hong Kong government's response was prompt and decisive. Starting from the third wave, the government sought ways to resume normal travel with Mainland China, as ongoing trade essential to the local economy was being disrupted. It launched the "Return2hk" and "Come2hk" schemes, allowing travelers to reserve entry and exit with detailed health reports. The Mainland and Hong Kong governments issued daily quotas for each available port. Under these schemes, individuals with special duties were exempted from mandatory quarantine starting in the fourth wave. While Hong Kong maintained "zero local cases" for most of the year, repeated attempts to fully reopen the border were delayed due to the dire situation in Mainland China. The mainland government ultimately held the decisive role, postponing travel resumption until December 2022, when it eased epidemic controls and accelerated reopening.

The term "zero local cases" refers to the absence of community transmission while allowing for imported cases, e.g., individuals tested positive during quarantine are not included in local transmission counts. In contrast, "zero cases" means no new cases at all. This standard is difficult to achieve in Hong Kong as an international hub with continuous risk of imported cases.

The adoption of definition "zero local cases" helps HK to align policy goals with practical feasibility and the core objective to resume quarantine-free travel with mainland China, which was also pursuing a "dynamic zero-COVID" strategy. Differentiating between local and imported cases was essential for harmonizing control standards while maintaining international air connectivity without a full border closure to achieve zero cases that might hurt its economy and international status. In addition, emphasizing "zero local transmission" conveyed community safety, bolstering economic recovery in sectors like retail and dining. In 2021, organizations like ASIFMA (Asia Securities Industry & Financial Markets Association) urged the government to shift from the zero-COVID strategy to "living with the virus" [44]. The government responded, stating that "zero local cases" would not hinder international engagement, demonstrating the strategic inflexibility.

Entry restrictions for foreign travelers were similarly stringent to prevent imported cases. For instance, when the UK was severely affected in 2021, the Hong Kong government banned all UK flights until the situation improved. To meet quarantine demands, hotels and industrial buildings were converted into centralized quarantine centers, reflecting the government's adaptability.

Home quarantine relied on travelers' self-discipline and was supported by "StayHomeSafe" electronic wristbands to track mobility and ensure compliance. Violations triggered alerts to the Department of Health and police, with penalties of up to 6 months in jail. However, full compliance was not achieved, as repeated violations were reported. While the wristbands effectively monitored individuals at low cost, their design allowed users to self-remove them after quarantine, leaving room for non-compliance. Though the government emphasized that electronic monitoring did not collect personal data, encrypted information was stored locally and deleted after quarantine, defiance by users undermined the system's effectiveness.

These long-standing border policies led to a major recession in Hong Kong's key industries, including tourism, aviation, and retail. During the pandemic, Hong Kong's GDP fell into an "L-shaped" recession, experiencing the largest single-quarter decline in its history (−9.1% in Q1 2020) [45] as well as the largest annual contraction on record (−6.1% in 2020) [46]. GDP continued to decline until August 2022, when international travel resumed with reduced quarantine times. However, business groups warned that Hong Kong's strict regulations caused a talent exodus and threatened its reputation as a global financial hub [47]. The differential quarantine periods for Mainland China ("7 + 7") and international travelers ("3 + 4") received public support, addressing economic concerns and business pressure. Yet, negative sentiments arose among Mainland visitors, particularly business travelers and those visiting family, who criticized the perceived "double standard" in quarantine durations. In response, the government improved transparency and diversified communication channels, reaffirming its commitment to increase cross-border quotas and restore normal travel flows [48].

Mandatory Testing, Contact Tracing, and Group Isolation

People entering Hong Kong and those in designated industries were required to undergo mandatory testing. Starting in January 2021, the government published a daily list of buildings with positive cases (or virus-positive sewage samples) for mandatory testing. Residents and visitors who stayed at these locations for more than 2 hours had to undergo nucleic acid testing within 2 days. Sampling was conducted using combined throat and nasal swabs at designated centers, with food and sanitary items provided to residents free by the government. This measure was costly, e.g., HKD 183 million just from July to September 2020 alone [49]. Test results were used for contact tracing, case confirmation, isolation, and medical care.

While testing at nearby neighborhood centers was voluntary, compliance fell short of 100%, reducing the policy's effectiveness. Concerns about cross-infection during travel and crowded queues at testing sites were raised. Additionally, test results took 3–10 days, during which infected individuals might no longer be infectious, or already recovered, further fueling public skepticism.

The criteria for determining which buildings required mandatory testing were also questioned. In densely packed residential areas with shared communal facilities, targeting specific buildings rather than entire complexes was challenging. This approach, while reducing the number of affected residents and the strain on medical resources, risked overlooking potential intra-community transmission, weakening the policy's impact.

Criticism of mandatory testing spread across social media and news outlets, eroding public trust. The threshold for triggering mandatory testing was also adjusted dynamically based on the epidemic's progression [50], fueling further public doubts. In response, the government shifted the focus to strengthen follow-up efforts, actively tracking and testing high-risk individuals.

Social Distancing

Social distancing was a key measure in the city's response to the second wave of COVID-19. It was the most dynamic policy, frequently adjusted and impacting daily life. As the pandemic worsened, gathering group sizes were reduced, late-night dine-in services in restaurants were banned, and more public venues were closed. Businesses and individuals breaching regulations faced fines and imprisonment [51]. When conditions improved, social distancing was often the first policy relaxed, with citizens urged to exercise self-discipline to minimize virus spread.

The government also focused on close contacts in schools. Schools were closed for face-to-face classes, with instruction, digital materials, and consultations provided online. While some schools developed interactive platforms and instructional videos, most relied on one-way lectures with limited interaction [52]. After social distancing was relaxed, teachers and students were required to self-test at home each morning before attending school, reporting positive results, and switching to online classes. However, reliance on self-discipline led to cases of forged negative results, making schools hotspots for outbreaks in later stages. Nonetheless, schools emphasized social responsibility and distributed free rapid antigen test kits to students in need [53].

Social distancing also caused significant social and economic challenges. Elderly and marginalized groups experienced heightened loneliness and psychological stress due to isolation [54]. Workers in service sectors like food, retail, and hospitality, often from low-income households, faced job losses and income reductions due to gathering restrictions and venue closures. In parallel, the business sector did not escape the impact of reduced mobility and consumption-constrained economic activities. Unemployment increased in the dining, tourism, and entertainment sectors. To ease the burden, the government provided loans, tax exemptions, and E-coupons to boost purchasing power and transaction volume. Remote work was encouraged to minimize travel and clustering. A Temporary Unemployment Support Scheme offered financial aid to affected individuals [55], and the Labor Department organized online job fairs [56]. These measures were critical during the fifth wave, when unemployment rose from 3.9% to 4.5%, and underemployment increased from 1.8% to 2.3%, figures considered high in the Hong Kong context [57].

Although experts predicted infection cases would rise after relaxing social distancing, the government prioritized economic recovery over virus control at the end. Relaxation reflected growing reliance on economic pressures and risk assessments from big data-driven public health modeling after a prolonged period of restricted social activities.

Mask Mandates

Starting from the third wave of the epidemic, a mask mandate was introduced. Initially, masking was required only in indoor public places during the third and fourth waves, but it became mandatory in all public areas and transportation during the fifth wave. Active promotion, peer pressure, and high fines for non-compliance ensured widespread adherence. Even after the mandate was lifted, most citizens continued wearing masks, making it a norm whenever infection risks arose.

This compliance was particularly strong among Asian [58] and older populations, who prioritized collective interests and social order. These groups showed higher trust in traditional media (e.g., television, newspapers) and government authority. The elderly, perceiving a higher risk of infection, exhibited stronger self-protection behaviors and policy compliance [59]. In contrast, defiance was more common among younger people and expatriates [60, 61], who valued personal freedom and were influenced by diverse narratives on social media. To address skepticism, the government emphasized credibility through frequent pandemic briefings and expert endorsements. However, individuals with differing political views or unfamiliarity with local systems remained unconvinced [62]. Some protested by wearing "fake" masks made of thin fabrics with no protective capability, a form of defiance difficult for authorities to monitor, apart from occasional media reports [63].

The mask mandate was one of Hong Kong's most thoroughly implemented public health measures. However, the city faced supply shortages, as its sole domestic mask production facility, run by 120 incarcerated women working overtime, could not meet the demand [64]. The city relied heavily on external procurement and donations. The government redirected public-sector mask supplies to frontline workers and vulnerable groups, such as those in care homes. Subsidies and appeals to industrial and commercial sectors helped establish local production lines [65], but global supply constraints caused frequent overnight queues at pharmacies. The local mask market developed a dual structure: large retailers sold masks at low prices but with limited supply, while smaller shops offered unlimited stock at higher prices, often of questionable quality. To combat counterfeit products, the government increased inspections but achieved limited success. Unable to regulate pricing under Hong Kong's free-market system, the government faced criticism from media for its rigid adherence to laissez-faire policies [65].

Information Dissemination

Throughout the pandemic, the government used the official website "Together, We Fight the Virus" [66] to manage information dissemination. Relevant details, such as confirmed cases and patient visit logs with specific times and locations, were published while protecting personal identities. This reflected the government's careful balance between transparency and privacy. The centralized platform provided comprehensive information, enabling residents and visitors to assess exposure risks, debunk rumors, track school closures, and understand policy rationales with real-time updates. Residents also used various media platforms to voice opinions and participate in public consultations. Specific interest groups and NGOs provided input through social media and interviews with official media [67]. The gradual relaxation of social distancing and immigration controls reflected the government's efforts to incorporate feedback from stakeholders [68].

A notable feature of the website was the "Clarification" section, which addressed rumors that could incite panic or distrust. For instance, when doubts arose about the quality of masks distributed by government offices, the Logistics Department disclosed batch numbers, origins, production dates, and inspection results. It explained that masks sourced from various suppliers might look different but maintained consistent quality. Such transparent and timely updates not only boosted public confidence but also provided valuable data for researchers tracking the pandemic. These efforts enhanced the government's image as a trustworthy authority, grounded in scientific principles. This platform fulfilled its role and could be further strengthened for future crisis management.

Vaccine Pass

The Hong Kong government launched a mobile App "LeaveHomeSafe" in November 2020, which served as a COVID-19 risk notification tool. Individuals could scan a QR code and record their visits upon their entry to a venue. When there was a COVID-19 positive case reported there, visitors would be notified of the potential risk of transmission. Concerns about personal privacy were raised regarding access to privileged information by the government or third parties [69, 70]. These concerns were addressed by assuring limited access to personal data which were stored only on personal devices [71]. Even when the adoption was voluntary, inconvenience over its role in pandemic management was raised, and it was addressed primarily via media briefings and extended hotline and technical support.

A vaccine passport was introduced in early 2022 during the fifth wave with notification made 2 months prior to its implementation [72]. The passport is required for access to designated venues for all citizens aged 12 and above with sufficient vaccination doses. The function was integrated into the "LeaveHomeSafe" App, with status updates automatically available for staff verification. Non-compliance resulted in fines and imprisonment, and there was mandatory venue closure and restricted access for unvaccinated individuals [73].

The abusive use of vaccine passes exemptions also created gray areas in implementation. Widespread misuse of medical exemption certificates, or "vaccination waivers," for groups such as pregnant women, children, the elderly, and those with health conditions was reported. While individuals could obtain exemptions from designated clinics and submit them through a platform for approval, fraudulent exemptions were provided by some private clinics for payment. This malpractice allowed vaccine-hesitant individuals to bypass restrictions in accessing facilities, e.g., restaurants. It prompted the government to tighten reviews and invalidate fraudulent certificates. This malpractice undermined vaccination coverage and damaged social cohesion and trust in both government and healthcare institutions.

The government used platforms like "Tamar Talk" to explain policies and held regular press conferences where officials clarified the rationale behind vaccine mandates. Medical experts debunked misinformation in mainstream outlets such as *Hong Kong Economic Times*, addressing concerns about vaccine safety and efficacy. New media strategies included short videos on Facebook and YouTube to explain vaccine pass requirements and vaccination intervals in accessible, multilingual materials [74], with promotional content featuring local celebrities [75]. Despite these efforts, dissenting media voices spread rumors about vaccine side effects, App-based location tracking, and promoted traditional medicine as a substitute. The Department of Health and experts countered misinformation via daily press briefings, Q&A sections on official websites, and transparent data disclosures to emphasize vaccination necessity and the vaccine pass policy's validity. During this period, media coverage had significantly shaped public perceptions and sentiment.

Influencing Factors of Public Health Policies

We have identified seven key elements that heavily influenced the adoption of public health policy in Hong Kong during this pandemic period, listed as (a) medical resources, (b) economic conditions, (c) public attitudes, (d) expert opinions, (e) mainland China, (f) political constraints, and (g) foreign government policies. They are depicted in Fig. 3.1 and are elaborated as follows.

Medical Resources

Hong Kong's medical resources heavily depend on imports. After the 2003 SARS epidemic, the Hong Kong government maintained a stockpile of essential items, like masks, which proved vital during the early stages of COVID-19. Lessons from SARS, where over 50% of healthcare workers were infected, prompted hospitals to adopt strict infection control measures to protect staff. During the first three waves, medical resources remained stable, as shown by the midnight inpatient bed utilization rate (Fig. 3.1, yellow line), internal medicine bed occupancy, emergency room usage, isolation ward availability, and quarantine facility activation [76]. The system handled the small-scale outbreak effectively.

However, this success led to government complacency and poor preparation for worsening conditions. In the fourth wave, medical utilization rates surged sharply. Strain on the system grew as non-emergent patients with mild symptoms occupied hospital beds. Over 10,000 public hospital staff became infected, further burdening the overstretched system.

By the fifth wave, the Health Authority underestimated the epidemic's scale, adopting a conservative triage protocol. Infected individuals, even asymptomatic cases, were urged to visit emergency rooms [77], while suspected cases were advised to seek medical care [78], leading to a collapse of emergency services. Quarantine centers were overwhelmed, and recovered individuals who still tested positive were hospitalized. Resource shortages prompted policy changes: patients with a CT value below 45 (later reduced to 33 and 30) were discharged based on physician judgment. These shifting criteria exposed hospital weaknesses [79], confused staff and the public, and eroded government credibility.

As the crisis deepened, the government began building and expanding temporary hospitals (Fangcang shelters), but delays in site selection and implementation slowed progress. By March 2022, before cases peaked, policies shifted away from mandatory nucleic acid testing. Mild or asymptomatic patients were encouraged to self-quarantine at home, freeing resources for critical cases. The government also secured oral antiviral drugs, like Paxlovid, offered free at some hospitals to reduce inpatient demand. In hindsight, this collapse of Hong Kong's medical system during the fifth wave was rooted in early government complacency and underestimating demand.

The home quarantine policy faced widespread criticism. Hong Kong's cramped housing made effective isolation difficult, leading to virus transmission among family members. Without timely medical intervention, public confidence in the healthcare system and government policies eroded. Delegating isolation and care to families without adequate support was seen as unfair and irresponsible [80], even though similar practices were used in other countries facing similar challenges.

Another shortcoming was poor integration between public and private healthcare sectors. Modeled after the UK system, Hong Kong lacked sufficient primary care physicians and had weak coordination between public and private hospitals. During the pandemic, public hospitals bore most of the burden due to affordability. Community-based or family medical care was limited, and the high cost of private care pushed most residents to public facilities. Many private hospitals lacked the capacity to admit COVID-positive patients. Despite efforts to recruit retired and private-sector medical staff, manpower shortages persisted [81, 82]. The government's delayed adoption of a tiered triage approach during the fifth wave worsened the situation. Public hospitals, community treatment centers, and temporary hospitals were eventually designated based on symptom severity to reduce mortality by focusing resources on critical cases [83]. The business sector, including hotels and buses, was also tapped for isolation facilities and patient transport [84].

The death rate during the fifth wave was high, especially in nursing homes. Over 1000 nursing homes experienced outbreaks, and treatment for elderly patients was strained as transfer services to isolation centers were overwhelmed. The lack of an aggressive elderly vaccination program was seen as the main cause. Later, medical personnel from the central government and local teams visited nursing homes to encourage vaccination, implemented closed-loop management, and transferred some elderly residents to temporary centers. Designated hospitals would focus on COVID-positive patients. Overall, neglecting vulnerable populations exposed flaws in emergency handling and prompted a review of resource allocation across the community.

Economic Conditions

Economic status had always been a focus for the Hong Kong government during this pandemic. Despite the low GDP growth rate and consumer confidence in this period, the overall economic data continued to present stable, gradual growth. However, macroscopically, Hong Kong's economy was hit hard by the pandemic in two aspects: (a) Economic inequality became more pronounced with a widened wealth gap. The strict border control and delayed customs clearance constituted a major blow to international trade, which heavily depends on physical interactions. When some selective sectors such as finance could circumvent the negative impact through online workflow, the widened social gap between sectors exposed the more vulnerable groups, such as the blue-collar workers. (b) The economic downturn also led to a net outflow of talent and capital from the city [85]. About 150,000 net outflows of residents were recorded in a European Chamber of Commerce report [86, 87] and about half of European companies were considering regional office relocation. Such circumstances put pressure on the government to relax the border control.

Balancing public health concerns, economic growth, and social stability has always been on the Government's priority, and it posed an ethical dilemma for policy makers. In these 3 years, relaxing social distancing, when the situation improved, opening special entry channels for business personnel or for those with special tasks were introduced. A fiscal plan spending over HK$100 billion to speed up economic recovery was put in place. At the end, the cancellation of many restrictive measures in March 2022 reflected the government's great emphasis on the economic recovery over purely health concern.

Public Attitudes

Public opinion played a significant role in shaping government policy. The Public Sentiment Index (PSI) (Fig. 3.1, pink line) from the HK Public Opinion Research Institute (HKPORI) reflected changing sentiments, showing an upward trend during the period reported. The low value in 2019 was likely tied to social unrest but rose steadily, except for a temporary dip during the fifth wave, indicating a generally positive attitude towards the government. This was attributed to the government's emphasis on public communication, including timely responses by the Hong Kong Information Services Department and the Government Press Office. Effective communication with media, community organizations, businesses, public services, and grassroots residents minimized confusion and built trust. The government's sensitivity to public feedback was evident, for example, in the adoption of the "zero cases" policy with strict border control and social distancing, explicitly communicated as essential for reopening borders with Mainland China. The public largely supported this rationale, accepting short-term inconvenience for lower health risks and faster economic recovery. Over time, as the virus became less virulent, citizen cooperation remained strong.

Nonetheless, large group gatherings, quarantine evasion, and fake test results all contributed to the rebound of cases in the fifth wave. These reflected the impatience of citizens with the lengthy negotiation with the central government on border opening, the strained medical system, and the high death toll caused by the Omicron strain. The support for the "zero cases" policy faded, and 57% of respondents saw the "coexistence" strategy more fitting for Hong Kong in an online poll of 7000 citizens by the HKPORI in February 2022. This public endorsement of an alternative strategy prompted the government to transition toward a "coexistence" policy. The political beliefs of Hong Kong residents also played a key part in the decision-making process. As a special administrative region of China for over 20 years, most Hong Kong residents had been more closely aligned with the Mainland Chinese government's strategy, but there remained many who did not share the same view. They preferred an independent position in crisis management. A low vaccination rate during the first three waves of the pandemic resulted when residents put personal freedom above community safety, a more liberal view on individuals' rights. The government actively sought support from corporations to encourage vaccination and ease public opinion with communication via medical experts, transparent data dissemination, and incentives [88]. The vaccination rate went up only when the government enforced a "LeaveHomeSafe" mandate. Liberal media skepticism toward vaccines further complicated efforts. Many residents preferred the BioNTech vaccine over Sinovac at a 1.75:1 ratio [89], despite both offering similar protection against severe illness and death [90, 91]. Resistance from healthcare unions to cooperate with Mainland medical personnel also initially affected service quality, requiring legal measures under the National Security Law and the Prevention and Control of Disease Regulation to curb misinformation.

Finally, policies like vaccine passports, mandatory testing, social distancing, and mask mandates required partial sacrifices of privacy and freedom—core values non-negotiable for some residents. To reduce ideological confrontations, enforcement was often lenient, which weakened policy effectiveness in certain communities.

Expert Opinions

Medical professionals could claim expertise in fields as well as being independent of the government [92]. Policy makers rely on external experts because they need effective solutions and help resolve disputes [93]. By consulting them in finding solutions, the government often follows the recommendation and may even delegate power to expert groups, thereby negating social conflicts [94]. Yet, disagreements between experts and the government may undermine public trust and compliance with public health policies. Under the notion of freedom of speech, many experts expressed their views through media interviews, published papers, and opinion articles (Fig. 3.1, lower panel). Most of these experts are in the medical profession with little to offer from the social science perspective. Their inputs were largely technical in nature and thus carried the professional limitations they entailed. With the urgency of policy decision-making, social and administrative issues such as community mobilization and logistics management were often taken as secondary. Data mostly related to medical risks and health resource management were emphasized, while the social behavioral needs of residents were often ignored.

The expertise and public opinions fell into three categories: (a) general assessment of the situation at the time; (b) expressing opinions on government policy; (c) reminder on specific behaviors. The divergence also stemmed from differing values between certain experts and segments of society, for example, the relative importance of individual freedom versus the right to life and health. Their opinions were often contradictory without a constant holistic perspective, which only caused confusion and perplexity for the residents. Experts used selected data as evidence to support their views with no clear balance of contributing factors [95]. In media reports, their conclusions were often taken as piece meal instead of being accompanied by elaborate analyses and reasoning. More importantly, these experts' lack of media literacy and communication skills often compromised the goal of bringing clarity and offering positive social impacts [96]. Likewise, journalists' reporting sometimes focused on sensational issues and reduced the effectiveness of public health information dissemination, given the education of lay people. Given these existing cognitive gaps, citizens lacking sufficient scientific literacy were more confused. Any public health policy would often invoke immediate support or opposition based on their political positions or personal interests, and the opinions would fall in line with camps even when some views were not scientifically based nor constructive in practice [97, 98]. The fierce debate among experts in the media was more like political disputes that helped little in formulating productive public health policy.

Mainland China

Mainland China's epidemic control strategy was critical in shaping Hong Kong's policy and its implementation. The Hong Kong government had to consider not only the mainland's overarching policy objectives but also the region's socio-economic interdependence of the city and mainland, such as population mobility and economic integration. The unique governance characteristics of Hong Kong also resulted in complex challenges in the local application of the "Dynamic Zero" strategy [99], necessitating a good balance of mainland expectations with local constraints and needs.

Mainland China's pandemic responses evolved through three distinct phases: (1) Emergency Containment Phase (Dec 2019–Apr 2020): Characterized by rapid emergency responses, such as the lockdown of Wuhan and nationwide mobility restrictions, aiming to halt transmission through travel bans for internal containment and external exportation control [100]; (2) Regularized Prevention and Control Phase (Apr 2020–Dec 2021): This stage emphasized the prevention of inbound cases and domestic resurgence and established the "Four Earlys" mechanism: early detection, early reporting, early isolation, and early treatment. Mass nucleic acid testing was a cornerstone of the surveillance system, and sporadic outbreaks were contained within two to three incubation periods. Vaccine development was accelerated, and emergent-use approvals were granted to vaccines such as Sinopharm and Sinovac by the end of 2020; and (3) Full-Chain Dynamic Zero Phase (Dec 2021–Nov 2022): In response to the Delta variant, the mainland implemented a full-chain dynamic zero-COVID approach, utilizing big data to trace close contacts and emphasizing rapid, precise interventions [101, 102].

In alignment with the change, the Hong Kong government adopted Dynamic Zero measures during its first four waves of outbreaks until early 2022. These policies effectively suppressed outbreaks, eventually leading to economic stagnation and dampening consumer sentiment. Throughout this time, the government continually sought to reopen its border with the mainland to revive business and tourism [103]. With the city's health code system ready by the end of 2021, the emergence of the Omicron variant in early 2022, accompanied by a surge of infections and overstretched medical support services, derailed preparations for reopening.

At the end, the central government dispatched medical teams [104] and initiated infrastructure projects [105], prompting the Hong Kong government to relax its containment policies. By late 2022, the mainland government introduced the "Ten New Measures" [106] and removed the term "Dynamic Zero" from official discourse. Despite reports of over 1000 cases daily, the government aligned its policies with the new national direction: shortening the isolation period for confirmed cases and fully eliminating isolation requirements by January 2023. The mask-mandate policy was officially lifted in March, and full border reopening in June. Even with positive cases rebounding significantly, this adjusted strategy was maintained to ensure stability in policy.

China's influence on Hong Kong's pandemic strategy stemmed from the constitutional constraints inherent in the "One Country, Two Systems" framework. According to Article 18 of the *Basic Law*, public health is considered a matter under the central government's authority, while the Hong Kong and Macau Affairs Office and state-run media are there to support Hong Kong's anti-pandemic efforts affirming the legitimacy of this strategy.

The most immediate driver of Hong Kong's alignment with mainland policies was its economic dependence on cross-border integration [107]. Given Hong Kong's reliance on China for trade, logistics, and tourism, prolonged border closures would have exacerbated economic recession. Thus, adherence to Beijing's strategy was perceived as a necessary condition for reopening social ties and economic recovery. This interdependence meant that shifts in mainland policy directly shaped Hong Kong's pandemic framework. For example, Hong Kong's vaccine passes systems and its policy adjustments following the "Ten New Measures" demonstrated close synchronization. China's support on essential medical equipment, food, expert teams, and healthcare personnel also played a key role in shaping policy execution in Hong Kong [108]. The "tiered treatment model" adopted during the fifth wave was indeed based on recommendations from mainland experts.

Political Constraints

Institutionalism endows and constrains the rights of an organization to influence public policy decisions [109]. Public health policy addresses how limited resources can be distributed to achieve a justifiable goal for the government and the operation of a civic society with the benefit and respect of its citizens on its agenda [110, 111]. The Hong Kong Special Administrative Region (HKSAR) government has relatively weak policy-making capability but more robust administrative/executive capabilities [112], making it a "weak government" [113] that can take care of routine functions but is slow in agenda setting, policy formulation, and critical decision in demand. The government was often compromised with limited authority [114]. The diverse views of legislators from different parties in the Legislative Council, legal feasibility of the policies, support from the civil service and grassroots communities, execution and implementation by qualified professionals, and budget of the policies had to be considered. To the community in need, this lack of a swift response would mean ineffective governance to develop a coping strategy.

Foreign Government Policies

As a free port of international trade, public health policy in Hong Kong is also dictated by many external factors imposed by other nations. When a stable condition was attained around April 2021 while resurgence of outbreaks popped up in nearby countries, the merit of HK's early containment strategy was praised. Unfortunately, the vaccination rate in HK stood at a low 12%, offering a low vaccine protection barrier in time. Any relaxation of such containment effort would lead to a sudden burst of symptomatic cases and mortality rates.

In contrast, nations of similar economic status but with higher infection rates were poised to open their border sooner. For example, a joint declaration "Living with Covid 19, Living as Usual" was made by the Joint Anti-Pandemic Working

Group in Singapore in June 2021. It outlined an alternative path to cope with the pandemic. "The Shift in Singapore's Approach: Moving Away from dynamic zero COVID-19 strategy" was announced in October 2021, 1 year ahead of HK. As a result, Singapore was able to attract a significant amount of capital in that year and quickly re-establish international connectivity. It allowed Singapore to surpass Hong Kong and become the third largest financial center globally [115]. The pressure was on Hong Kong's ability to expedite the re-establishment of its global connections. A video conference attended by 130 representatives from the International Chamber of Commerce (ICC) and Consuls-General and Honorary Consuls in Hong Kong was held [116]. The International Financial Leaders' Investment Summit also attracted 200 delegates from 120+ financial institutions around the world in November 2022 to reconfirm HK's status as a financial hub. In the wake of this recovery, political and business sectors were actively making overseas visits along the Belt and Road regions to promote HK as a hub for sustainable finance, innovation and technology, and energy, among others [117].

Policy Implications

The SARS outbreak in 2003 taught the Hong Kong government valuable lessons. From a public health perspective, Hong Kong's pandemic response encompassed a range of actions, including the establishment of early warning systems, preparedness for infection control among frontline workers, widespread use of surgical masks, early implementation of visitation restrictions, vaccine development and advocacy, as well as timely school closures [118, 119]. In parallel, the Hong Kong government also considered the city's socio-demographic and infrastructural conditions with improvement of drainage and ventilation systems in buildings [120], mitigating housing-related transmission risks through stratified quarantine and wastewater surveillance [121], implementing closed-loop management and financial subsidies for key sectors and essential occupations [122], and providing targeted care and interventions for the vulnerable groups. Furthermore, the government emphasized transparency in communicating the rationale behind public health interventions and restrictions. These experiences paved the road toward a fast response of the government to curb COVID-19 in early 2020. All these measures successfully got the city through the first three waves of outbreaks. The barely successful containment of the fourth wave and the tragic response to the fifth wave exposed more weaknesses of the Hong Kong government and provided insights for the authority to combat large-scale public health crises. We believe that inputs from five important stakeholders should be carefully evaluated by the municipal government and should be understood by the community at large (see Fig. 3.2).

Improving Mechanisms to Manage Major Public Health Crisis

Firstly, the government should enhance its ability to coordinate with various departments and stakeholders for fast policy formulation and implementation. This

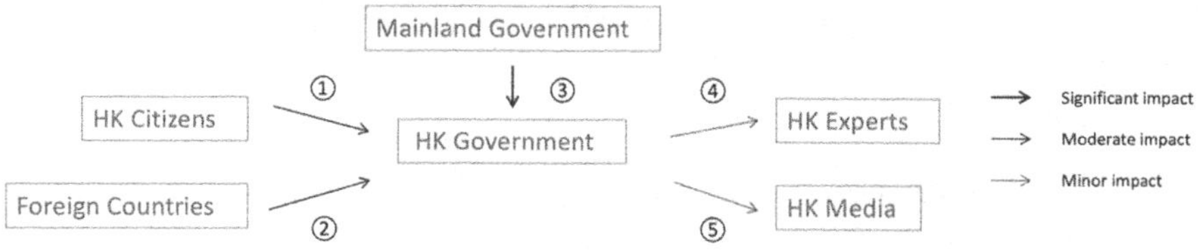

Fig. 3.2 Five policy implications in the relationship between key policy actors

requires effective communication of scientifically reliable information with a persuasive narrative. A clear set of priorities should be established within the government in any decision-making process. Secondly, there is an urgent need to increase its medical capacity during peacetime. Improvement in mobilizing personnel, emergency medical resources, their production, and imports during crises must be well in place ahead of time. To alleviate the strain on public healthcare institutions during the pandemic, a territory-wide healthcare network, including the Department of Health, the Hospital Authority, and the Social Welfare Department must be established [123]. This system will integrate healthcare services by region, encompassing public and private hospitals, community hospitals, and clinics. Primary healthcare services and tele-medicine initiatives require further strengthening. Moreover, there should be increased collaboration between public and private healthcare providers, including non-governmental organizations, to better coordinate resources and facilitate community support and mobilization. Thirdly, protection mechanisms for vulnerable groups and susceptible communities, e.g., the elderly and children, need better attention. Routine provision of adequate medical supplies and services for these groups should be practiced but not as an emergency response. Fourthly, ongoing investment in epidemic prevention-related technologies and research like digital tracking and positioning technology would be the best preparation for future public health crises.

Enhancing Local Production Lines for Essential Supplies

During the high time of the outbreak, the entire production line of medical supplies was strained because of our weak manufacturing sector. The government faced difficulties in procuring, purchasing, and transporting foreign-produced materials. An orchestrated supply chain network would need to be set up locally for necessary materials, mobilization, and production plans during emergencies. Cross-border transportation between Hong Kong and Mainland China needs better management to ensure the swift transfer of materials of desired quality and safety. The government is also advised to use tough price control policy and law enforcement to regulate the market to control the product price during emergencies and avoid gouging.

The Administrative Relationship Between Hong Kong and Mainland China

Hong Kong residents generally consider themselves independent and do not want to be overly constrained by the central government policy. This constitutional and administrative relationship between the two needs to be clarified through everyday practices, such that a priority in the public health policy that serves both the local and national needs is set early on. When public health policy is not an area of high political sensitivity, its economic and social implications would take priority. Hence, good mutual understanding and feasible working protocols should have been established ahead of time, rather than being set on an ad hoc basis.

Building and Improving the Government Think Tank

Experts are valuable resources for governments to make policy recommendations. The Special Advisory Group and the Center for China and Globalization have currently taken up such functions in Hong Kong. They are composed of well-known figures from various domains such as politics, business, and academia to advise on the government's strategic development. Yet, they lack technical expertise in specific areas, e.g., public health. Neither are they capable of advising a fast response to crisis management. In the past few years, only a handful of professors have continually spoken out on public health issues, airing individual opinions without much coordination and consensus of a clear objective. This led to more confusion and only eroded the public trust in both the experts and the government.

To circumvent the problem, expert committees in the field of public health, other key technical fields, including ethical practices, are expected to help the government formulate public health policy, solve technical problems, and promote technological innovation with long- and short-term strategies that align with community priorities. This unified and official expert organization can improve the government's credibility by helping the public to filter and validate relevant information. Authoritative information accompanied by professional interpretation and analysis will help the community have a better understanding of the rationale behind any proposed initiative. In turn, trust in the government can be raised.

Optimization of Information Dissemination

The official information dissemination channels for the government were press conferences, press releases, and the "Together, We Fight the Virus" website. The promotion through these channels could be more aggressive, a weak link at this point. The platform could have been more user-friendly to provide authentic and verified information. It helps combat the social media that usually provoke sensational responses through their reports that often bias readers' views by repeated bombardment with eye-catching headlines. Such an official central platform with credibility would allow citizens to have a better grasp of the facts. During public health crises, conflicting and unsubstantiated reports circulated will only tarnish the effectiveness of any measure to curb health risks. A credible platform will be essential to forge trusted exchanges between the public and the management authority.

Conclusion

From the experience of the epidemics management in previous years, it is evident that measures that align well with local needs and adhere to the principle that integrates all local and external inputs would work best. A proactive, flexible, and adaptive approach to the evolving situation is critical. While science does not provide a more concrete answer in time and polarized opinions spread, a clear message of communication from a trusted authority is of paramount importance. Policies may be subject to frequent minor modifications, sometimes leading to a state of bewilderment among the public. The confusion can be attributed to the following: (a) the policy appears fragmented in handling issues in a piece meal manner, and lack of comprehensive formulation with a coherent strategy would only create more follow-up problems; (b) inadequate transparency of communication between the government and the public regarding policy and amendments disseminated to citizens without providing elaborate scientific bases, implementation details, and rationales; (c) the absence of a coherent and credible platform for experts to express their viewpoints and to engage the community at large through public discussion. Poor comprehension of valid information will only cause misinterpretation, fueling more discontentment on social media instead of bringing about coherent efforts in addressing prioritized concerns.

Considering these issues, the following suggestions are made to enhance the efficacy of scientific decision-making processes, prioritize the objectives, ensure proper dissemination and utilization of scientific data, and implement a policy framework that guarantees policy coherence and durability. It is imperative to (a) adopt an open approach in communicating precise information regarding the virus or agents, its transmission dynamics, and the necessary preventive measures implemented; (b) maintain honesty while discussing the difficulties and constraints associated with the response efforts; (c) establish confidence and credibility among the public through strategic utilization of reliable platforms for consistent messages; (d) use clear and succinct language in all forms of communication materials; (e) include public feedback and input via specific avenues in the decision-making process; and (f) reduce the influence of politics and special interest groups in policy communication, and keep the focus more on facts and an expected outcome with clear priority. After all, only when citizens have a clear understanding of the priority of the government on the general well-being of the community, collective support and tolerance would be effectively harnessed to overcome challenges the society faces.

Appendix 3.1 Summary of Key Changes of Hong Kong COVID-19 Policy

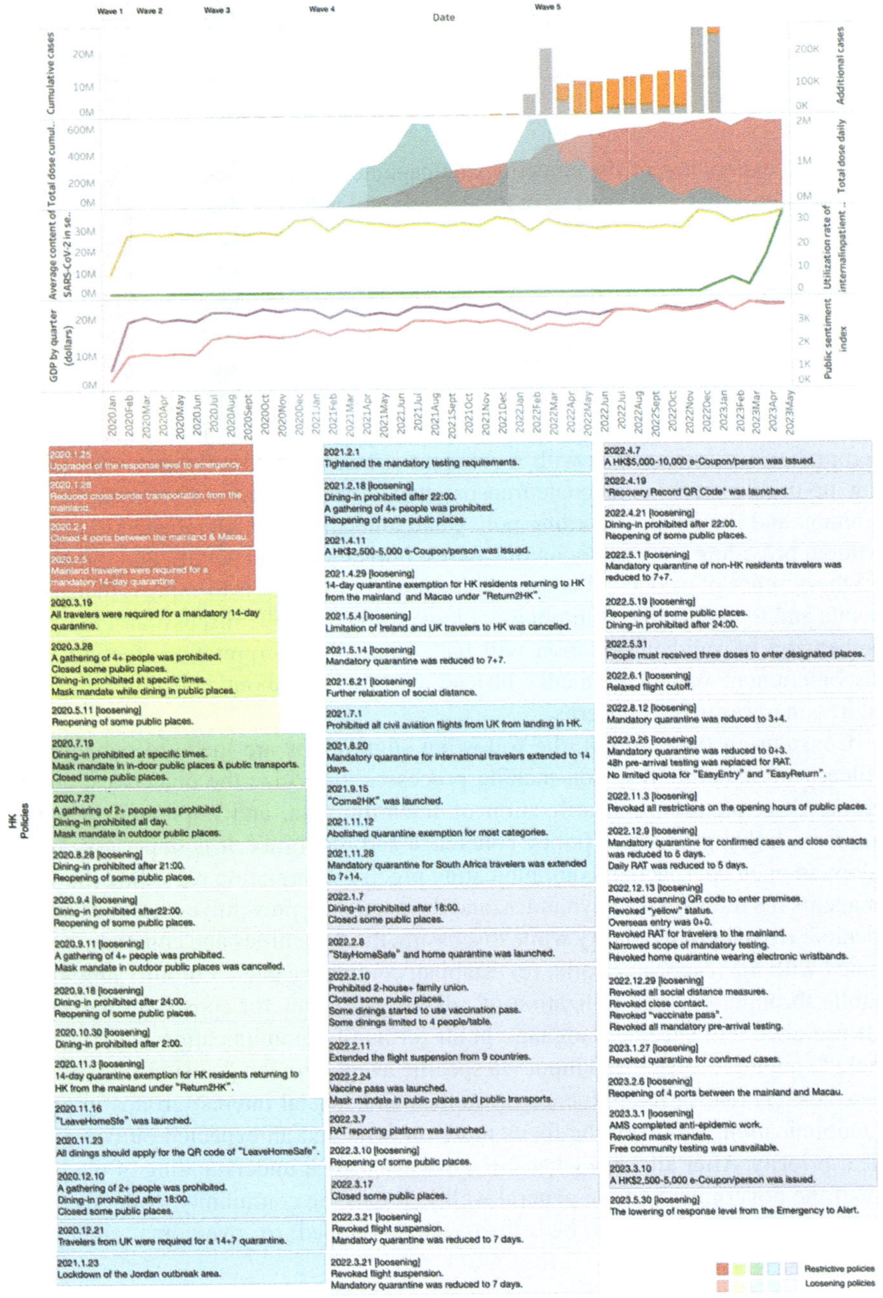

References

1. Worldometer (2023 October 6) Hong Kong COVID - Coronavirus Statistics - Worldometer. WORLD / COUNTRIES / CHINA, HONG KONG SAR. https://www.worldometers.info/coronavirus//country/china-hong-kong-sar/
2. Hui M (2023 January 5) Hong Kong has the world's highest Covid death rate—again. QUARTZ. https://qz.com/hong-kong-has-the-world-s-highest-covid-death-rate-agai-1849947517
3. HK Gov (2020 January 25) No delay in fight against virus: CE. News.Gov.Hk. https://sc.news.gov.hk/TuniS/www.news.gov.hk/chi/2020/01/20200125/20200125_191553_178.html
4. IndexMundi (2023 January 31) Country comparison > population density. IndexMundi. https://www.indexmundi.com/g/r.aspx?v=21000&l=zh
5. BlueNetNews (2023 January 17) Ranking of the world's top 10 ports (latest ranking of international maritime ports in 2022). Shipping China. http://info.shippingchina.com/bluenews/index/detail/id/202148.html
6. GovHK (2020 January 25) Hospital authority activates emergency response level. The Government of the Hong Kong Special Administrative Region Press Releases. https://www.info.gov.hk/gia/general/202001/25/P2020012500689.htm
7. GovHK (2020 January 28) Special arrangements for cross boundary transport. The Government of the Hong Kong Special Administrative Region Press Releases. https://www.info.gov.hk/gia/general/202001/28/P2020012800727.htm
8. GovHK (2020 February 4) Land-based cross-boundary transport services. The Government of the Hong Kong Special Administrative Region Press Releases. https://www.info.gov.hk/gia/general/202002/04/P2020020400718.htm
9. GovHK (2020 February 5) Government to impose mandatory quarantine on people entering Hong Kong from mainland. The Government of the Hong Kong Special Administrative Region Press Releases. https://www.info.gov.hk/gia/general/202002/05/P2020020500793.htm
10. CCTV News (2021 January 4) Hong Kong 2020: facing difficulties amidst changes in the epidemic. Liaison Office of the Central People's Government in the Hong Kong S.A.R. http://www.locpg.gov.cn/jsdt/2021-01/04/c_1210963175.htm
11. GovHK (2020 March 19) Temporary test centres speed up tests for people upon arrival. The Government of the Hong Kong Special Administrative Region Press Releases. https://www.info.gov.hk/gia/general/202003/19/P2020031900664.htm
12. GovHK (2020 March 28) Prevention and control of disease (Prohibition on Group Gathering) regulation. The Government of the Hong Kong Special Administrative Region Press Releases. https://www.info.gov.hk/gia/general/202003/28/P2020032800720.htm
13. Marlow I, Lung N (2020 May 5) Hong Kong inches toward normalcy with move to open schools, bars. Bloomberg. https://www.bloomberg.com/news/articles/2020-05-05/hong-kong-will-relax-virus-restrictions-soon-as-cases-stay-low
14. Tingand V, Cheung E (2020 July 21) Hong Kong third wave: how did City's scariest surge in Covid-19 cases start, and what did authorities miss? South China Morning Post. https://www.scmp.com/news/hong-kong/health-environment/article/3093978/hong-kong-third-wave-how-did-citys-scariest-surge
15. GovHK (2020 July 27) Government further tightens social Distancing measures. The Government of the Hong Kong Special Administrative Region Press Releases. https://www.info.gov.hk/gia/general/202007/27/P2020072700650.htm
16. Yu W (2020 August 4) The mainland aid team has started to design the Hong Kong Shelter Hospital. Guanchezhe. https://www.guancha.cn/politics/2020_08_04_560026.shtml
17. News.gov.hk (2020 November 17) Travel Scheme Bookings to Open. https://www.news.gov.hk/eng/2020/11/20201117/20201117_172601_515.html
18. GovHK (2020 November 3) Government extends social distancing measures under prevention and control of disease ordinance. The Government of the Hong Kong Special

Administrative Region Press Releases. https://www.info.gov.hk/gia/general/202011/03/P2020110300750.htm
19. GovHK (2020 November 23) Over 2 140 people returned to Hong Kong on first day of implementation of Return2hk scheme. The Government of the Hong Kong Special Administrative Region Press Releases. https://www.info.gov.hk/gia/general/202011/23/P2020112300869.htm
20. GovHK (2020 November 16) OGCIO explains access right issues on "LeaveHomeSafe" COVID-19 exposure notification mobile app. The Government of the Hong Kong Special Administrative Region Press Releases. https://www.info.gov.hk/gia/general/202011/16/P2020111600904.htm
21. GovHK (2021 January 23) Update on testing at specified "restricted area" in Jordan (as at 1pm). The Government of the Hong Kong Special Administrative Region Press Releases. https://www.info.gov.hk/gia/general/202101/23/P2021012300507.htm
22. GovHK (2021 June 18) Government announces details of consumption voucher scheme. The Government of the Hong Kong Special Administrative Region Press Releases. https://www.info.gov.hk/gia/general/202106/18/P2021061800842.htm?fontSize=1
23. GovHK (2021 September 14) Government announces arrangements for Come2hk scheme. The Government of the Hong Kong Special Administrative Region Press Releases. https://www.info.gov.hk/gia/general/202109/14/P2021091400592.htm?fontSize=1
24. GovHK (2021 November 12). CHP provides update on latest investigations on two COVID-19 imported cases. The Government of the Hong Kong Special Administrative Region Press Releases. https://www.info.gov.hk/gia/general/202111/12.htm
25. Centre for Health Protection of the Department of Health (2022 March 5) The 5th Wave Data of Coronavirus Disease in 2019. Together, We Fight the Virus! https://www.coronavirus.gov.hk/pdf/5th_wave_statistics/5th_wave_statistics_20220305.pdf
26. GovHK (2022 January 5) Government tightens social distancing measures in view of latest epidemic situation. The Government of the Hong Kong Special Administrative Region Press Releases. https://www.info.gov.hk/gia/general/202201/05/P2022010500859.htm
27. News.gov.hk (2022 February 20) Home Quarantine Scheme Launched. https://www.news.gov.hk/eng/2022/02/20220208/20220208_180204_696.html?type=category&name=covid19&tl=t
28. News.gov.hk (2022 April 19) New QR Code for Recovered Residents. https://www.news.gov.hk/eng/2022/04/20220419/20220419_175727_552.html?type=category&name=covid19&tl=t
29. News.gov.hk (2022) RAT Reporting Platform Launched. https://www.news.gov.hk/eng/2022/03/20220307/20220307_180254_157.html?type=category&name=covid19&tl=t
30. GovHK. Consumer Voucher Scheme Website. https://www.consumptionvoucher.gov.hk/tc/#
31. News.gov.hk (2022 May 30) Vaccine Pass Reminder Issued. https://www.news.gov.hk/eng/2022/05/20220530/20220530_174815_614.html?type=category&name=covid19&tl=t
32. News.gov.hk (2022 February 8) Home Quarantine Scheme Launched. https://www.news.gov.hk/eng/2022/02/20220208/20220208_180204_696.html?type=category&name=covid19&tl=t
33. GovHK (2022 April 19) Government announces arrangement to download recovery record QR code and implementation details of second stage of vaccine pass. The Government of the Hong Kong Special Administrative Region Press Releases. https://www.info.gov.hk/gia/general/202204/19/P2022041900477.htm?fontSize=1
34. Qianjinhuyaofang (2022 February 17) Take Orders to Go out! The Central Government Sent an Epidemiological Expert Group to Hong Kong to Fight the Epidemic! Sohu. https://www.sohu.com/a/523496660_121118850
35. Ye L (2022 February 21) The second batch of mainland experts went to Hong Kong for support, and the critical illness expert group participated in the treatment. Renminzhengxiewang. https://www.rmzxw.com.cn/c/2022-02-21/3053121.shtml

36. Shenzhentequbao (2022 March 14) The Guangdong provincial group sent 75 mainland medical teams to Hong Kong to carry out work. Smartjx. http://gdp.smartjx.com/yw/2022/0314/2530834.html
37. GovHK (2022 September 30) Government relaxes certain social distancing measures. The Government of the Hong Kong Special Administrative Region Press Releases. https://www.info.gov.hk/gia/general/202209/30/P2022093000818.htm
38. GovHK (2023 February 3) Full resumption of normal travel between Hong Kong and Mainland. The Government of the Hong Kong Special Administrative Region Press Releases. https://www.info.gov.hk/gia/general/202302/03/P2023020300649.htm
39. GovHK (2023 February 28) Government lifts all mandatory mask-wearing requirements. The Government of the Hong Kong Special Administrative Region Press Releases. https://www.info.gov.hk/gia/general/202302/28/P2023022800677.htm
40. GovHK (2023 May 30) Government lowers response level in relation to COVID-19 epidemic to alert level. The Government of the Hong Kong Special Administrative Region Press Releases. https://www.info.gov.hk/gia/general/202305/30/P2023053000552.htm?fontSize=1
41. Oriental Daily News (2022) Gov't admits difficulties in clearing out the epidemic press conference was suspended. Available at: https://orientaldaily.on.cc/content/%E8%A6%81%E8%81%9E%E6%B8%AF%E8%81%9E/odn-20220925-0925_00176_022/%E6%94%BF%E5%BA%9C%E8%AA%8D%E6%B8%85%E9%9B%B6%E6%9C%89%E5%9B%B0%E9%9B%A3%2D%2D%E7%96%AB%E6%83%85%E8%A8%98%E8%80%85%-E6%9C%83%E5%81%9C%E8%BE%A6 (accessed 25 Sept 2022)
42. Chan HY, Chen A, Ma W, Sze NN, Liu X (2021) COVID-19, community response, public policy, and travel patterns: a tale of Hong Kong. Transp Policy 106:173–184
43. Roziqin A, Mas' udi SY, Sihidi IT (2021) An analysis of Indonesian government policies against COVID-19. Public Adm Policy 24(1):92–107
44. Youlieryoumianer (2021 November 8) Foreign Organizations Actually Ask Hong Kong to "Coexist with the Virus"! ? Guanchazhe. https://user.guancha.cn/main/content?id=629252&page=0
45. Liu Y (2020 July 29) Hong Kong's GDP fell by 9% in the second quarter, and the signs of economic recovery were suppressed by the epidemic. Caixin. https://economy.caixin.com/2020-07-29/101586160.html
46. Zhou W (2020 January 29) Hong Kong's economy will shrink by 6.1% in 2020 and decline for two consecutive years. Caixin. https://economy.caixin.com/2021-01-29/101657980.html
47. Bloomberg (2022 June 29) Hong Kong mulls easing quarantine rules to 5 days, with 2 more in home isolation, amid China's shift in Covid-19 policy. South China Morning Post. https://www.scmp.com/yp/discover/news/hong-kong/article/3183453/hong-kong-mulls-easing-quarantine-rules-5-days-2-more
48. Rufeng (2022 July 27) Mainland Customs Clearance Is Indefinite, Hong Kong or International Customs Clearance First! It Is Planned to Shorten the Quarantine and Implement the "4+3" Arrangement!... Guangtandawanwang. https://www.dww11.com/article/207671.html
49. Zhu L (2020 September 30) Interview with Lu Kaijun, CFO of Prenetics, the largest private Laboratory in Hong Kong: there is still room for the government to consider subsidies for nucleic acid testing in Hong Kong. Sina. https://finance.sina.com.cn/tech/2020-09-30/doc-iivhvpwy9617076.shtml
50. Chen Y (2021 February 2) Hong Kong SAR government further strengthens mandatory building inspection. CCTV.. https://m.news.cctv.com/2021/02/02/ARTI1obsKOvHaz89eJrVbLs7210202.shtml
51. GovHK (2022 August 23) Government largely extends social distancing measures. The Government of the Hong Kong Special Administrative Region Press Releases. https://www.info.gov.hk/gia/general/202208/23/P2022082300649.htm?fontSize=1
52. Han X (2020 February 12) Hong Kong under the epidemic situation: from Kindergarten to Middle School, "Suspension of Classes and Non-Stop Learning." HxNews. http://www.hxnews.com/news/jypd/jyzx/202002/12/1859142.shtml

53. News.gov.hk (2022 April 11) Daily Testing for Teachers, Pupils. https://www.news.gov.hk/eng/2022/04/20220411/20220411_120620_439.html
54. Ho GW, Leung SF, Mak YW, Wong KH, Kwok RK, Loke AY (2025) Social distancing and mental health two years into a global pandemic: a sequential exploratory mixed-methods investigation in Hong Kong. BMJ Open 15(3):e092160
55. News.gov.hk (2022 March 20) Temporary unemployment support scheme accepts applications next Wednesday. https://sc.news.gov.hk/TuniS/www.news.gov.hk/chi/2022/03/20220318/20220318_121258_459.html
56. Liang S, Chang Y, Wu X (2022 May 20) Affected by the 5th wave, the unemployment rate in Hong Kong soared to 5.4%, and 200,000 people were unemployed. TaKungPao. https://www.takungpao.com/news/232109/2022/0520/721594.html
57. Office of the Financial Secretary of the Hong Kong Special Administrative Region Government Office of the Government Economic Adviser (2022 December 15) Economic Report for the Third Quarter of 2022. https://www.hkeconomy.gov.hk/tc/pdf/er_c_22q3.pdf
58. Zheng Y (2020 March 17) Cultural differences in masks Hong Kong people call it civic duty. Europe and the United States advocate leaving it to medical care. Hong Kong Economic Journal. https://inews.hket.com/article/2592766/口罩文化差異%20港人稱公民義務%20歐美主張留給醫護
59. Kwan RYC, Lee PH, Cheung DSK, Lam SC (2021) Face mask wearing behaviors, depressive symptoms, and health beliefs among older people during the COVID-19 pandemic. Front Med 8:590936
60. Zhu J (2022 March 14) Hong Kong Media: some Hong Kong citizens and foreigners ignored the gathering restriction order and gathered on the beach without wearing masks. Sina. https://news.sina.com.cn/c/2022-03-14/doc-imcwipih8429975.shtml?cre=tianyi&mod=pcpager_news&loc=12&r=0&rfunc=39&tj=cxvertical_pc_pager_news&tr=174&wm=#!/ranking/1?hasPlayedTime=5.110536
61. ExploreHK (2022 February 21) Strict law enforcement! More than 600 people were punished for violating the mask order and the gathering restriction order on the 2 Nd! Lin Zheng: No More Kindness. Sohu. https://www.sohu.com/a/524429694_121124403
62. Chen L (2020 February 24) Mask Politics in Hong Kong. ChinaTimes. https://www.chinatimes.com/newspapers/20200224000861-260502?chdtv
63. Zhu K (2020 February 20) Zhu Kaidi, an opposition member of Hong Kong, vigorously promoted cloth masks. Netizen: a little knowledge kills people (Hong Kong News Network). Guanchazhe. https://user.guancha.cn/main/content?id=246625
64. Takungpao (2020 January 31) Prisoners in Luohu Volunteered to Work Overtime to Make Masks. TaKungPao. https://www.takungpao.com/news/232109/2020/0131/410727.html
65. Jiao J (2020 February 13) Hong Kong anti-epidemic record: it is difficult to buy masks at high prices, but should the price of masks be regulated? QQNews. https://news.qq.com/rain/a/20200213A0OMPN00
66. The Government of the Hong Kong Special Administrative Region. Together, We Fight the Virus! https://www.coronavirus.gov.hk/sim/
67. Liu M (2022 February 22) Qi Xin from all walks of life in Hong Kong supports the fight against epidemic. Renminwang. http://hm.people.com.cn/n1/2022/0222/c42272-32357180.html
68. Xinhuawang (2022 March 3) All walks of life in Hong Kong have taken active actions to fight the epidemic. Chinese Government Network. https://www.gov.cn/xinwen/2022-03/03/content_5676840.htm
69. The Efficacy and Privacy Protection of Hong Kong's Imitation of China's "Health Code" Are Questioned (2022 July 12) Deutsche Welle Chinese Network. https://www.dw.com/zh/香港拟仿中国健康码-功效及隐私保护遭质疑/a-62438501
70. Zheng M (2021 December 3) Hong Kong government pushes health code to register political parties in real name, worrying about citizens' privacy. SoundofHope. https://www.soundofhope.org/post/571433

71. Yang C (2020 April 24) Hong Kong privacy department: a series of responses and guidelines on privacy issues caused by the COVID-19 epidemic. Safety Internal Reference. https://www.secrss.com/articles/18959
72. Wen S (2022 July 9) How to Implement Vaccine Pass in Hong Kong. Caixin. https://www.caixin.com/2022-07-09/101910543.html
73. Wen S (2022 September 8) The Hong Kong government lowered the age of vaccine pass to 5 years old. Children Need Injections to Eat in Class. Caixin. https://www.caixin.com/2022-09-08/101937605.html
74. Centre for Health Protection (2025 July 2) Education & Media Resources. COVID-19 Vaccination Programme. https://www.chp.gov.hk/sc/features/106990.html
75. Shou P (2021 August 4) Joey Yung Sang "Da Da Da City" to promote vaccination. Toutiao. https://www.toutiao.com/article/6992545212895592973/?upstream_biz=doubao&source=m_redirect
76. Zhang X (2022 December 26) How can Hong Kong, where the number of new cases has exceeded 20,000 for three consecutive days, prevent a run on medical resources? Jiemian. https://m.jiemian.com/article/8634962.html
77. Wen S (2022 March 8) Why are private hospitals in Hong Kong "absent" in the fight against the epidemic? Caixin. https://www.caixin.com/2022-03-08/101852464.html
78. Laoziliao S (2022) COVID-19: why are Hong Kong hospitals "full"? LaoZiLiao. https://www.laoziliao.net/health/info/56405070
79. Stennett J, Hou R, Traverson L, Ridde V, Zinszer K, Chabrol F (2022) Lessons learned from the resilience of Chinese hospitals to the COVID-19 pandemic: scoping review. JMIRx med 3(2):e31272. https://doi.org/10.2196/31272
80. Xiangganghao (2022 February 18) Central Hand! What Is the Most Urgent Need to Solve the "War Epidemic" in Hong Kong Now? ThePaper. https://www.thepaper.cn/newsDetail_forward_16750044
81. Li, K. (2022 February 19) [Exclusive] Hong Kong Hospital Authority responded with three major measures to deal with the shortage of medical resources under the surge in diagnosis. Jiemian. https://www.jiemian.com/article/7120424.html
82. Li K (2022 February 23) What is the imbalance between public and private medical resources in Hong Kong behind the implementation of universal testing? Jiemian. https://www.jiemian.com/article/7130822.html
83. Tian F (2022 March 11) Tian Feilong: the difficult time of Hong Kong's anti-epidemic and the way out. Aisixiang https://www.aisixiang.com/data/131951.html
84. China Resources Macroeconomic and Policy Research Institute (2022 March 1) Some thoughts and suggestions on the anti-epidemic work carried out by the SAR Government. Bauhinia Megazine. https://zijing.com.cn/article/2022-03/01/content_948283915094073344.html
85. GovHK (2022 August 11) Mid-year population for 2022. The Government of the Hong Kong Special Administrative Region Press Releases. https://www.info.gov.hk/gia/general/202208/11/P2022081100393.htm
86. Hong Kong: EU Citizens Head for the Exit (2022 February 28) Finews.Asia. https://www.finews.asia/finance/36399-hong-kong-eu-citizens-head-for-the-exit
87. Zhang X (2023 February 20) What are the reasons for the negative growth of Hong Kong's population for three consecutive years and the loss of 187,000 people? Jiemian. https://www.jiemian.com/article/8928635.html
88. Li M (2021 June 7) More and more "interested" Hong Kong vaccine. Tongshuo. https://mp.weixin.qq.com/s?__biz=MzIyNTA5Mzg3Nw==&mid=2651599832&idx=1&sn=fb54907834995d0a49a1958abef97f30&chksm=f21bb95d265a169d606fe962a99fa67b073408209f82435b647f2f99db04fc38e08563e25800#rd
89. News.gov.hk (2021 October 20) 8.93m Vaccine Jabs Given. Health & Community. https://www.news.gov.hk/eng/2021/10/20211020/20211020_213756_778.html
90. McMenamin ME, Nealon J, Lin Y, Wong JY, Cheung JK, Lau EH et al (2022) Vaccine effectiveness of one, two, and three doses of BNT162b2 and CoronaVac against COVID-19 in Hong Kong: a population-based observational study. Lancet Infect Dis 22(10):1435–1443

91. Ranzani OT, Hitchings MD, de Melo RL, de França GV, Fernandes CDFR, Lind ML et al (2022) Effectiveness of an inactivated Covid-19 vaccine with homologous and heterologous boosters against the Omicron (B. 1.1. 529) variant. medRxiv, 2022–03
92. Lundin M, Öberg P (2014) Expert knowledge use and deliberation in local policy making. Policy Sci 47:25–49. https://doi.org/10.1007/s11077-013-9182-1
93. Lavazza A, Farina M (2020) The role of experts in the Covid-19 pandemic and the limits of their epistemic authority in democracy. Front Public Health 8:356
94. Flinders M, Dimova G (2020) Bringing in the experts: blame deflection and the COVID-19 crisis. British Policy and Politics at LSE
95. Armstrong R, Waters E, Dobbins M, Anderson L, Moore L, Petticrew M et al (2013) Knowledge translation strategies to improve the use of evidence in public health decision making in local government: intervention design and implementation plan. Implement Sci 8:1–10
96. Gao Y, Lau JTF, Zhang Q, Ming WK, Shen F, Huang YHC, Jiang LC (2025) The impact of expert opinions in COVID-19 News coverage on vaccination during the fifth outbreak in Hong Kong: a time series analysis. Health Commun 40(2):305–320
97. Yuen VWH (2022) Political attitudes and efficacy of health expert communication on the support for COVID-19 vaccination program: findings from a survey in Hong Kong. Vaccine 40(15):2282–2291
98. Yuen VW (2023) The efficacy of health experts' communication in inducing support for COVID-19 measures and effect on trustworthiness: a survey in Hong Kong. Soc Sci Med 317:115602
99. Wang Y, Wang X (2022 February 17) Hong Kong is facing a new epidemic crisis: it cannot coexist with the virus, nor can it be "cleared". NYTimes. https://web.archive.org/web/20220503093410/https://cn.nytimes.com/china/20220217/hong-kong-covid-omicron-wave/zh-hant/
100. The Information Office of the State Council (2020) Fighting COVID-19: China in Action. Available at: http://www.scio.gov.cn/ztk/dtzt/42313/43142/index.htm (accessed 7 June 2020)
101. The Information Office of the State Council (2022) Transcript of the press conference held by the Information Office of the State Council on 29 April 2022. Available at: https://wsjk.ln.gov.cn/wsjk/ztzl/qlzhxxgzbdfyyqfkgz/fkdt/DDA64C5BFFF14209AFA01BF1702A7BE9/index.shtml (accessed 29 Apr 2022)
102. The Xinhua News Agency (2023) China's war on "epidemic" enters a new stage - China optimizes epidemic prevention and control measures according to the times and the situation. http://www.nhc.gov.cn/xcs/yqfkdt/202301/53e4bdbadd0f4a3384c0af7e38f16427.shtml (accessed 8 Jan 2023)
103. Shao S (2022 February 19) Why did the the 5th wave outbreak occur in Hong Kong, which also implemented the "zero clearing" policy? Kunlunce. http://www.kunlunce.com/ssjj/fl11/2022-02-19/158984.html
104. China News Service (2022) Overview: all-round support from the central authorities helps Hong Kong stabilize and control the epidemic. Available at: http://www.chinanews.com.cn/ga/2022/04-11/9726029.shtml (accessed 11 Apr 2022)
105. Architectural Service Department (2022) Community Isolation and Therapeutic Facilities at Penny's Bay (Phases 5 and 6), Kai Tak, Tam Mei, Yuen Long, Hung Shui Kiu, Ma Shi Road, Fanling, Hong Kong-Zhuhai-Macao Bridge Hong Kong Boundary Crossing Facilities Artificial Island, San Tin, and Tsing Yi. Available at: (accessed 2022)
106. National Health Commission of the People's Republic of China (2022) On further optimizing the implementation of preventive and control measures for the new crown pneumonia epidemic. Available at: http://www.nhc.gov.cn/xcs/zhengcwj/202212/8278e7a7aee34e5bb378f0e0fc94e0f0.shtml (accessed 7 Dec 2022)
107. Ruo R (2022 October 4) Review think tank: what is the way to link up the epidemic prevention policies between the mainland and Hong Kong? Zhongpingshe. https://mp.weixin.qq.com/s?__biz=MzA3MTAyMzAxMg==&mid=2650538800&idx=1&sn=f9562f17766dc

05d9d6ac82901c24f3e&chksm=873b43fbb04ccaed502b597da4dcd4faa19e7fb2967d727d30eae8b15a363b0a3af2e3626e68#rd
108. Ma X, Wang F, Hu N (2022 February 19) Starry night to help fight the epidemic-the mainland fully supports Hong Kong's first-line anti-epidemic scan. Baidu. http://mbd.baidu.com/newspage/data/landingsuper?third=baijiahao&baijiahao_id=1725120483275951108
109. Greer SL, Fonseca EM, Raj M, Willison CE (2022) Institutions and the politics of agency in COVID-19 response: federalism, executive power, and public health policy in Brazil, India, and the US. J Soc Policy:1–19
110. Falkenbach M, Greer SL (2021) Denial and distraction: how the populist radical right responds to COVID-19 comment on "a scoping review of PRR parties' influence on welfare policy and its implication for population health in Europe". Int J Health Policy Manag 10(9):578
111. Kavakli KC (2020) Did populist leaders respond to the COVID-19 pandemic more slowly? Evidence from a global sample. Unpublished manuscript, Covid crisis Lab, Bocconi University, Milano
112. Cheung AB (2004) Strong executive, weak policy capacity: the changing environment of policymaking in Hong Kong. Asian J Political Sci 12(1):1–30
113. Kwong KM, Yew CP (2013) Hong Kong politics: weak leadership, frustrated society and prospect. East Asian Policy 5(01):55–65
114. Hui VTB (2020) Beijing's hard and soft repression in Hong Kong. Orbis 64(2):289–311
115. Mike Wardle, Professor Michael Mainelli (2022) The Global Financial Centres Index 32. Available at: https://www.longfinance.net/media/documents/GFCI_32_Report_2022.09.22_v1.0_.pdf (accessed Sept 2022)
116. Lin Z (2022 March 30) Carrie Lam Cheng Yuet-Ngor: inviting representatives of the International Chamber of Commerce to a Meeting. ChinaTimes. https://www.chinatimes.com/cn/realtimenews/20220330002323-260409?chdtv
117. China News Service (2023) Overview: Hong Kong strengthens "connectivity" after quarantine, consolidates position as international financial center. Available at: https://m.chinanews.com/wap/detail/chs/zw/10032260.shtml (accessed 27 June 2023)
118. Cheung PHH, Chan CP, Jin DY (2022) Lessons learned from the fifth wave of COVID-19 in Hong Kong in early 2022. Emerg Microbes Infect 11(1):1072–1078
119. Chow L (2021) Care homes and COVID-19 in Hong Kong: how the lessons from SARS were used to good effect. Age Ageing 50(1):21–24
120. Xing G, Liu X (2020 April 18) Fighting "epidemic": public health safety and planning response. Nanyue Planning. https://mp.weixin.qq.com/s?__biz=MzU2MzAwODQ4Nw==&mid=2247492734&idx=1&sn=964ff1658e88796b2c1ce5d98ab32670&chksm=fd8aa9836c7711b5df725dd496ea058ce978fc12b0a3c6b103964f97af038ea8e500e8f7cb1d#rd
121. Wang X (2020 March 11) From over 50% of medical staff infected with SARS to zero infection in COVID-19, what has Hong Kong done? QQNews. https://news.qq.com/rain/a/20200311A0ORGL00
122. Zhao L (2008 February 7) Hong Kong Way of Public Epidemic Prevention. Sohu. http://business.sohu.com/20080207/n255098455.shtml

Health Bureau. Establish a community primary medical system. Basic Medical Health Blueprint. https://www.primaryhealthcare.gov.hk/bp/sc/supplementary-documents/develop-a-community-based-system/

Overview of COVID-19 Situation, Challenges, and Countermeasures Taken in Japan

4

Junko Kiriya and Akira Shibanuma

Abstract

Japan faced various challenges during the COVID-19 pandemic since January 2020. The Japanese government led all the responses to the pandemic in collaboration with local governments with the help of the advisory group consisting of experts. Though Japan did not introduce strong legal measures such as national lockdown, the numbers of infected cases and deaths were smaller compared to other countries at the same economic level. The following shortcomings became apparent during the pandemic: insufficient capacity of testing and medical care, delay in data collection and data sharing, and lack of legal framework for ethical issues such as discrimination. The first two affected the government's risk assessment and, consequently, health communication with the public. This may have caused public's distrust of advice from the government and the advisory group. *Jishuku* (self-restraint) by the society may have contributed to infection control without strong legal measures. *Jishuku*, however, has also led to discrimination against COVID-19 patients, their families, and healthcare providers because *jishuku* originates from social pressure, and getting infected was stigmatized as a breach of social rules. Regulation of these prejudicial acts was difficult as Japan lacks a legal basis to protect human rights. For future pandemics, capacity building and legal reform for ethical issues are required. At the same time, evaluation of all the countermeasures introduced during the pandemic will provide lessons for planning new policies in case of next health-related national emergencies.

J. Kiriya · A. Shibanuma (✉)
Department of Community and Global Health, Graduate School of Medicine, The University of Tokyo, Tokyo, Japan
e-mail: shibanuma@m.u-tokyo.ac.jp

L. P. Marquez et al. (eds.), *COVID-19 in the Pacific Rim and Beyond*,
Collaborative Bioethics 6, https://doi.org/10.1007/978-3-032-21218-4_4

Keywords

COVID-19 · Pandemic preparedness · Health communication · Public health infrastructure · Japan

Introduction

In January 2020, Japan first confronted a confirmed coronavirus disease 2019 (COVID-19) case [1]. Soon after that, a large cruise ship, Diamond Princess, stopped by Yokohama port with hundreds of passengers on board and susceptible to Severe Acute Respiratory Syndrome Coronavirus 2 (SARS-CoV-2), the cause of COVID-19, as one old man who got off the ship at Hong Kong had tested positive for the virus [2]. This case gathered international attention as it could have been a good or bad example of infection control.

Starting with this difficult case, Japan has gone through about 3 years of pandemic and restrictions. Well-established public health organizations have been known as a key player of achieving universal health coverage in Japan [3]. Regarding infectious disease control, Japan has developed a decentralized system consisting of public health centers and laboratories at the prefectural and city levels. These organizations provide surveillance, early warning, and contact tracing for key infectious diseases, such as tuberculosis and influenza [4, 5]. As Japan was not heavily affected by Severe Acute Respiratory Syndrome (SARS) and Middle East Respiratory Syndrome (MERS), unlike other East Asian countries, these organizations had limited experience in handling emerging infectious diseases [6]. This chapter aims to overview the main events during the pandemic and discuss the responses to the challenges by the Japanese society.

Summary of COVID-19 Situation in Japan and Key Events

Figure 4.1 shows the number of COVID-19 confirmed cases and deaths over the pandemic period with related key events in Japan (the data are publicly available on the website of the Ministry of Health, Labour, and Welfare [MHLW]). The state of emergency was declared four times between April 2020 and September 2021 [7]. In the early stage of the pandemic, the number of deaths did not seem to correspond with that of confirmed cases, whereas in the latter half of the pandemic, they mostly overlapped. This could be because the Japanese healthcare system fell short of case detection due to the insufficient testing capacity in the early stage of the pandemic [6, 8]. There were several spikes and a sharp increase in the number of cases and deaths in 2022 compared to the previous years. This surge may be partially attributed to the fact that major regulations were lifted after 2 years of the pandemic and people were going back to the pre-COVID lifestyle, as seen in the most governmental responses taking place in the first 2 years (2020 and 2021). Mass vaccination and closing the Tokyo Olympic Games may have led people to be less cautious about COVID-19.

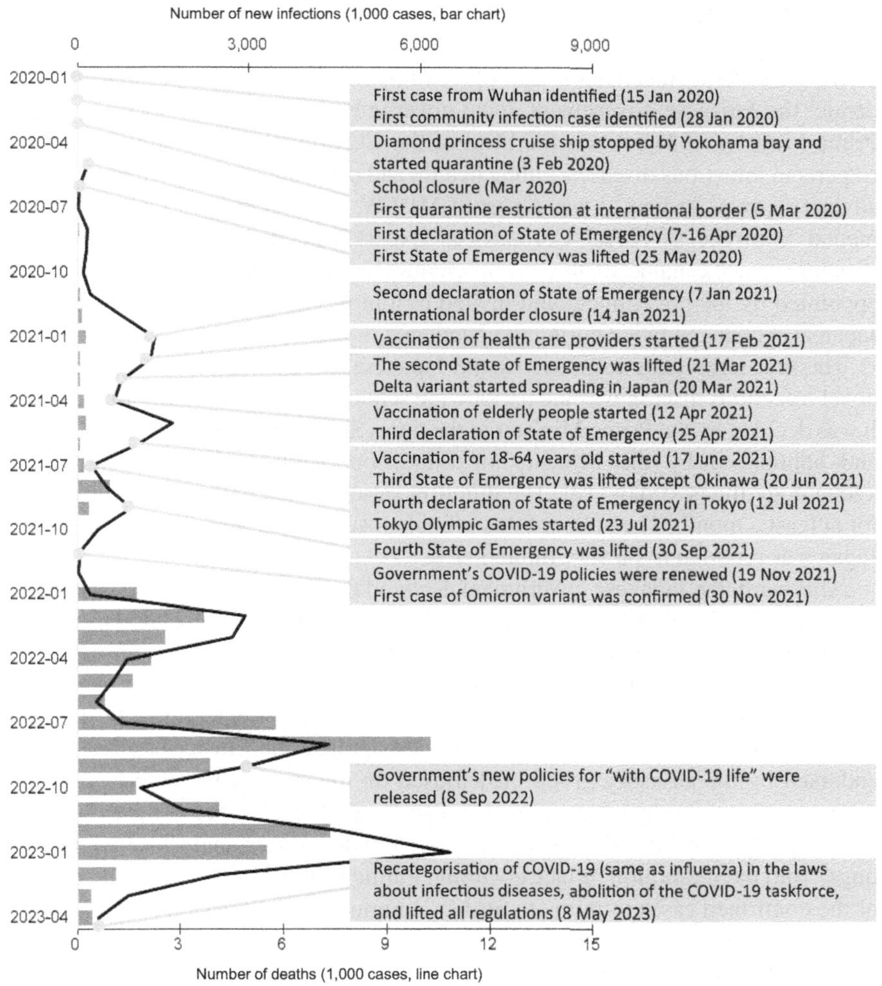

Fig. 4.1 Numbers of COVID-19 confirmed cases and deaths in Japan from January 2020 to April 2023. (*The nature and the quality of the data changed after May 8, 2023, due to the alteration in the data collection methods. The figure shows the data up to the end of April 2023, before the recategorization of COVID-19 in laws)

COVID-19 was recategorized into the same level as influenza in the laws related to infectious diseases on May 8, 2023 [9]. Based on this recategorization, data collection methods of COVID-19 cases switched from population-based surveillance to sentinel site surveillance [10], and the number of COVID-19 deaths was counted through vital statistics, in which COVID-19 deaths were identified based on the cause of death that a doctor diagnosed [11].

Measures Taken in Japan in Response to the Pandemic

There are several stakeholders that played important roles in response to the pandemic: the Japanese government, local governments, and the advisory group. The Prime Minister of Japan was the head of the COVID-19 governmental taskforce. Experts in infectious diseases, public health, and other areas were invited to join the advisory group. At first, it was a part of MHLW, but was merged into the governmental taskforce as Novel Coronavirus Expert Meeting group afterwards. The structure and the name of this group changed several times. The members were appointed by the government and invited to join the group. The Prime Minister of Japan also changed twice during the pandemic.

The first large-scale response by the Japanese government to the spread of the novel infectious disease was the nationwide school closure from early March 2020. It was decided by the Prime Minister at the time (Shinzo Abe) on February 27, 2020, and announced by the Ministry of Education, Culture, Sports, Science and Technology the next day [12]. The duration varied across prefectures, but it lasted for at least a month. The effect of this school closure on the spread of COVID-19 in Japan was reported to be very limited [13].

Afterward, the Japanese government declared the State of Emergency starting on April 7, 2020. As there was no legal basis to implement a strong lockdown in Japan, it was a request-based control over human mobility, unlike in other countries where strict lockdowns were introduced. Still, these non-compulsory measures were effective enough to reduce human mobility in the first state of emergency [14]. Public health centers and surveillance laboratories applied the cluster-based approach endorsed by the advisory group [15]. Based on this approach, several preventive measures were introduced at national and prefectural levels. One of the examples is the campaign to avoid '3Cs (Closed spaces, Crowded spaces, and Close-contact settings)' and the promotion of face covering. Public health centers traced the contacts of the confirmed cases to mitigate further community transmission [16]. They also launched health promotion programs for behavioral change (such as avoiding 3Cs) while operating testing centers and telephone-based consultation centers for triaging suspected cases. Public health surveillance centers provided polymerase chain reaction (PCR) testing that supported hospital- and community-based COVID-19 testing, and reported newly confirmed cases daily at the prefectural level. With the limited testing capacity, the Japanese government tried to minimize the infection risks among the population by active epidemiological investigations and border control along with the health promotion programs.

COVID-19 vaccines were introduced first among healthcare providers in February 2021, followed by prioritized vaccination among the elderly people in April, a mass vaccination campaign by the Ministry of Defense in May, and finally vaccination at general workplaces in June [7]. Vaccination was not compulsory, but was encouraged by the vaccine and testing package campaign. Although the roll-out of the COVID-19 vaccination in Japan was delayed [17] and vaccination was not mandated, its coverage (people with a complete initial protocol) became higher, by the end of October 2021, than those countries where vaccination was compulsory such as Italy [18]. As the vaccine coverage increased, most prefectures started to decrease active epidemiological investigations. They also had to limit the investigations due to work overload in public health centers as the number of newly infected cases showed a rapid increase in August 2021 [19].

In March 2022, when the majority (over 80%) of its population was fully vaccinated (three doses), Japan re-opened its borders to international visitors for non-touristic purposes [20]. Pre-travel testing was exempted from September 2022 for those who were fully vaccinated and the borders were re-opened to all international travelers including tourists in October 2022 [21, 22].

Following the deregulation, Japan saw a rapid increase in the numbers of COVID-19 cases and deaths in 2022. However, despite the non-strict measures, the death rate of COVID-19 in Japan (57.7 per 100,000 population) overall remained lower, as of March 16, 2023, than those countries where strong measures were implemented such as the USA, France, and Germany (341.1, 254.7, and 203.2 per 100,000 population, respectively) [23].

Issues the Japanese Society Faced During the Pandemic

Insufficient Capacity for Testing, Public Health Service, and Medical Care

While the cluster-based approach may be one of the successful factors in preventing the surge of infections in the early stage of the pandemic in Japan, it also placed an excessive workload on public health workers [24, 25]. The number of public health centers in Japan had decreased by half over the last 30 years [26]. Public health centers took advantage of their prior experience of infectious disease control. However, this approach required intensive input from public health workers as the number of infections dramatically increased. They received officers from other municipal departments and part-time workers, which might not lead to sustainable supply of workforce [27, 28].

As for the treatment of COVID-19 patients, Japan faced crucial issues. Japan seemed to have a sufficient number of hospital beds to face the pandemic [29]. The number of hospital beds in Japan was 12.6 per 1,000 population, and this was higher than the Organisation for Economic Co-operation and Development (OECD) average of 4.3 per 1,000 population in 2021. Regarding hospital beds for adult acute intensive care including COVID-19 in Japan, the number was a little lower than the OECD average (Japan: 14.4 vs. OECD: 16.9 per 100,000 population) [30]. However, Japan's medical system nearly collapsed in the midst of the pandemic [31]. This crisis may be attributed to several bottlenecks in Japan's healthcare system. Firstly, the proportion of hospital beds allocated to COVID-19 was small such as 4.2% for general COVID-19 cases and 0.6% for severe cases [29]. Behind this fact was the transition of function of hospital beds from dealing with acute to chronic diseases [32]. With its aging population, the Japanese medical system had shifted the allocation of hospital beds, forecasting a decrease in the resource-demanding medical care.

Another issue is the small number of clinicians for the population. In 2021, the number of clinicians was 2.6 per 1,000 population, whereas the OECD average was 3.7 per 1,000 [30]. The shortage of clinicians was also an issue for vaccination as this kind of medical practice is strictly limited to medical doctors and nurses by law. When human resource shortage became crucial for vaccination, MHLW made an exception that this invasive practice by dentists, paramedics, and clinical laboratory technicians would not be illegal without an amendment to the current medical laws [33, 34]. Whereas, many other countries adopted emergency legislation to reform the health systems for improving the capacity in the health workforce [35].

The third is that the majority of healthcare facilities in Japan are private and primary/secondary hospitals. The government had limited legal basis for ordering these hospitals to accept COVID-19 patients [36]. Taking care of COVID-19 patients is resource- and cost-intensive. Hospitals that accepted those patients were not financially well supported. Although local public health centers were supposed to coordinate with hospitals for the hospitalization of COVID-19 patients under the law applied to various infectious diseases, they faced excessive workloads for coordination during the pandemic. Moreover, good relationship between hospitals or municipalities was not established enough to send patients to each other or adequately allocate them to the right place in case of shortage in medical care capacity [37].

Data Collection

The National Epidemiological Surveillance of Infectious Diseases (NESID) was established in Japan in 1981, and the diseases included in this surveillance extended along with newly introduced laws to undefined diseases such as COVID-19. Healthcare providers must report all confirmed cases to their local public health center. The information on the reported cases is eventually communicated to MHLW. As the conventional case reporting process was not planned for reporting on a daily basis, it was overwhelmed with the dramatically increasing number of newly confirmed COVID-19 cases. MHLW, therefore, asked local governments to report the number of cases directly back to them. However, there were often delays and errors in the reports. As Mr. Katsunobu Kato, then Minister of Health, Labour, and Welfare, pointed out, the biggest challenge in Japan during the pandemic was digital transformation (DX), since the country lacked digital infrastructure to share the data on the numbers of cases, deaths, beds, and in-hospital patients [38]. This issue in the surveillance and the communication between the Japanese government and local governments made it difficult for the national COVID-19 cluster taskforce to keep up-to-date with the COVID-19 situation in Japan for risk assessment and provide timely advice to the central government and the public. Therefore, a new database was set up to collect COVID-19-related data quickly on a daily basis [39].

Whereas, it was not clear between the Novel Coronavirus Meeting group and the central government who should play what roles. The Expert Meeting group was initially expected to analyze the data and understand the situation regarding COVID-19 to advise the government on possible countermeasures. However, depending on the pandemic situation, the role they were expected to play expanded, and they appealed to the public to raise awareness on the disease, which may have been seen as the Expert Meeting group did beyond what they should have done. This may have caused the public's distrust of the Expert Meeting group and consequently, the mistrust of the information and advice from them [40].

In terms of patient monitoring, patients had to answer phone calls from the local clinic or public health center to report their health condition. Whereas, clinicians had to report the identified cases to the local public health center by sending a handwritten report using a facsimile. Local public health centers, then, elevated all the data to higher institutions such as local governments. As this data collection method required efforts and extra work, MHLW has developed a new system to collect COVID-19-related data immediately. It is called HER-SYS (Health Center Real-time Information-sharing System on COVID-19). These efforts were made to help with timely data collection and reduce work overload in public health and healthcare sectors.

Local governments were also struggling with information disclosure of cases. Though MHLW presented a basic guideline for that matter [41], the actual practice in some areas (prefectures, cities, towns, etc.) appeared not to follow it [42]. One of the reasons for this issue may be that the suggested rules on information disclosure were not clearly or sufficiently communicated to the local governments, let alone the public.

Discrimination Against COVID-19 Patients, Healthcare Providers, and Their Families

In the early stage of the pandemic, COVID-19 patients and healthcare providers as well as their families experienced discrimination and stigmatization at work or in their communities [43, 44]. In a study on COVID-19 survivors, more than 40% of the participants reported to have experienced COVID-19-related stigma and discrimination such as being treated as contagious after recovery and harmful rumors [45]. Mass media are said to have played a crucial role in encouraging prejudice in the society by labeling certain occupational groups and sending misleading messages to the audience. In addition, several local governments unnecessarily disclosed patients' personal information and activity records under the name of "infection risk management," which became controversial [46].

One of the Novel Coronavirus Expert Meeting members (Muto) has expertise in ethical, legal, and social issues. She set up a working group on Prejudice, Discrimination, and Privacy in September 2020. Muto points out that it was and still is difficult to set up anti-discrimination policies in Japan, as it lacks a legal basis to protect human rights, which other countries have. The working group meetings have not been held after the fourth meeting in November 2020 [47].

The Japanese society was also divided by the discussion over vaccination like in other countries. As vaccination is often connected to or considered to be a part of social norms [48, 49], those unvaccinated might be seen as undermining the public benefit. An experimental study showed that COVID-19-vaccinated people in Japan had strong ingroup favoritism probably generated by their outgroup bias, whereas unvaccinated group did not present ingroup favoritism [50]. Based on the concerns and negative experiences reported, the Ministry of Justice in Japan launched a campaign to abolish discrimination against those who were infected with and/or unvaccinated against COVID-19 [51].

Public Health Ethics Discussion

As described in the previous sections, the COVID-19 pandemic revealed the vulnerability of the Japanese organizational system and healthcare system. For future preparedness against novel infectious diseases, Japanese society must overcome some challenges.

Capacity Building in Risk Assessment and the Healthcare System

To respond to public health emergencies appropriately without delay, real-time risk assessment using up-to-date data is crucial. For that, efficient data collection methods and databases for recording information on cases need to be established. In terms of organizational reform, frequent communication and cooperation at prefecture and ministry levels may be the key to a swift response in case of a national emergency.

Japan is a super-aging society, and the whole healthcare system has been transformed to meet the needs in this kind of demographic structure, which includes reduction of public health centers and more emphasis on treatment for chronic diseases. This kind of healthcare system is not always ready for the epidemic of acute infectious diseases. To build a flexible and robust healthcare system in such a complex situation, amendments to the current law may be required, for example, enabling task shifting and task sharing among healthcare providers when necessary [35].

To tackle these challenges, in November 2023, MHLW and the main associations of healthcare professionals (physicians, nurses, pharmacists, dentists, etc.) agreed on the declaration of improvement in the healthcare system for the post-COVID age [52]. In this declaration, they plan to ensure 19,000 beds and 1,500 healthcare facilities to deal with novel infectious diseases for the first 3 months of the outbreak. In the meantime, they pursue DX in the healthcare system. DX enables universal management of medical records of individuals as well as seamless health care, which will benefit patients.

In April 2025, aiming to achieve required improvement in research on infectious diseases and the healthcare system, two representative health institutions in Japan—National Institute of Infectious Diseases and National Center for Global Health and Medicine—merged to establish a new institute, Japan Institute for Health Security (JIHS) [53]. JIHS will perform the unique function as a research institute and a top-tier hospital, being designated as one of the national focal points for clinical care at the onset of the next pandemic. It mainly aims to create a network with research institutes such as universities inside and outside of the country for clinical trials and collaborative research. At the same time, its mission is to become a hub for the central and local governments while encouraging DX for data collection in case of an outbreak of a new infectious disease.

Legal Reform to Address Ethical Issues

The pandemic also evoked discussions over the ethical, legal, and social issues in Japan. Strong public health measures sometimes require compromise between public benefit and individual's basic human rights such as restriction of movement, vaccination mandate, and contact tracing. Depending on the custom and tradition in the country, a balance between the local values and public priorities may differ [54]. The WHO guidelines suggest the following ethical considerations especially for public health surveillance: common good, equity, respect for persons, and good governance.

Muto points out that there was not enough discussion over the ethical issues. When the demand for COVID-19 treatment went beyond the capacity of healthcare facilities, the central government did not provide guidelines on prioritization of patients, and the decision was left to clinicians [55]. The absence of clear guidance on the allocation of limited resources and compromising the care for patients without COVID-19 made the healthcare professionals and civil servants suffer in the overwhelming situation [56]. The same concern is true of the discrimination against COVID-19 patients and their relatives. Nakazawa also emphasizes the need for ethical consideration on problems of vulnerable populations [57].

Comprehensive analyses of smartphone geolocation data of the Japanese population showed that while the declaration of state of emergency affected people's mobility to some extent, the reduction in human mobility can be mainly attributed to people's "voluntary lockdown" observing the increase in the number of new cases and deaths as well as the crises in the Japanese healthcare system [58]. This suggests providing appropriate information and advice in a timely manner to the public is effective enough to cause behavioral change and will consequently lead to successful infection control rather than imposing strong legal countermeasures.

Compared to many other countries, Japan was unique in that it managed to control the COVID-19 situation without introducing strong legal measures especially on human mobility and close contact. The Japanese government chose to request *jishuku* (self-restraint) to the public instead of forcing people to stay home by law, which was successful in a way as seen in the aforementioned "voluntary lockdown"

[59]. Sato points out *jishuku*-based requests worked out in Japan because of its unique social rules [60]. He argues that Western countries (including North American and European countries) rely on "rule of law," whereas Japan and its people rely on "rule of society," where peer pressure controls the public. Under such rules, people "voluntarily" refrain from going out or having close contact with other people. This rule, however, is a double-edged sword because on one hand, the government does not require strong restriction or legal reform on human mobility in a national emergency. On the other hand, the peer pressure could cause ethical issues such as discrimination and stigmatization as the whole society can act like police, and the peers can impeach those who disobeyed the rules of society, which was called "*Jishuku* police." COVID patients were blamed as it was due to their lack of responsibility and a result of breaking the *jishuku* rule even when that was not the case [60]. Therefore, even though the non-compulsory measures were effective, they should be implemented with caution as people may comply with such measures affected by the influence of stigma and discriminations stemming from the fear of infection, excessive social norms, and peer pressure.

In February 2021, an amendment was made to the Act on Special Countermeasures against New-type Flu and Other Novel Infections, one of the legal bases for the policies and countermeasures against COVID-19 in Japan, to add prevention of discriminatory actions. It added the obligation of the central and local governments to examine and identify any discriminatory actions as well as raise awareness in the public of the related issues [47]. In addition, there are ways to legally punish those threatening actions based on prejudice or apply codes of the local communities [46]. However, appealing for criminal punishment is a big hurdle especially for patients or healthcare providers, and awareness raising may not be enough as a deterrent to discrimination when the whole society is reacting with panic to a pandemic.

Evaluation of the Countermeasures

Unlike the majority of public health measures in Japan, school closure was imposed nationwide as compulsory in 2020. After re-opening schools, three measures were widely imposed in the school setting: face mask mandate, silence during lunch, and the termination of extracurricular activities. Further research is required to investigate how these measures affected the daily life of students and if a high rate of suicides and mental disorders among adolescents could be attributed to these public health measures at schools [61].

In addition, like in other countries, the government needed to address potential injustice caused by inequity in public health measures. Although Japan achieved a high coverage of COVID-19 vaccination, the coverage might be lower among migrants, partly because of limited vaccine-related information in various languages [62]. Addressing language and cultural differences may be the key for a diverse society to prepare for a global health emergency.

These are only a few examples of questionable countermeasures introduced during the pandemic. There are quite a few COVID-19 policies and countermeasures awaiting evaluation that will determine if they worked or even did more harm than good. Nakazawa claims "procedural justice" for transparency of decision-making and post-evaluation of the policies introduced during the pandemic [57]. To allow fair and genuine evaluation from different disciplines, relevant data should be open to public.

Conclusion

In this chapter, we overviewed the situation and challenges regarding COVID-19 in Japan as well as the countermeasures taken by the Japanese government. We described several issues related to the capacity of the Japanese healthcare system and unethical behavior in Japanese society. To prepare for the next national emergency such as pandemic, the country needs reform in its healthcare system and legislation as well as improvement in the communication between the stakeholders (e.g., local governments, ministries, etc.).

For these improvements, evaluating and understanding what countermeasures worked and what did not is required. Future pandemic preparedness is unachievable without lessons learned from the experiences of the COVID-19 pandemic.

References

1. National Institute of Infectious Diseases (2020) Nihon kokunai no shingata korona uirusu kansensho dai ichirei wo keiki ni kenchi sareta chugoku bukan shi ni okeru sichu kansen no hassei [Community acquired infection in Wu-Han, China, detected based on the first case of COVID-19 in Japan]. Infect Agents Surveill Rep (IASR) 41(8):143–144
2. National Institute of Infectious Diseases (2020) Daiyamondo purinsesu gou shingata korona uirusu kansensho jirei ni okeru jirei hassei shoki no ekigaku [Epidemiological analysis of the early stage of the COVID-19 cases on the cruise ship, diamond princess]. Infect Agents Surveill Rep (IASR) 41(7):106–108
3. Ikegami N (2014) Improving population health through public health centers in Japan. In: Universal health coverage for inclusive and sustainable development: lessons from Japan. The World Bank, Washington DC, pp 163–180
4. Izumi K, Ohkado A, Uchimura K, Kawatsu L, Suenaga M, Urakawa M et al (2017) Evaluation of tuberculosis contact investigations in Japan. Int J Tuberc Lung Dis 21(2):188–195
5. Ohshiro H, Kawamoto K, Nose T (1996) Surveillance system of infectious diseases in Japan. J Epidemiol 6(3 Suppl):81–85
6. An BY, Tang SY (2020) Lessons from COVID-19 responses in East Asia: institutional infrastructure and enduring policy instruments. Am Rev Public Adm 50(6–7):790–800
7. Machida M, Wada K (2022) Public health responses to COVID-19 in Japan. Glob Health Med 4(2):78–82
8. Shimizu K, Negita M (2020) Lessons learned from Japan's response to the first wave of COVID-19: a content analysis. Healthcare 8(4):426
9. Omi S, Wakita T, Sahara Y, Wada K (2024) COVID-19 Taiou Kiroku <2022–2024> report no: 2 [Record of COVID-19 countermeasures <2022–2024>]. Tokyo
10. Ministry of Health, Labour, and Welfare (2023) COVID-19 ni Kansuru Hasseidoukou no Teitenhaaku ni tsuite [Sentinel site surveillance to capture the level of infection spread of COVID-19]. https://www.mhlw.go.jp/content/001065724.pdf. Accessed 30 Jan 2025

11. Ministry of Health, Labour, and Welfare (2023) Dai 74 kai Kousei Kagaku Shingikai Kansensho bukai Gijiroku [Minutes of the 74th meeting of Health Sciences Division] [Internet]. https://www.mhlw.go.jp/stf/newpage_32940.html. Accessed 30 Jan 2025
12. Ministry of Education, Culture, Sports, Science and Technology (2020) Shingata korona uirusu kansenshou taisaku no tameno shougakkou, chugakkou, koutougakkou oyobi tokubetusien gakkou niokeru issei rinji kyugyou ni tsuite [Nation-wide school closure in response to the COVID-19 pandemic]. https://www.mext.go.jp/content/202002228-mxt_kouhou01-000004520_1.pdf. Accessed 30 Jan 2025
13. Fukumoto K, McClean CT, Nakagawa K (2021) No causal effect of school closures in Japan on the spread of COVID-19 in spring 2020. Nat Med 7(12):2111–2119
14. Yabe T, Tsubouchi K, Fujiwara N, Wada T, Sekimoto Y, Ukkusuri SV (2020) Non-compulsory measures sufficiently reduced human mobility in Tokyo during the COVID-19 epidemic. Sci Rep 10(1):18053
15. Oshitani H (2020) Cluster-based approach to coronavirus disease 2019 (COVID-19) response in Japan, from February to April 2020. Jpn J Infect Dis 73:491–493
16. Imamura T, Saito T, Oshitani H (2021) Roles of public health centers and cluster-based approach for COVID-19 response in Japan. Health Secur 19:229–231
17. Kosaka M, Hashimoto T, Ozaki A, Tanimoto T, Kami M (2021) Delayed COVID-19 vaccine roll-out in Japan. Lancet 397(10292):2334–2335
18. Our World in Data (2024) Coronavirus (COVID-19) vaccinations. https://ourworldindata.org/covid-vaccinations. Accessed 21 Feb 2024
19. KYODO NEWS (2021) Active epidemiological investigations decrease in Tokyo in response to the rapid increase in COVID-19 cases. The Nikkei 15 Aug
20. Ministry of Health, Labour, and Welfare (2022) Deregulation in border control after March 2022, Tokyo
21. NHK (2022) Deregulation in border control from 11 October. https://www3.nhk.or.jp/news/special/coronavirus/restrictions/detail/detail_94.html. Accessed 27 Feb 2024
22. Ministry of Health, Labour, and Welfare/FORTH (2022) COVID-19 quarantine (After 7 September, 2022). https://www.forth.go.jp/news/20220903_00001.html. Accessed on 27 Feb 2024
23. Johns Hopkins Coronavirus Resource Center (2023) Mortality analyses. https://coronavirus.jhu.edu/data/mortality. Accessed 21 Feb 2024
24. Zenroren (2020) What's New from Zenroren. Jichiroren exposes excessive workloads at public health centers amid COVID-19 pandemic. https://zenroren.gr.jp/jp/english/2020/10/english201021_01.html. Accessed 3 March 2026
25. Tsuruda K, Hatono Y (2023) Mental health status of public health nurses and its related factors under the coronavirus disease 2019 pandemic in Japan. J Int Nurs Res 2(1):e2022-0007-e2022-0007
26. Takemura S, Ohmi K, Sone T (2020) Public health center (Hokenjo) as the frontline authority of public health in Japan contribution of the National Institute of Public Health to its development. J Natl Inst Public Health 69(1):2–13
27. Taira K, Hosokawa R, Shiomi M (2022) Changes in the number of public health nurses employed in local governments in Japan during the Covid-19 pandemic: a cross-sectional study. SAGE Open Nurs 8. https://doi.org/10.1177/23779608221140718
28. Osawa E, Okuda H, Koto-Shimada K, Shibanuma A, Saito T (2022) The environment encouraging COVID-19 response at public health centers and future challenges in Japan. Int J Environ Res Public Health 19(6):3343
29. Suzuki W (2021) Koronaka to iryou teikyou taisei no kadai [COVID-19 pandemic and the challenges in the health care system in Japan]. Gakushuin daigaku keizai ronshu 58(3):203–226
30. OECD (2023) Health at a glance 2023: OECD indicators. OECD Publishing, Paris. https://doi.org/10.1787/7a7afb35-en
31. NHK World – Japan (2021) Why Japan's medical system has come close to breaking point https://www3.nhk.or.jp/nhkworld/en/news/backstories/1524/. Accessed 21 Feb 2024

32. Nakamura H (2021) Tekisetsu na Yoboukanri nakushite Kikikanri wa Jubun ni Kinou sezu [Risk management does not fully function without appropriate preventive control]. J Health Care Soc 31(3):333–335
33. Ministry of Health, Labour, and Welfare (2021) Shingata korona uirusu kansenshou ni kakaru wakuchin sesshu no tameno kinnikunai chusha no shika ishi ni yoru jisshi ni tsuite [COVID-19 vaccination by dentists]. https://www.mhlw.go.jp/content/000773564.pdf. Accessed 21 Feb 2024
34. Ministry of Health, Labour, and Welfare (2021) Shingata korona uirusu kansenshou no wakuchin sesshu wo suishin surutame no kaku iryou kankei shokushu no senmonsei wo fumaeta taiou no arikatatou ni tsuite [Responses to promote vaccination against COVID-19 based on expertise of each type of medical profession]. https://www.mhlw.go.jp/content/000788723.pdf. Accessed 21 Feb 2024
35. Williams GA, Maier CB, Scarpetti G, de Belvis AG, Fattore G, Morsella A et al (2020) What strategies are countries using to expand health workforce surge capacity during the COVID-19 pandemic? (Special Issue: COVID-19 health system response.). Eurohealth (Lond) 26(2):51–57
36. Kadomatsu N (2022) Legal countermeasures against COVID-19 in Japan: effectiveness and limits of non-coercive measures. China-EU Law J 8(1):11–32
37. Nakahara S, Inada H, Ichikawa M, Tomio J (2021) Japan's slow response to improve access to inpatient care for COVID-19 patients. Front Public Health 9:791182
38. Asia Pacific Initiative (2020) Shingata Korona Taiou Minkan Rinji Chousakai – Chosa Kensho Houkokusho [COVID-19 response – evaluation and report by a temporary examination committee of private sectors]. Discover 21, Tokyo
39. Ninomiya K, Kanamori M, Ikeda N, Jindai K, Ko YK, Otani K et al (2022) Integration of publicly available case-based data for real-time coronavirus disease 2019 risk assessment. Jpn West Pac Surveill Response J 13(1):1–6
40. Novel Coronavirus Expert Meeting (2020) Tsuginaru Nami ni Sonaeta Senmonka Jogensoshiki no Arikata nitsuite [What an expert advisory committee should be for the next wave of COVID-19 cases]. https://note.stopcovid19.jp/n/nc45d46870c25. Accessed 14 May 2025
41. Ministry of Health, Labour, and Welfare (2020) Ichirui Kansensho ga Kokunai de Hassei shita Baai niokeru Jouhou no Kouhyou nikakaru Kihon Houshin [Basic guideline for information disclosure on cases of Category I Infectious Diseases]. https://www.mhlw.go.jp/content/000652973.pdf. Accessed 14 May 2025
42. Nagai A, Ri I, Fujisawa K, Muto K (2022) Chihou Jichitai niokeru COVID-19 Kansensha nikansuru Jouhoukouhyou no Jittai: 2020-nen 1-gatsu ~ 8-gatsu no Kouhyounaiyou no Bunseki [Facts on information disclosure on COVID-19 cases by local governments: analysis of data disclosed January–August, 2020]. Jpn Soc Public Health 69(7):554–567
43. Asaoka H, Sasaki N, Kuroda R, Tsuno K, Kawakami N (2021) Workplace bullying and patient aggression related to COVID-19 and its association with psychological distress among health care professionals during the COVID-19 pandemic in Japan. Tohoku J Exp Med 255(4):283–289
44. COVID-19 taskforce working group on prejudice discrimination and privacy (2020) Summary of discussions. https://www.cas.go.jp/jp/seisaku/ful/henkensabetsu_houkokusyo.pdf. Accessed 28 Feb 2024
45. Sugiyama A, Miwata K, Kitahara Y, Okimoto M, Abe K, E B et al (2022) Long COVID occurrence in COVID-19 survivors. Sci Rep 12(1):6039
46. Tsujimura-Ito T (2021) Shingata korona uirusu kansenshou to hou no kakawari -Nihon ni okeru koudouseigen sochi to henken sabetsu wo megutte [COVID-19 and the law: movement restriction measures in relations to prejudice]. IASS Rev 46(1):22–31
47. Muto K (2022) COVID-19 ni kansuru sabetsuteki gendou no boushi ni kansuru torikumi wo furikaette [Efforts to prevent stigmatizing and discriminatory behaviors related to COVID-19 in Japan]. J Health Care Soc 32(1):32–83
48. Chevallier C, Hacquin AS, Mercier H (2021) COVID-19 vaccine hesitancy: shortening the last mile. Trends Cogn Sci 25(5):331–333

49. Vriens E, Tummolini L, Andrighetto G (2023) Vaccine-hesitant people misperceive the social norm of vaccination. PNAS Nexus 2(5):132
50. Sasaki S, Kurokawa H (2022) Vaccination and discrimination: experimental evidence under the COVID-19 pandemic. SSRN. https://doi.org/10.2139/ssrn.4246154
51. Ministry of Justice (2021) Abolition of prejudice and discrimination related to infectious diseases. https://www.moj.go.jp/JINKEN/stop_coronasabetsu.html. Accessed 28 Feb 2024
52. Takemi K, Matsumoto K, Takahashi H, Yamamoto N, Takahashi H, Aizawa T, et al (2023) Posuto Korona Iryo Taisei Jujitsu Sengen [Declaration of improvement in medical system for post-COVID age]. https://www.nisseikyo.or.jp/about/teigen/images/20231106_postcorona.pdf. Accessed 28 Feb 2024
53. Ministry of Health, Labour, and Welfare (2024) Kokuritsu Kenko Kiki Kanri Kenkyu Kiko no Sousetsu ni Mukete [Establishment of Japan Institute for Health Security]. https://www.mhlw.go.jp/content/10600000/001248419.pdf. Accessed 16 May 2025
54. World Health Organization (2017) WHO guidelines on ethical issues in public health surveillance. https://www.who.int/publications/i/item/who-guidelines-on-ethical-issues-in-public-health-surveillance. Accessed 16 May 2025
55. Muto K (2023) COVID-19 wo Meguru Rinriteki Shokadai [Ethical issues regarding COVID-19]. J Jpn Soc Intern Med 112(9):1553–1556
56. Muto K (2023) COVID-19 ni Kansuru Kantougen wo Furikaeru [Overview of past forewords of the journal regarding COVID-19]. J Health Care Soc 33(1):1–3
57. Nakazawa E (2023) COVID-19 pandemic and public health ethics in Japan. JMA J 6(1):76–80
58. Watanabe T, Yabutomo Y (2020) Nihon no jihatsuteki rokkudaun ni kansuru kousatsu report no.: 26. ['Voluntary lockdown' in Japan during the COVID-19 pandemic]. Tokyo
59. Iijima W (2021) Jishuku as a Japanese way for anti-COVID-19. Some basic reflections. Hist Soc Res Suppl 33:284–301
60. Sato N (2023) Naze, Jishuku-Keisatsu wa Nihon dake nanoka: Douchou-Atsuryoku to 'Seken' [Why only Japan experienced 'Jishuku (self-restraint) police': peer pressure and the society] [Internet]. Gendai Shokan, Tokyo
61. Goto R, Okubo Y, Skokauskas N (2022) Reasons and trends in youth's suicide rates during the COVID-19 pandemic. Lancet Reg Health West Pac 27:100567
62. Fujita M, Kanda M, Kiyohara H, Ikeda S, Iwamoto A, Sudo K et al (2023) Migrants' access to COVID-19 vaccination in Japan: progress and challenges. J Migr Health 7:100169

South Korea's COVID-19 Response: A Critical Analysis of Public Health Measures and Their Implications

5

Ilhak Lee and Cheol Kang

Abstract

South Korea's COVID-19 response, known as the K-Quarantine model, effectively controlled the virus through large-scale testing, digital contact tracing, technology-enabled surveillance, and targeted social distancing. While the response achieved low direct mortality, it overlooked basic public health ethics principles, raising questions about state authority and individual liberties. The pandemic also disproportionately affected vulnerable populations, highlighting the need for future responses to integrate ethical equity, economic support, and mental health strategies. South Korea's pandemic response, anchored in the Infectious Disease Control and Prevention Act, enabled rapid and coordinated measures. The K-quarantine model, while effective, raised privacy concerns due to extensive digital surveillance. Despite universal healthcare, equity gaps emerged in testing, vaccination, and economic support, highlighting the need for more inclusive strategies. The COVID-19 pandemic in South Korea exposed existing social inequities and highlighted the need for more inclusive emergency planning. Political dynamics, including resistance to healthcare reforms and populist appeals, influenced the response, sometimes undermining effective crisis management. The government's transparency strategy, while enhancing public trust, compromised privacy and led to social stigmatization, underscoring the need for ethical frameworks that balance transparency with privacy protection and ensure equitable resource allocation. South Korea's COVID-19 response, while effective in controlling the disease, highlighted ethical challenges related

I. Lee (✉)
Department of Medical Humanities and Social Sciences, College of Medicine, Yonsei University, Seoul, South Korea
e-mail: arete2@yuhs.ac

C. Kang
Graduate School of Public Health, Yonsei University, Seoul, South Korea

L. P. Marquez et al. (eds.), *COVID-19 in the Pacific Rim and Beyond*,
Collaborative Bioethics 6, https://doi.org/10.1007/978-3-032-21218-4_5

to privacy, equity, and democratic governance. The extensive digital surveillance system, while epidemiologically effective, raised concerns about privacy rights and government overreach. Additionally, resource allocation disparities and the politicization of crisis management underscored the need for ethical frameworks that prioritize individual rights and collective health goals.

Keywords

COVID-19 · Public health · Trust · South Korea · Public health ethics

Introduction

Pandemics like COVID-19 reveal the foundations upon which a society stands by exposing how its members understand and treat one another. On the other hand, the ethical issues discovered during the course of a pandemic are as important as the appropriateness of public health measures and should form the basis for future public health responses. Therefore, examining South Korea's public health countermeasures to COVID-19 from February 2020 to May 2023, along with the ethical considerations surrounding these responses, can be considered a valuable endeavor.

South Korea's response to the pandemic, though widely acclaimed for its efficacy in curbing viral transmission, constitutes a distinctive and instructive case study in public health ethics. Public health ethics principles, such as respect for individual rights, protecting social justice, and collective goods, were overlooked in the process of formulating and implementing public health measures. Therefore, it is necessary to reexamine South Korea's COVID-19 response from an ethical perspective.

South Korea's strategy, characterized by its extensive deployment of technology and data-driven surveillance mechanisms, raises critical questions about the limits of state authority and the protection of individual liberties during public health emergencies. These inquiries take on heightened relevance as societies worldwide confront similar challenges in preparing for future health crises.

This chapter provides an overview of the impact of COVID-19 on South Korean society and introduces the country's response measures, including the so-called K-Quarantine model and relevant legal actions. Based on these policy materials, it examines the ethical challenges inherent in South Korea's pandemic response. Finally, it concludes by summarizing the key lessons learned and proposing a future ethical framework for public health preparedness and response.

Impact of COVID-19 of Korean Society

From the onset of the pandemic through mid-2024, South Korea recorded approximately 34.57 million confirmed COVID-19 cases and 35,934 deaths, yielding a case fatality rate (CFR) of 0.10%, among the lowest globally. The country's cumulative mortality reached about 69 deaths per 100,000 population, significantly lower than figures in many Western countries. However, the COVID-19 toll extended beyond reported cases: excess mortality between 2020 and June 2022 increased by ~8.7% (~61,000 deaths) compared with expected levels, with a pronounced rise during the Omicron wave [1]. Modelling data further highlights a substantial excess of ~39,000 deaths among those aged 80 and over in 2022 [2].

Specific populations were disproportionately affected. Elderly individuals with a Charlson Comorbidity Index ≥2 had a 1.5-fold increased risk of death, and those in Medical Aid (low-income) programs experienced a 1.7-fold higher mortality risk. These disparities underscore the heightened vulnerability of older and socioeconomically disadvantaged groups [3].

On the economic front, while South Korea avoided comprehensive lockdowns, the second quarter of 2020 saw national GDP contract by approximately 2.9%, marking a steep economic decline. The initial OECD forecast projected a modest −0.8% contraction for the year, compared to an average −7.6% among OECD countries. Labor markets were heavily impacted: overall unemployment surged, and youth unemployment rose from ~11.3% to 16.7% in 2020; the employment rate among those aged 15–29 dropped by 1.3 percentage points, and twice the decline was seen in older cohorts [4].

Mental health consequences were profound. Surveys indicate that 12–34% of adults experienced moderate to severe depression, and 7–29% reported similar levels of anxiety during 2020–2021. Among adolescents, one national survey reported 19% with anxiety and 17.5% with depression. Notably, young individuals facing economic hardship exhibited nearly three times higher suicidal ideation (26%) compared to their counterparts (10.8%) [5].

Collateral damage extended to disrupted healthcare, with deferred chronic disease management and preventive services disproportionately affecting rural and low-income citizens. Education gaps widened as remote learning underscored existing technology disparities, potentially worsening long-term inequality. Migrant and undocumented communities often refrained from testing or treatment due to fear of legal repercussions, deepening health inequities [6].

In summary, while South Korea maintained low direct mortality from COVID-19, the pandemic's broader impact included substantial excess mortality, severe socioeconomic stress, and mental health deterioration—especially among vulnerable and marginalized populations. These outcomes underscore the need for future public health responses to integrate ethical equity, economic support, and mental health strategies alongside epidemiological control measures.

Public Health Strategies

The integrated approach of South Korea's COVID-19 response achieved remarkable outcomes, including over 80% full vaccination coverage by the end of 2021, effective surge capacity management during multiple waves of infection, maintenance of essential healthcare services throughout the pandemic, and successful transition from emergency response to endemic management. The strategy demonstrated the effectiveness of combining technological solutions with healthcare system flexibility, showcasing how rapid innovation, coordinated policy implementation, and adaptive healthcare delivery models could work together to create a comprehensive pandemic response framework that balanced immediate crisis management with long-term system sustainability.

Public Health Strategy: The K-Quarantine Model

South Korea developed a comprehensive pandemic control strategy known as the K-quarantine model (Trace-Test-Treat), built on four integrated components that operated synergistically to establish an effective containment framework. The strategy centered on large-scale diagnostic testing through drive-through testing sites and community-based screening centers, enabling rapid, high-volume screening with scalable infrastructure that prioritized speed and accessibility. Digital contact tracing formed the second pillar, integrating credit card transactions, mobile phone location data, and CCTV footage to enable real-time tracking of transmission pathways and exposure networks through mandatory participation with comprehensive data collection [7].

Technology-enabled surveillance provided the third component, utilizing advanced monitoring systems for quarantine compliance, digital health certificates, movement tracking, and automated alert systems for exposure notifications. The fourth element involved targeted social distancing measures that implemented risk-based restrictions, strategic closure of high-risk venues and activities, and flexible adjustment based on transmission patterns. These core strategies were supported by comprehensive border control measures including mandatory quarantine for international arrivals, entry restrictions, health screening protocols, and coordinated efforts between health authorities and immigration services.

Healthcare Innovation and Adaptive Responses

South Korea's pandemic response was characterized by rapid implementation of innovative healthcare delivery models that fundamentally transformed the landscape of medical services. The drive-through testing infrastructure enabled high-volume screening with minimal exposure risk and efficient patient flow, reduced facility contamination, and provided a scalable model adaptable to local transmission patterns. Community Treatment Centers emerged as specialized facilities for mild COVID-19 cases, providing alternative care settings to reduce hospital burden through standardized isolation and monitoring protocols, resource optimization during peak transmission periods, and systematic patient stratification and triage [8, 9].

The accelerated expansion of telemedicine services represented a pivotal transformation in healthcare delivery, encompassing remote consultations for routine care and follow-up, digital health monitoring and chronic disease management, mental health support and counseling services, and prescription management with medication delivery coordination. Digital health integration played a crucial role, incorporating electronic health records for contact tracing, mobile health applications for symptom monitoring, digital vaccination certificates and health passes, and real-time data sharing between healthcare facilities to create a comprehensive digital health ecosystem.

Healthcare institutions underwent extensive structural reconfigurations to accommodate surge capacity while maintaining essential services. These adaptations included establishment of specialized COVID-19 treatment wards, implementation of segregated patient flow pathways, creation of negative pressure isolation units, and conversion of existing facilities for pandemic response. Resource reallocation strategies involved rapid redistribution of medical equipment and supplies through centralized procurement and distribution systems, strategic stockpiling of critical care resources, and flexible staffing models across healthcare facilities [10].

Strategic healthcare workforce management became essential for maintaining system functionality during the crisis. This involved redeployment of staff to high-need clinical areas, extended work schedules and surge capacity planning, implementation of comprehensive safety protocols, training programs for COVID-19 specific care requirements, and development of mental health support systems for healthcare workers. Capacity management strategies included dynamic bed allocation across hospital networks, establishment of surge capacity protocols, integration of public and private healthcare resources, and regional coordination for patient transfer and care distribution to optimize resource utilization.

Vaccination Strategy

South Korea's vaccination strategy evolved through multiple phases, beginning with initial procurement challenges due to limited global supply and the country's emphasis on thorough safety evaluation processes. The government participated in the COVAX Facility while pursuing bilateral agreements and strategic partnerships with multiple pharmaceutical companies. Domestic vaccine development efforts led to SK Bioscience's SKYCovione development and approval in June 2022, representing investment in domestic vaccine manufacturing capacity, though production was later suspended in November 2022 due to market saturation.

Implementation followed a carefully planned phased rollout approach, beginning with healthcare workers and high-risk populations, followed by elderly populations and individuals with underlying conditions, then expanding to the general population through age-based prioritization before entering a mass vaccination phase. Delivery infrastructure included mass vaccination centers in major population centers, mobile vaccination clinics for rural and underserved areas, integration with local health facilities and clinics, and workplace vaccination programs for essential workers [11, 12].

Digital integration and monitoring systems supported the vaccination campaign through online booking systems for appointment scheduling, digital vaccination certificates and health passes, real-time monitoring of vaccination coverage and adverse events, and integration with existing health information systems. The vaccine pass implementation required vaccination proof for entry to public spaces using digital verification systems and QR code technology, with phased implementation across different venue types and integration with existing digital health infrastructure [13].

Legal Framework and Institutional Structure

The legal foundation of South Korea's pandemic response was anchored in the Infectious Disease Control and Prevention Act (IDCPA), which underwent substantial revisions following the 2015 MERS outbreak. These amendments fundamentally transformed the scope of public health authority during health emergencies, establishing a comprehensive legal architecture that enabled rapid and coordinated response measures [14].

Core Legal Provisions and Powers

The revised IDCPA granted health authorities extensive emergency powers including mandatory access to personal data for epidemiological investigations, authority to require diagnostic testing and impose quarantine measures on individuals and groups, power to restrict or prohibit public gatherings and events, and implementation of movement controls and border restrictions. These provisions established the legal basis for contact tracing operations that could compel disclosure of personal information including location data, financial transaction records, and communication patterns.

The Act also empowered local and national health authorities to requisition private facilities for public health purposes, mandate cooperation from telecommunications companies and financial institutions for data sharing, impose penalties for non-compliance with health orders including fines and imprisonment, and override certain privacy protections during declared health emergencies. Administrative procedures were streamlined to enable rapid decision-making, with reduced judicial oversight during emergency periods and expanded executive authority to implement health measures without legislative approval.

Institutional Coordination Mechanisms

The legal framework established clear hierarchical structures for pandemic response, with the Central Disease Control Headquarters serving as the primary coordinating body under the Ministry of Health and Welfare. Regional response centers were granted authority to implement national directives while adapting to local conditions, and local governments received enhanced powers to enforce health measures within their jurisdictions. Inter-agency coordination protocols facilitated collaboration between health authorities, law enforcement, immigration services, and telecommunications companies.

The institutional architecture included mandatory reporting requirements for healthcare facilities and laboratories, standardized protocols for information sharing between government agencies, legal immunity provisions for public health officials acting within their authority during emergencies, and mechanisms for requisitioning medical supplies and healthcare workers. Emergency procurement procedures bypassed normal administrative processes, enabling rapid acquisition of medical equipment and supplies.

Enforcement Mechanisms and Compliance Systems

The legal framework established comprehensive enforcement mechanisms including administrative fines for quarantine violations, criminal penalties for knowingly exposing others to infection, civil liability for damages caused by non-compliance with health orders, and immediate detention powers for individuals posing public health risks. Digital monitoring systems were legally mandated for quarantine enforcement, with electronic tracking required for all quarantined individuals and automated violation detection through mobile applications and location monitoring.

Public-private partnerships were formalized through legal agreements requiring cooperation from technology companies in developing and maintaining digital health infrastructure, telecommunications providers in sharing location and communication data, and financial institutions in providing transaction records for contact tracing. Legal protections were established for data sharing between agencies, with specific provisions allowing health authorities to access normally protected personal information during health emergencies.

Social Impacts and Population Effects of COVID-19 Policy

Loss of Social Trust and COVID-19 Policy

South Korea initially succeeded in its early COVID-19 response, leading to a significant surge in public trust in the government. However, this trust declined sharply over time, returning to pre-pandemic levels and even falling below them [15]. The prolonged nature of quarantine measures and strict regulations triggered conflict and division within communities. While national unity was initially formed through early quarantine achievements, subsequent policy side effects and discrimination controversies revealed backlash from specific groups and social fractures.

Particularly, questions were raised about fairness and equity in the application and enforcement of quarantine policies. For example, some religious groups were identified as sources of cluster infections and faced strong sanctions. In the case of Shincheonji, all gatherings were banned, and they received stricter measures than other religions, leading to complaints [16]. Meanwhile, other churches also caused cluster infections during worship services but did not receive equivalent sanctions, creating equity controversies. This response damaged trust between religious communities and the government, and some conservative Protestant forces strongly opposed government quarantine measures, viewing them as oppression.

Additionally, far-right conservative groups expressed their dissatisfaction with quarantine policies through anti-government protests. In August 2020, conservative pastor Jun Kwang-hoon forced a rally attended by thousands and was subsequently confirmed with COVID-19. His megachurch became the center of over 300 confirmed cases, raising concerns about spread in the metropolitan area [17]. Despite requests from the government and quarantine authorities to refrain from gatherings, these protests continued, leading to conflicts between quarantine and freedom. Government supporters and quarantine compliance advocates criticized these groups, but the conservative camp argued that the government was suppressing legitimate political freedom of expression, escalating social tensions.

During the vaccination phase, vaccine skeptics and anti-vaccination movements known as anti-vaxxers also emerged. They primarily used online platforms to exaggerate information about vaccine side effects or held protests opposing the government's introduction of "vaccine passes." For instance, when the government required vaccine passes even for minors, some parent groups protested against this, holding demonstrations opposing forced vaccination of children [18]. While these movements were not large in scale, they clearly demonstrated distrust in quarantine policies and resistance to government control. Even in South Korea, where vaccination rates were high, these vaccine refusal movements caused division in public opinion, creating conflicts over quarantine cooperation.

Self-employed individuals and small business owners expressed significant dissatisfaction while enduring economic sacrifices due to quarantine regulations. As social distancing repeatedly imposed business hour restrictions and gathering bans, those feeling threatened in their livelihoods demanded substantial compensation and regulatory relaxation from the government [19]. In late 2021, hundreds of self-employed individuals gathered for collective action, chanting slogans like "Provide proper loss compensation" and "Withdraw vaccine passes." They questioned policy equity, arguing that large retailers and corporations operated normally while only small business owners were forced to sacrifice. This dissatisfaction spread nationwide through online communities, and some business owners even attempted extreme measures, creating social shock. Class conflicts surrounding quarantine measures came to the surface.

The Itaewon club cluster infection case in May 2020 highlighted issues of hatred and stigmatization against the LGBTQ+ community. When initial reports revealed that the club was a gay club, homophobic criticism poured out online with messages like "homosexuals spread COVID." One netizen commented that "[faggots] deserve to be hated," and after media emphasized patients' sexual identity, hate messages flooded online like a dam burst [20]. Anti-gay hatred videos spread, and the incident was labeled the "Itaewon gay club incident," deepening social stigma against sexual minorities. This atmosphere was so serious that health authorities had to arrange for anonymous testing for club visitors. Since South Korea still lacks comprehensive anti-discrimination laws, sexual minorities hesitated to cooperate with testing or epidemiological investigations due to fear of having no legal protection if their identities were revealed. Eventually, some confirmed cases and contacts went into hiding, creating counterproductive effects that made quarantine cooperation difficult. This exemplifies how quarantine policies threatened the safety of the entire community by stigmatizing specific groups.

In this way, South Korea's quarantine policies created conflicts with various groups including migrants, religious groups, self-employed individuals, and sexual minorities. Some local governments imposed comprehensive COVID testing only on foreign workers, creating the perception that "foreigners = potential infection sources." In 2021, Seoul attempted to mandate COVID testing for all foreign workers and impose fines for non-compliance, but faced strong domestic and international backlash and withdrew the policy [21]. The National Human Rights Commission also warned that such foreigner-targeted quarantine measures could incite hate crimes. However, some regions (including Gyeonggi Province) maintained these orders for some time, creating migrant discrimination controversies. When discriminatory regulations occur during quarantine processes, it deepens government distrust among affected groups and weakens solidarity within Korean society.

Privacy and Surveillance Trade-Offs

The K-quarantine model's extensive digital surveillance created unprecedented tensions between public health effectiveness and individual privacy rights. The integration of credit card data, mobile location tracking, and CCTV footage enabled highly effective contact tracing but raised fundamental questions about data privacy and government overreach. Public disclosure of confirmed cases' detailed movement histories, while intended to enhance community awareness and contact tracing, resulted in unintended consequences including social stigmatization, harassment, and privacy violations that often made individuals identifiable within their communities.

The mandatory nature of surveillance measures, with minimal opt-out provisions, challenged core principles of informed consent and individual autonomy. The absence of clear sunset clauses for emergency surveillance powers raised concerns about the potential entrenchment of pervasive state monitoring beyond the pandemic period, highlighting the need for robust regulatory frameworks governing surveillance technologies in public health emergencies.

Equity and Resource Allocation Challenges

Despite South Korea's universal healthcare system, the practical implementation of pandemic strategies revealed significant equity gaps. Testing infrastructure, while comprehensive in urban centers, often left rural populations underserved, exposing persistent geographic and socioeconomic disparities in healthcare access. The online booking systems for vaccination appointments created barriers for elderly populations and individuals with limited digital literacy, highlighting how digital solutions can inadvertently exclude vulnerable groups.

Economic support measures during the pandemic exposed limitations in social protection systems. Relief programs frequently required formal employment status or official business registration, creating substantial coverage gaps that disproportionately affected informal sector workers, freelancers, and undocumented migrants. These eligibility frameworks revealed the need for more inclusive economic protection strategies during public health emergencies.

Healthcare System Adaptations and Worker Impact

The rapid reallocation of healthcare resources toward COVID-19 management, while necessary for crisis response, resulted in significant disruptions to routine medical care and preventive services. Community treatment centers effectively managed mild cases and reduced hospital overcrowding but raised concerns about care standards and patient autonomy, particularly regarding enforced separation from family support networks and restricted access to specialized medical expertise.

Healthcare workers faced unprecedented pressures including heightened workloads, elevated exposure risks, and morally challenging decisions about patient triage and resource allocation. The pandemic exposed inadequate support infrastructure for healthcare workers' mental health, with limited access to psychological services and professional stigma around seeking support contributing to increased rates of burnout, anxiety, and depression among medical personnel.

Vulnerable Population Impacts and Systemic Exclusions

The pandemic response disproportionately affected vulnerable populations, exposing and often exacerbating existing social inequities. Low-income households struggled with social distancing requirements due to inadequate living spaces and limited work-from-home options. The transition to remote work and online education created additional burdens for families lacking technological infrastructure or digital literacy.

Migrant workers and foreign residents faced distinct challenges including language barriers that limited access to health information, increased discrimination, and fear of legal consequences that deterred undocumented workers from seeking testing or treatment. These groups were often excluded from policy design and public information channels, revealing gaps in inclusive emergency planning.

Political Dynamics and the "Poli-demic" Phenomenon

The intersection of political dynamics with pandemic response created what researchers termed the "poli-demic" phenomenon. Professional medical organizations sometimes prioritized institutional interests over public health goals, resisting healthcare reforms designed to strengthen pandemic preparedness. The alignment of certain religious organizations with political opposition groups contributed to mass infection clusters, particularly following the Shincheonji Church outbreak in February 2020.

The distinction between constructive political engagement and destructive politicization became apparent during crisis management. While constructive discourse involved logical reasoning and public accountability, destructive politicization relied on contradictory rhetoric and populist appeals that undermined effective crisis response.

A growing concern is the weakening of public cooperation in subsequent vaccination campaigns. Vaccination willingness in South Korea evolved markedly during the COVID-19 pandemic. Through a combination of public education, policy incentives, and mandates such as the vaccine pass system, vaccination numbers increased substantially, with full vaccination coverage reaching 82.2% by January 2022 [22]. However, booster uptake declined sharply thereafter: by late 2022, only 36% expressed willingness to receive a third dose. While this trend is difficult to interpret in purely epidemiological terms, it likely reflects more than just concerns about vaccine efficacy or side effects. Diminished trust in policy-making processes and the perceived politicization of public health—a phenomenon often described as "*poli*-demic"—may have eroded public confidence [23]. The interplay between biomedical information and political narrative likely shaped public perceptions, underscoring the need for transparent, participatory, and depoliticized public health communication in future immunization efforts.

Stigmatization and Social Discrimination

Public health transparency measures had significant unintended social consequences. The Shincheonji Church outbreak highlighted how public health crises can amplify existing social biases, with members facing widespread discrimination that extended beyond health concerns to broader cultural and religious prejudices. Regional stigmatization affected communities associated with outbreak clusters, creating lasting economic and social repercussions including employment and housing discrimination.

These stigmatization patterns potentially deterred honest disclosure of symptoms and contact histories, undermining public health surveillance effectiveness [24]. The government's failure to implement proactive strategies for public education or social integration revealed significant gaps in crisis communication planning and the need for more comprehensive approaches to managing social cohesion during health emergencies.

Policy Critique and Recommendations for Future Pandemic Preparedness

Decisions Cost Social Trust

South Korea poured almost every available resource into suppressing viral spread and reducing direct morbidity and mortality. That single-minded focus sidelined investments in social resilience—mental-health support, income safeguards, and community cohesion—so secondary harms accumulated while epidemiological indicators improved.

Because choices were driven mainly by technical models and epidemiological targets, social and ethical dimensions entered late, if at all. Policy design stayed expert-centric; opportunities for public deliberation or explicit value balancing were rare, leaving the needs of vulnerable groups largely invisible.

As the emergency dragged on, this imbalance eroded public trust. Initial confidence in government collapsed once economic burdens, privacy intrusions, and perceived inequities became clear. Conflicts deepened between compliance advocates and groups who felt singled out—self-employed workers, certain religious communities, migrants—fueling social polarization at a moment when solidarity was most needed.

Criticism Based on "Immunity Paradigm"

Roberto Esposito [25] presents the "immunity paradigm" as a key concept for understanding contemporary politics. According to his concept of immunity, communities exclude or isolate certain elements like an immune response to protect themselves from external threats. This logic of exclusion and protection is also found in COVID-19 quarantine measures. The Korean government took bold measures to protect the community from viral risks, but in the process defined certain groups as risk factors and excluded them, thereby alienating them from the community. For instance, Shincheonji members, Sarang Jeil Church congregants, Itaewon club visitors, and foreign workers were stigmatized as viral "spreaders" and subjected to strong control and surveillance. While such immunity strategies are justified in the short term as protecting the safety of the majority, in the long term they can create internal fractures within the community and trigger autoimmune adverse effects.

Esposito points out that when the immunity paradigm reaches extremes, autoimmune phenomena occur where society begins to attack itself [25]. Indeed, when excessive control is implemented or exclusion of specific groups is justified under the pretext of quarantine, communities end up damaging their own internal trust and solidarity. For example, excessive exposure of confirmed patients' movements and personal information under the guise of quarantine information disclosure led to privacy invasion controversies, creating distrust and stigmatization among community members. Additionally, discriminatory measures against sexual minorities and migrants pushed these community members outside the public safety net, ultimately creating a dilemma that made community protection—the original purpose of quarantine—more difficult. Viewed through Esposito's conceptual lens, this contradictory phenomenon represents immune actions intended to protect the community becoming autoimmune responses that sicken the community.

Esposito's perspective aligns with Agamben's critique of the state of exception. Giorgio Agamben has criticized the tendency of modern governments to abuse the "state of exception" during crisis situations. He viewed the COVID-19 crisis as governments worldwide normalizing emergency measures and justifying severe restrictions on freedom [26]. Indeed, in South Korea, strong administrative orders and surveillance systems that would normally be unacceptable were deployed under the justification of infectious disease crisis. Personal movement information was disclosed through emergency disaster text messages, and private information such as mobile phone locations and credit card usage records was extensively collected and utilized for epidemiological investigation purposes. Additionally, fundamental rights such as freedom of assembly and religious activity were restricted under quarantine purposes, making some feel as though they were experiencing a constitutional state of exception.

From Agamben's perspective, such measures carry the risk of making emergency situations a normal means of governance. In other words, there is concern that public health crises could become opportunities to perpetuate expanded government authority and strengthened control. Indeed, when governments restrict freedom citing public desire for safety, citizens come to accept those restrictions out of fear. This can lead to erosion of freedom and rights in democratic societies. While most quarantine measures in South Korea were based on law, citizens experienced intensive restrictions for extended periods, creating areas difficult for democratic control. For instance, violating self-isolation orders could result in electronic ankle monitors or criminal punishment—measures that would be controversial in normal times but received widespread support during the pandemic. The problem lies in the fact that while such exceptional governance methods should be limited to temporary emergency responses, the point of return to normalcy becomes unclear. As quarantine measures prolonged, social consensus on "exactly how long should freedom be restricted" wavered, and growing voices of opposition can be seen as warnings against the normalization of the state of exception.

In essence, Esposito's immunity concept critically illuminates the duality of exclusion and protection in Korean quarantine policy, while Agamben's state of exception concept points to phenomena where democratic values are suspended under the justification of quarantine. Korean quarantine policy simultaneously demonstrated positive effects of protecting the community from the virus and negative effects of pushing some members outside the community or restricting freedom and rights. This shows that quarantine policy is more than simple health administration—it is policy with socio-philosophical implications.

Transparency and Information Disclosure Challenges

The South Korean government's transparency strategy during COVID-19 exposed fundamental tensions between public information needs and individual privacy protection. While proactive disclosure of epidemiological data enhanced public trust and risk communication, the granular detail of certain disclosures compromised personal privacy and led to social stigmatization. The lack of predefined standards for data disclosure contributed to public unease and privacy infringements. The government's information-centric, top-down communication model, while effective for rapid dissemination, provided limited avenues for public engagement and participatory input in policy decisions.

Political dynamics significantly influenced pandemic response policies, with the entanglement of public health decision-making and political calculations occasionally resulting in delayed responses, policy reversals, and enforcement inconsistencies. Opposition scrutiny of scientific rationales and socio-economic consequences of restrictions highlighted challenges in sustaining coherent, evidence-based strategies within democratic political systems.

Critical Gaps in Crisis Management

The pandemic revealed that ethical considerations were often reactive rather than embedded from the outset of emergency planning. Vulnerable groups including undocumented migrants, informal workers, the digitally marginalized, and older adults were inadequately identified in preparedness plans, lacking tailored strategies for outreach, healthcare access, and financial support. The absence of independent oversight over emergency policies, particularly regarding surveillance and vaccine access, compromised public trust and democratic accountability [27].

The exclusive focus on disease-specific control metrics neglected collateral health consequences of emergency interventions. Widespread postponement of routine healthcare services resulted in significant care backlogs, disruptions in chronic disease management, and exacerbation of pre-existing conditions for vulnerable populations. Mental health infrastructure proved inadequate for addressing the psychological toll of prolonged social distancing, economic instability, and health-related anxieties across diverse population groups.

Recommendations for Ethical Governance Frameworks

Toward an Accountable and Inclusive Decision Framework

Future pandemic governance must weave together public-health effectiveness, social resilience, and inclusiveness. That requires formal channels for civil society, community representatives, and multidisciplinary experts to shape strategy from the outset, ensuring that trade-offs reflect a full spectrum of values, not just infection metrics.

Decision grounds must be public, reviewable, and open to revision, because pandemic choices inevitably "select" among population groups for protection or sacrifice. Transparent reasoning plus a guaranteed right to contest and adjust policies are essential to procedural legitimacy, especially when rights restrictions or priority setting are unavoidable. We could consider accountability for reasonableness frame for future decision-making process [28].

Comprehensive Ethical Decision-Making Structures

Future emergency governance requires robust ethical frameworks that explicitly address individual privacy protection, equitable resource allocation, and proportional rights restrictions. These frameworks should establish clear criteria for initiating, continuing, and de-escalating emergency measures alongside transparent mechanisms for public accountability and independent review. Emergency measures including surveillance, lockdowns, and public disclosures must be subject to rigorous ethical scrutiny with defined boundaries for data collection, legal oversight of emergency powers, and sunset clauses to prevent normalization of rights restrictions.

Participatory Governance and Democratic Oversight

Governments should institutionalize mechanisms for structured input from civil society, healthcare professionals, and affected communities through standing citizen advisory committees, deliberative forums on risk communication, and formal consultations with public health experts. These measures increase public understanding and compliance while safeguarding against the politicization of crisis management. Enhanced accountability mechanisms should include routine evaluation of policy effectiveness and proportionality, transparent publication of findings, and legal avenues for redress when rights are infringed.

Equity-Centered Response Strategies

Future frameworks must incorporate explicit provisions protecting vulnerable populations, promoting equitable distribution of healthcare resources, and addressing social determinants of health. This includes designing targeted support programs for marginalized communities and ensuring cultural competence, social justice, and equity principles inform emergency response protocols. Data anonymization protocols and predetermined thresholds for information release must balance transparency with privacy protection.

Sustainable Healthcare System Adaptations

Healthcare systems must be redesigned to maintain continuity of essential services during acute crises through hybrid care models that transition between emergency and routine care, flexible workforce deployment systems, and equitable infrastructure distribution across urban and rural regions. Comprehensive workforce planning should extend beyond immediate staffing to encompass long-term sustainability, including robust training programs, standardized emergency deployment protocols, and integrated support systems prioritizing healthcare professional safety, mental health, and well-being.

Holistic Health Metrics and Long-Term Planning

Effective preparedness frameworks must incorporate multidimensional health metrics including mental health, chronic care continuity, and social well-being, not merely epidemiological indicators. Strategies should ensure continuity of essential healthcare services during crises, including preventive care, chronic disease management, and mental health support. Long-term capacity building should prioritize sustainable, resilient healthcare infrastructures through strategic investments in adaptable infrastructure, comprehensive training programs, and robust public health surveillance systems.

Global Solidarity and International Cooperation

Future frameworks must embed global solidarity from the outset, viewing engagement in international vaccine-sharing initiatives, bilateral health diplomacy rooted in equity, and investment in regional public health capacity as integral components of ethical pandemic governance rather than charitable acts [27]. This requires proactive participation in global health security mechanisms and consideration of international responsibilities alongside national priorities.

Implementation Principles for Future Preparedness

Moving forward, pandemic preparedness must move beyond efficiency metrics toward a principled foundation embedding ethics centered on rights protection, equity, participation, and transparency at the heart of public health planning. This approach requires balancing immediate infection control with broader long-term health outcomes, establishing flexible resource allocation mechanisms with regular stress testing and evaluation, and creating integrated models of routine and emergency care that maintain service continuity.

Success in future pandemic responses necessitates comprehensive approaches that consider interrelated dimensions of public health preparedness while maintaining adaptability to diverse circumstances. The COVID-19 experience demonstrates that effective preparedness extends beyond technical proficiency to encompass robust ethical governance frameworks, sustainable health system adaptations, and comprehensive support structures safeguarding both individual and community well-being. Embedding these principles will not only improve crisis management effectiveness but also foster public trust and social cohesion in facing future global health threats.

Conclusion

South Korea's COVID-19 response worked—but at what cost? The K-quarantine model controlled the virus through mass testing, digital tracking, and swift interventions. Yet this success came with serious ethical problems. The government tracked people's every move using phone data, credit cards, and security cameras. When officials published detailed movement histories, it led to harassment and discrimination, especially against religious groups. People had no say in these surveillance measures, raising fundamental questions about consent and privacy.

The pandemic also exposed deep inequalities. Elderly people couldn't navigate digital vaccine booking systems. Migrant workers and informal laborers were left out of support programs. Healthcare workers burned out without proper mental health resources. Politics infected public health decisions, with partisan concerns often trumping scientific evidence. Emergency powers bypassed normal democratic checks, weakening public trust and accountability.

This study has clear limitations. The pandemic kept changing, so some observations may already be outdated. The analysis relies mostly on public information, which probably misses important details about how decisions were made and how marginalized communities really experienced these policies. Focusing only on South Korea also limits how much can be generalized to other countries. Important questions remain about healthcare workers' long-term trauma, surveillance overreach, and how politics will keep interfering with public health.

The lesson is clear: technical competence isn't enough. Good pandemic response needs ethical frameworks that protect individual rights while protecting collective health. It needs healthcare systems that can adapt without breaking. Most importantly, it needs to include diverse voices in decision-making, not just expert committees and government officials. Future pandemic planning must care about social cohesion and democratic values, not just case numbers and hospital capacity.

References

1. Mathieu E, Ritchie H, Rodés-Guirao L, Appel C et al (2020) Coronavirus pandemic (COVID-19). Published online at OurWorldinData.org. Retrieved from: https://ourworldindata.org/coronavirus [Online Resource]
2. Jang M, Kim S, Choi S, Ryu B, Choi SY, Choi S, An M, Kim SS (2024) Estimating excess mortality during the COVID-19 pandemic between 2020–2022 in Korea. J Korean Med Sci 39(40):e267. https://doi.org/10.3346/jkms.2024.39.e267
3. Im SJ, Shin JY, Lee DH (2024) Excess deaths in Korea during the COVID-19 pandemic: 2020-2022. J Prev Med Public Health 57(5):480–489

4. World Economic Forum (2020) How South Korea is suppressing COVID-19 through innovative technology and public trust. https://www.weforum.org/stories/2020/08/south-korea-covid19-government-pandemic-response/
5. Lee KS, Sung HK, Lee SH, Hyun J, Kim H, Lee JS, Paik JW, Kim SJ, Sohn S, Choi YK (2022) Factors related to anxiety and depression among adolescents during COVID-19: a web-based cross-sectional survey. J Korean Med Sci 37(25):e199. https://doi.org/10.3346/jkms.2022.37.e199
6. Jeong H, Park S, Kim J, Oh K, Yim HW (2022) Mental health of Korean adults before and during the COVID-19 pandemic: a special report of the 2020 Korea National Health and Nutrition Examination Survey. Epidemiol Health 44:e2022042. https://doi.org/10.4178/epih.e2022042
7. Kim H (2022) The Korean 3T practice: new biosurveillance model utilizing new information technology and digital tools. JMIR Format Res 6(5):e34284
8. Choi S, Han C, Lee J, Kim S-I, Kim IB (2020) Innovative screening tests for COVID-19 in South Korea. Clin Exp Emerg Med 7:73–77
9. Kang J et al (2020) South Korea's responses to stop the COVID-19 pandemic. Am J Infect Control 48:1080–1086
10. Lee D, Lee DH (2020) Healthcare service justice and community engagement in crisis situation: focusing on failure cases in response to COVID-19. J Korea Serv Manag Soc 21:293–312
11. Ha T (2023) The effects of COVID-19 and response measures on poverty, household income and household consumption expenditure in South Korea. J Econ Stud. https://doi.org/10.1108/jes-09-2021-0453
12. Park WB, Hwang YH, Cheong HJ (2023) COVID-19 vaccination in Korea. Infect Chemother 55(1):135–149. https://doi.org/10.3947/ic.2023.0023
13. Ministry of Health and Welfare (2021) White paper on health and welfare. Ministry of Health and Welfare, Republic of Korea, Seoul
14. Republic of Korea (2020) Infectious Disease Control and Prevention Act, Act No. 16725. Ministry of Government Legislation. https://www.law.go.kr/LSW/eng/engLsSc.do?menuId=2&query=infectious%20disease#liBgcolor0
15. Edelman (2021) 2021 Edelman Trust Barometer. Edelman. Retrieved from https://www.edelman.com/trust/2021-trust-barometer
16. Kasulis K (2020) South Korea's coronavirus scapegoat. Foreign Policy. https://foreignpolicy.com/2020/08/19/south-korea-coronavirus-scapegoat-shincheonji/
17. The Associated Press (2020) South Korea pastor tests positive amid COVID-19 spike at Seoul church. The Associated Press
18. Chang K, Lee SY (2019) Why do some Korean parents hesitate to vaccinate their children? Epidemiol Health 41:e2019031. https://doi.org/10.4178/epih.e2019031
19. Reuters (2021). South Korean businesses protest against return of strict COVID rules. The Star. https://www.thestar.com.my/news/world/2021/12/22/south-korean-businesses-protest-against-return-of-strict-covid-rules
20. Rashid R, Kasulis K (2020) Coronavirus resurgence in South Korea reignites homophobia. Foreign Policy. https://foreignpolicy.com/2020/05/13/coronavirus-resurgence-south-korea-reignites-homophobia/
21. Rashid R (2021) South Korea's foreigner-targeted COVID testing will hurt its New Southern Policy. The Diplomat. https://thediplomat.com/2021/03/south-koreas-foreigner-targeted-covid-testing-will-hurt-its-new-southern-policy/
22. Korea Disease Control and Prevention Agency (2022) National vaccination progress report. Korea Disease Control and Prevention Agency
23. Kang C, Lee I (2021) COVID-19 pandemic, transparency, and "Polidemic" in the Republic of Korea. Asian Bioeth Rev:1–12. https://doi.org/10.1007/s41649-021-00164-4
24. Hwang IK (2024) The politics of life in the post-COVID-19 era: an intersectional perspective on immunity and vulnerability focusing on healthcare and public health. Sookmyung Res Inst Hum 17:71–100. https://doi.org/10.37123/th.2024.17.71
25. Calleja C (2021) Gregory of Nyssa's "Reverse Contagion" and Roberto Esposito's "Immunity": which way forward in the aftermath of the pandemic? J Moral Theol 10(1):26–45

26. Agamben G (2020) The state of exception provoked by an unmotivated emergency. Positions Politics. https://positionspolitics.org/giorgio-agamben-the-state-of-exception-provoked-by-an-unmotivated-emergency/
27. Ajana B (2021) Immunitarianism: defence and sacrifice in the politics of Covid-19. Hist Philos Life Sci 43(1):25. https://doi.org/10.1007/s40656-021-00384-9
28. Hasman A, Holm S (2005) Accountability for reasonableness: opening the black box of process. Health Care Anal 13:261–273. https://doi.org/10.1007/s10728-005-8124-2

Part III

Southeast Asia

Beyond Lockdowns: Equity, Trust, and Social Justice in Malaysia's Pandemic Strategy

6

Jeffrey Soon-Yit Lee, Varaalakshmy Gokilavanan, Kai Hong Ooi, and Mark Kiak Min Tan

Abstract

Malaysia's response to the COVID-19 pandemic unfolded across four waves, driven by variants like Delta and Omicron, resulting in 5.1 million cases and 37,200 deaths in 2023. This emphasized the value of harm prevention, equity, societal responsibility, and the common good. Movement Control Orders (MCOs) effectively reduced transmission but disproportionately impacted marginalized groups, such as migrants and rural communities, deepening unemployment and poverty. Vaccination efforts achieved 85% adult coverage by 2022, yet indigenous populations like the Orang Asli lagged at 32% due to logistical barriers and mistrust. The Delta wave overwhelmed ICU capacity, exposing healthcare disparities and ethical tensions as collective welfare often sidelined individual rights. The chapter highlights key lessons for future pandemics. Adaptive policies, like the Situational Matrix, must incorporate equity audits to address disparities, ensuring support like cash transfers for vulnerable groups. Health system resilience demands expanded ICU capacity and digital triage systems with offline alternatives for rural access. Communication strategies should leverage cultural leaders and multilingual campaigns to counter misinformation, as seen with fatwas boosting Muslim vaccination rates. Equity-centered aid,

J. S.-Y. Lee
Clinical Research Centre, Hospital Sibu, Sarawak, Malaysia

V. Gokilavanan
Hospital Kuala Lumpur, Kuala Lumpur, Malaysia

K. H. Ooi
Bioscience and Biomedical Engineering Thrust, System Hub, Hong Kong University of Science and Technology (Guangzhou), Guangzhou, China

M. K. Min Tan (✉)
Jeffrey Cheah School of Medicine and Health Sciences, Monash University Malaysia, Bandar Sunway, Selangor, Malaysia

L. P. Marquez et al. (eds.), *COVID-19 in the Pacific Rim and Beyond*, Collaborative Bioethics 6, https://doi.org/10.1007/978-3-032-21218-4_6

using mobile banking and grassroots networks like Kita Jaga Malaysia, proved essential for reaching underserved populations. Public-private partnerships, such as enlisting private hospitals, require pre-emptive agreements to prioritize equity. Malaysia's experience underscores the need to balance epidemiological control with social justice. Future strategies must institutionalize equity by decoupling healthcare from immigration status, fostering community trust, and embedding intersectional audits into policy frameworks. This approach ensures inclusive crisis management that leaves no population behind.

Keywords

COVID-19 · Malaysia · Pandemic response · Pandemic preparedness · Public health

Introduction

Origins and Spread

The COVID-19 pandemic reached Malaysia on January 25, 2020, when three Chinese nationals tested positive after arriving from Wuhan [1]. Over the next four years, the country endured four distinct waves driven by different variants: the original Wuhan strain (January–October 2020), Beta (November 2020–March 2021), Delta (April 2021–January 2022), and Omicron (February 2022 onward). The Delta wave proved catastrophic, contributing 55.4% of total cases and overwhelming intensive care unit (ICU) capacity; daily deaths peaked at 592 in August 2021 [2]. By December 2023, Malaysia had recorded 5.1 million cases and 37,200 deaths, revealing systemic vulnerabilities in healthcare infrastructure and socioeconomic equity [1].

Scope and Ethical Framework

This chapter evaluates Malaysia's pandemic response through the lens of *public health ethics*—a discipline prioritizing (1) prevention of harm (e.g., curbing disease transmission, morbidity, and mortality); (2) equity (fair distribution of health resources and protection from disproportionate burdens across socioeconomic, ethnic, and migrant populations); (3) societal responsibility (collective duty to safeguard vulnerable groups through policies like lockdowns and vaccine mandates); and (4) the common good (maximizing population health outcomes even when limiting individual freedoms). These principles guided crisis policies but also exposed tensions between collective welfare and individual rights.

Policies and Implementation

Movement Control Orders (MCOs) and Public Health Social Measures

In March 2020, Malaysia imposed a comprehensive lockdown strategy under the Movement Control Order (MCO), aligning with WHO guidelines to restrict non-essential travel, enforce school closures, and mandate masking [3]. While daily cases dropped from 2000 to single digits by June 2020, the economic fallout was severe: 2.4 million jobs were lost, disproportionately affecting blue-collar workers and refugees [3]. The Enhanced MCO (EMCO) further targeted high-risk areas like factories and migrant dormitories, where outbreaks surged due to overcrowding. An example of this saw 5000 cases in Selangor's Top Glove factories during November 2020, prompting total lockdowns that halted production but protected workers and prevented spread to the community. By 2021, the government adopted the COVID-19 Situational Matrix, dynamically adjusting restrictions based on community transmission rates and hospital capacity [3]. This framework reduced blanket lockdowns but struggled to address inequities as marginalized groups lacked savings to weather prolonged closures.

Even prior to the pandemic, it was well known that the government clinics and hospitals were already congested with limited resource and manpower [4]. Thus, the main purpose of the MCO was to temporarily halt the virus transmission to prevent overwhelming the healthcare system, particularly government healthcare. However, the authorities' decision needed to be balanced against the country's economic needs and the people's livelihood. While public health and safety took precedence at the beginning of the pandemic, MCOs should not have been the only measure taken or implemented for the long term. For a highly infectious virus like COVID-19, access to testing and effective vaccines was also crucial.

Testing, Isolation, and Hospitalization Strategy

Testing strategies evolved from polymerase chain reaction (PCR) dominance to rapid antigen kits during Omicron, with private labs processing most COVID-19 tests by 2022. Hospitalization policies prioritized severe cases, but the Delta wave exposed ICU shortages [23]. Occupancy reached a high amount in Selangor, forcing hospitals to ration ventilators [2]. Virtual assessment centers diverted mild cases to home isolation, yet rural areas lacked digital infrastructure, leaving vulnerable populations underserved.

Prior to the COVID-19 pandemic, molecular diagnostics such as the PCR were a luxury to most government healthcare systems in Malaysia. In the past, there were only a handful of laboratories offering PCR service, mainly in the bigger state hospitals or universities. As a result, other smaller hospitals or healthcare facilities that required PCR diagnostics would have to send the specimens to bigger hospitals or university hospitals. A major issue during handling specimens intended for PCR was enforcement of cold-chain transfer, which was sometimes broken during transport. Moreover, there were issues of specimens going missing or being spilled during transport. If the specimens made it to the reference lab, it took days to weeks for the results to be available due to the sheer volume of specimens the reference lab had to work on.

The COVID-19 outbreak highlighted the disparity of access to diagnostics within the government healthcare system. For a highly virulent disease, containment of infected individuals is vital. While Malaysia implemented a quarantine and test policy, the testing capacity at the beginning of the pandemic was inadequate, as the main reference labs (Institute of Medical Research and the National Public Health Laboratory) were overwhelmed with specimens from all over the country. As a result, some persons under observation (PUO) were quarantined for an extended period, while patients with influenza-like illness (ILI) and severe acute respiratory infection (SARI) had to wait for days, and sometimes weeks, for their results. This situation allowed for the potential infection of others during the waiting period [23].

The disparity was particularly felt in the East Malaysian states of Sarawak and Sabah, where specimens needed to be couriered by flight to the capital city of Kuala Lumpur. In Sarawak, where communal living in the longhouse is the norm, one positive case could rapidly lead to disaster. Thus, rapid turnaround time for the test results is exceptionally important to enable isolation. Yet, due to geographical factors and lack of assess, the most vulnerable population had least access to the diagnostics test. The COVID-19 outbreak was thus a wake-up call to the government and authorities to focus on the development and upgrading of healthcare facilities in all regions rather than in the capital city alone.

Vaccination Rollout: Equity and Access

Malaysia's vaccination drive began in February 2021, prioritizing frontline workers, the elderly, and those with comorbidities. By 2022, 85% of adults were fully vaccinated, supported by the *National People's Well-Being and Economic Recovery Package*, (PEMULIH), which allocated MYR150 billion for vaccine procurement and cash aid [3]. However, rural and migrant populations faced barriers: only 32% of Orang Asli (indigenous communities) were vaccinated by late 2021 due to logistical hurdles and mistrust [3]. In contrast, a significant number of populations who had ready access refused the vaccines, giving rise to the anti-vaccine movement which led to vaccine hesitancy among the population. This paradoxical distribution proved to be a challenge to policy makers whose main aim was to ensure equitable access to limited vaccines.

One strategy was addressing vaccine hesitancy. Vaccine hesitancy persisted nationwide and was fueled by misinformation on social media and political scandals, such as allegations of "VIP queue-jumping" for doses [5]. Vaccine mistrust was further fueled by various conspiracy theories which were mostly baseless but believed by many, even by some healthcare professionals. For instance, COVID-19 is only considered severe in susceptible people, hence vaccination is unnecessary for the healthy. There were suggestions that alternative coronavirus therapies would be more useful than vaccinations [8]. Despite various efforts by the Malaysian Ministry of Health and healthcare professional groups to debunk misinformation, unfortunately, there were still significant levels of mistrust regarding the strategy to make COVID-19 vaccine mandatory. By 2023, booster uptake consequently stagnated, signifying either increased complacency due to the perception that the pandemic was over or the increasing of subscribers to conspiracy theorists. This underscores the need for a community-based study to understand the people's perception toward pandemic prevention and regulatory policies which are crucial in the event of future pandemics.

Practices and Stakeholder Dynamics

Public-Private Collaboration

To cope with the exponential number of cases, the government enlisted 95 private hospitals to manage COVID-19 cases in January 2021, expanding ICU beds by 30%. Private labs processed almost a hundred thousand tests daily. Corporate initiatives, such as the Selangor Testing Consortium, conducted mass screenings in factories. However, critics argued these efforts favored economic productivity over worker safety.

Telemedicine platforms like DoctorOnCall provided remote consultations during the pandemic. While telemedicine proved advantageous during the pandemic where physical distancing was preferred and crisis standards of care were applied, concerns about the quality of care provided arose. The main issue was that physical examination, a vital component of diagnosis, was impossible with telemedicine. Other issues included data security and the need for protection of patients' privacy [24]. In order to ensure that telemedicine was conducted in an ethical manner, the Malaysian Medical Council issued an Advisory on Virtual Consultations (during the COVID-19 pandemic) in April 2020 [9].

Communication and Enforcement

Multilingual campaigns on TikTok and WhatsApp improved compliance with mask mandates, with surveys showing public approval of standard operating procedures. However, enforcement was often perceived as biased; for example, migrant workers faced fines up to MYR10,000 for protocol breaches, while those in the upper-class flouting rules received lighter penalties. Stigmatization of COVID-19 patients, particularly in-migrant dormitories, further eroded trust.

Community and Cultural Responses

Grassroots networks like *Kita Jaga Malaysia* mobilized volunteers to deliver food and personal protective equipment to low-income households. Indigenous communities, such as the Orang Asli, erected barricades and retreated into forests to avoid infection while blending traditional practices with pandemic precautions. However, gatherings during cultural festivals contributed to localized outbreaks, exacerbating transmission risks and precipitating surges in COVID-19 cases. This compelled authorities to navigate the delicate equilibrium between safeguarding public health and upholding cultural and religious freedoms [10].

Ethical Framework for Public Health Decision-Making

Public health ethics prioritizes the prevention of harm, promotion of population well-being, and equitable distribution of health resources. It balances collective welfare with individual rights through principles of *equity* (fair allocation of burdens and benefits), *societal responsibility* (shared duty to protect vulnerable groups), and the *common good* (maximizing health outcomes for all). These principles guide decisions during crises like COVID-19, where policies must address systemic inequities while maintaining public trust and justice.

Lives Versus Livelihood

For a new and unknown threat like COVID-19, the public health field may adapt the precautionary principle. The precautionary principle asserts that when faced with an unknown threat, the uncertainty must be resolved in favor of preventing exposure [11]. In the COVID-19 context, preventing exposure to the SARS-CoV-2 virus is the priority, manifesting as the MCOs in the early phase. In the utopian world, administrative hazard management measures like the MCO should be in place until the hazard is eliminated. In reality, containing the virus spread with MCO needed to be balanced against economic well-being. In other words, public health intervention has to save lives without causing too much disruption to the livelihood of the population.

Nevertheless, the economic impact of MCO on people's livelihood is alarming. While the MCO managed to avert about 500,000 COVID-19 deaths during the pandemic, it had deepened poverty for many more. Over 30,000 businesses were shut down and employees were dismissed from their permanent jobs, leading to a spike in unemployment rate to 4.5% by May 2021. The pandemic poverty phenomenon impacted mental health negatively, leading to a rise in mental issues such as depression and suicide [12]. The lower- and middle-income groups were particularly vulnerable to the negative economic and emotional impact of the MCO [13].

To cushion the negative economic impact, the government introduced multiple economic stimulus packages, which provided some relief to the lower- and middle-income group. Additionally, data-driven decisions made by the authorities enabled modifications to the MCOs based on risk stratifications. For example, the conditional MCO (CMCO) employed during the later phase of the pandemic allowed more economic activities, which improved livelihood to a certain extent.

The MCO highlights the ethical challenges of making public health decisions. While some of the policies made during the MCO were seen as infringing on freedom and human rights, the ultimate aim of any public health policy is to save lives [23]. Realistically, there was no perfect solution to address a crisis of a pandemic scale. Considering that, Malaysia has done reasonably well to mitigate the pandemic effect, both on the lives and livelihood of her people.

Societal Responsibility and Trust

For effective crisis management, stakeholder's trust is crucial [14]. In the context of the COVID-19 pandemic response, trust between the authorities and the public is paramount to shape public behavior in the right direction. To achieve a trusting relationship, effective communication is the key.

The Malaysian authorities had to put in much effort in ensuring important messages were communicated to the public. Daily reports of the Director General of the Malaysian Ministry of Health on mainstream and social media regarding the pandemic situation provided some confidence to the public in the authority's capability to manage the pandemic in the country. Nevertheless, conspiracy theorists continued to flourish despite the government's effort. The most common conspiracy theory was that the government had hidden important facts from the public, including the actual death toll. While the majority of the population, especially those working in the health sector, were aware that these claims were baseless, it certainly eroded the trust of the population toward the authorities.

By the time the COVID-19 vaccines became available, the trust in the government had already eroded. Misinformation was thriving on social media, where the authorities had little control. Due to the MCO, a large number of the population who previously did not use any form of social media became internet savvy. As a result, misinformation reached a significant number of the public. The most common misinformation was on the adverse effects of the vaccine, which discouraged people from accepting it. Unfortunately, although most of the claims made were not backed by scientific evidence, a significant number of people accepted them. While the authorities attempted to dispel misinformation with scientific explanations, these messages were less persuasive to the public because the vaccines were newly developed and conclusive scientific evidence was still limited. As there was a paucity of evidence of whether the vaccines were good or bad, the loudest group attracted more attention.

Although the authorities had attempted to improve communication and debunk misinformation, there were areas that could have been improved. Greater enhancement of the law to curb misinformation was needed. Although there were concerns that this would have impinged on individuals' freedom of speech, the COVID-19 misinformation menace had proven that freedom of speech should come with the responsibility to provide accurate facts. Irresponsible spread of misinformation masked under the guise of freedom of speech can be catastrophic, especially in times of crisis when all should unite [5].

Vulnerable Populations

While every Malaysian was affected by the pandemic response, vulnerable groups including migrants, refugees, stateless, and indigenous peoples were the most affected due to underlying untreated health conditions, pre-existing barriers to healthcare, and undetermined immigration or citizenship status. Many migrants, particularly those from developing countries, had difficulty accessing affordable healthcare. Healthcare costs for migrants can be many times higher compared to the costs for citizens. This could be due to a combination of government policies, structural differences in the healthcare system, and limited insurance options. Documented migrant workers are required to have health insurance through schemes such as SPIKPA (*Skim Perlindungan Insurans Kesihatan Pekerja Asing*). However, this coverage is restricted to public hospital admissions and includes a financial cap, which may fall short of covering severe medical conditions or prolonged treatments. Consequently, even insured migrants often face out-of-pocket expenses for outpatient care, medications, or services beyond the scheme's limits [15]. Undocumented migrants, who had no access to any insurance, had to bear the full cost of medical care themselves, further exacerbating their financial vulnerability. These limitations underscore the precarious healthcare situation faced by migrant workers, regardless of their legal status. As a result, many migrants either self-medicate or ignore underlying health conditions such as hypertension, which is a significant risk for COVID-19 mortality. Besides high healthcare costs, refugees, stateless people, and illegal immigrants feared arrest and deportation should they seek treatment at hospitals. In certain areas with a high number of migrants staying in crowded living conditions, outbreaks were inevitable. Outbreaks among migrants population triggered stigmatization and community isolation, which subsequently contributed to loss of employment and financial insecurity. This vicious cycle contributed to the rapid spread of the virus among migrants by May 2020 [16].

To prepare for a future pandemic, public health policies should recognize and include vulnerable groups, guided by the principle of equity and distributive justice.

Lessons for Future Pandemics

Adaptive Policymaking: The Situational Matrix Model and Equity Audits

Malaysia's COVID-19 Situational Matrix, introduced in 2021, marked a shift from blanket lockdowns to dynamic, data-driven restrictions. This framework categorized districts into color-coded tiers (green, yellow, red) based on two metrics: community transmission (CT) rates and healthcare response capacity (RC). For example, a district with high CT but sufficient RC might face targeted workplace closures, while one with overwhelmed hospitals would trigger stricter measures like school shutdowns. This approach reduced economic disruption compared to the 2020 MCO, which cost Malaysia MYR2.4 billion daily in GDP losses. However, the model's focus on aggregate data often overlooked disparities affecting marginalized groups. During the Delta wave, industrial zones like Selangor's Klang Valley, which is home to migrant worker dormitories, remained in perpetual lockdown due to persistent outbreaks. On the other hand, urban areas reopened faster despite similar case counts.

To address this, equity audits must be integrated into adaptive frameworks. These audits would evaluate policies through an intersectional lens which assesses impacts on vulnerable populations such as refugees, indigenous communities, and informal workers. For instance, the EMCO in migrant dormitories prioritized containment but failed to provide adequate food or wage protections, leading to hunger strikes and human rights criticisms [17]. Equity audits could mandate parallel support measures, such as conditional cash transfers or guaranteed sick pay, to ensure that containment policies do not exacerbate poverty. Additionally, decentralized decision-making by empowering local leaders in rural or indigenous areas to tailor restrictions could improve compliance and reduce mistrust. By embedding equity into adaptive models, Malaysia can balance public health goals with social justice to provide safeguards for marginalized groups so that these groups may not be compromised in future crises.

Health System Resilience: ICU Expansion and Digital Triage

The Delta wave exposed critical gaps in Malaysia's healthcare infrastructure, with ICU bed occupancy reaching a high percentage in Selangor by August 2021. Many hospitals resorted to rationing ventilators and prioritizing younger patients with higher survival odds—a triage policy that sparked ethical debates. To prevent such crises, Malaysia must invest in surge-ready ICU capacity that aims for at least 10 ICU beds per 100,000 people (up from 4.5 in 2021) [18]. This requires upgrading rural hospitals, which lacked ventilators entirely in states like Kelantan, and training specialized staff through partnerships with institutions like the National Institute of Health.

Equally vital is the adoption of digital triage systems to optimize resource allocation. During Omicron, Malaysia piloted virtual assessment centers where patients uploaded symptoms via apps like MySejahtera, allowing clinicians to prioritize severe cases for hospitalization. However, rural communities faced barriers due to poor internet access and digital literacy. Future systems must integrate offline options such as SMS-based symptom reporting while ensuring AI algorithms were audited for biases. For example, AI triage tools in Singapore were found to underestimate risks for elderly patients with comorbidities—a flaw Malaysia must avoid

[19]. Pairing technology with grassroots health workers such as training village nurses to use portable diagnostics can provide equitable care during surges.

Inclusive Communication: Cultural Leaders and Multilingual Campaigns

Malaysia's multicultural fabric necessitated tailored messaging to combat misinformation and vaccine hesitancy. While national campaigns used platforms like TikTok and WhatsApp, rural indigenous communities and migrant workers often relied on oral traditions and community leaders. For example, the *Orang Asli* in Pahang distrusted government posters but responded to outreach by tribal elders who participated in vaccine trials. Similarly, Islamic authorities issued *fatwas* declaring vaccines *halal*, boosting uptake among Malay-Muslims from 62% to 82% in 2021 [6].

However, one-size-fits-all approaches failed non-Malay speakers. Refugees from the Myanmar and Pakistan communities struggled with Malay-language guidelines which exacerbated their exclusion. Future campaigns must partner with cultural brokers—NGOs, religious figures, and migrant advocates—to co-design multilingual, faith-sensitive materials. During Ramadan, Islamic NGOs successfully framed vaccination as a *fardhu kifayah* (collective obligation), while radio stations debunked myths about infertility [7]. Leveraging these networks can bridge trust gaps, especially where historical marginalization fuels skepticism toward authorities.

Equity-Centered Aid: Mobile Banking and Grassroots Networks

Malaysia's pandemic aid programs, like the *Bantuan Prihatin Nasional* (BPN) cash transfers, faced criticism for excluding undocumented migrants and rural households without bank accounts [20]. By late 2021, only 23% of refugees had received aid, compared to 85% of citizens. Mobile banking companies offered a solution during the PEMULIH package rollout. E-wallets like Touch 'n Go dispersed MYR500 one-off payments to 8 million Malaysians. Expanding such systems to refugees via biometric IDs (e.g., the United Nation Refugee Agency (UNHCR) cards) and rural communities through partnerships with telecommunication companies (e.g., Celcom's village kiosks) can ensure direct, leakage-proof aid.

Grassroots networks also proved indispensable. The Kita Jaga Malaysia initiative, a volunteer coalition, delivered food to 200,000 urban poor households during lockdowns by collaborating with local grocers and ride-hailing platforms [21]. Formalizing partnerships such as state-funded microgrants for community kitchens can institutionalize equity-centered aid. For indigenous groups, mobile clinics staffed by tribal health advocates (trained via NGOs) can bypass bureaucratic delays in remote areas.

Public-Private Synergy: Pre-pandemic Resource-Sharing Agreements

Malaysia's reliance on 95 private hospitals during the pandemic highlighted the potential of public-private partnerships (PPPs). However, ad hoc negotiations delayed resource mobilization. Private ICU beds were only activated 6 months into the Delta wave. Legislating pre-pandemic agreements would mandate annual joint drills, standardized protocols, and predefined cost-sharing models. For example, Singapore's PPP framework guarantees private hospitals state subsidies for reserving ICU beds during emergencies—a model Malaysia could adapt.

To ensure accountability, agreements must include equity clauses. Mandating minimum allocations for low-income patients and penalizing non-compliance through additional taxes can align private interests with public health goals. Additionally, integrating private labs into a national testing registry, as done in South Korea's drive-through system [22], can prevent duplication and ensure equitable distribution of testing kits to underserved regions.

Conclusion

Malaysia's pandemic response achieved significant milestones in containment through stringent Movement Control Orders (MCOs) and high vaccination coverage, yet it revealed systemic inequities in safeguarding marginalized populations. The prioritization of collective welfare often overlooked the vulnerabilities of migrants, refugees, and indigenous communities, who faced barriers to healthcare access, vaccine hesitancy rooted in mistrust, and exclusion from economic relief programs. These gaps underscore the ethical tensions inherent in public health decision-making, where rapid crisis management inadvertently deepened social divides. The Delta wave's disproportionate toll on overcrowded migrant dormitories and rural regions with fragmented healthcare infrastructure exemplifies the consequences of policies that lacked intersectional equity audits. To uphold ethical imperatives, future strategies must reconcile public health efficacy with a commitment to protecting those most at risk.

Moving forward, Malaysia's experience offers critical lessons for aligning pandemic preparedness with social justice. Ethical frameworks must institutionalize equity-centered measures, such as integrating vulnerable groups into policy design through grassroots partnerships, decoupling healthcare access from immigration status, and legislating transparency in data-sharing to rebuild public trust. Pre-emptive reforms—such as reserving private-sector ICU beds for low-income populations, decentralizing communication through cultural brokers, and formalizing equity audits in crisis protocols—can transform reactive policies into proactive safeguards. By spotlighting equity, fostering community agency, and addressing historical exclusion, Malaysia can pioneer a model that balances epidemiological rigor with moral accountability, ensuring that no population is left behind in the next global health crisis.

References

1. Ministry of Health Malaysia (2025) KKMNOW COVID19. https://data.moh.gov.my/dashboard/covid-19
2. Jayaraj VJ, Rampal S, Ng CW, Chong DWQ (2021) The epidemiology of COVID-19 in Malaysia. Lancet Reg Health West Pac 17:100295. https://doi.org/10.1016/j.lanwpc.2021.100295
3. United Nations (2024) Embracing system-wide reforms for equitable, sustainable health services through strengthening primary health care delivery and enhanced governance in Malaysia. https://malaysia.un.org/en/268006-embracing-system-wide-reforms-equitable-sustainable-health-services-through-strengthening
4. World Health Organization, Regional Office for the Western Pacific (2012) Malaysia health system review (Health systems in transition, Vol. 2 No. 1) [PDF]. World Health Organization. https://iris.who.int/bitstream/handle/10665/206911/9789290615842_eng.pdf
5. Yusof A, Muuti M, Ariffin L, Tan MKM (2020) Sharing information on COVID-19: the ethical challenges in the Malaysian setting. ABR 12:349–361. https://doi.org/10.1007/s41649-020-00132-4
6. Mardian Y, Shaw-Shaliba K, Karyana M, Lau CY (2021) Sharia (Islamic law) perspectives of COVID-19 vaccines. Front Trop Dis 2:788188. https://doi.org/10.3389/fitd.2021.788188
7. Alsuwaidi AR, Hammad HAA, Elbarazi I, Sheek-Hussein M (2023) Vaccine hesitancy within the Muslim community: Islamic faith and public health perspectives. Hum Vaccin Immunother 19(1):2190716. https://doi.org/10.1080/21645515.2023.2190716
8. Wong LP, Alias H, Wong YL, Megat Hashim MMAA, Lin Y, Hu Z (2023) COVID-19 anti-vaccine sentiments in Malaysia: narratives of comments from Facebook post. Vaccine 11(4):834. https://doi.org/10.3390/vaccines11040834
9. Malaysian Medical Council (2024) Malaysian Medical Council Guideline on Telemedicine. Retrieved from, https://mmc.gov.my/wp-content/uploads/2024/01/MMC-Guideline-on-Telemedicine.pdf. Accessed 30 July 2025
10. Tan HS, Tan MKM (eds) (2020) Bioethics and COVID-19: guidance for clinicians, 1st edn. Malaysian Bioethics Community. Retrieved from, https://doi.org/10.5281/zenodo.3821126
11. Goldstein BD (2001) The precautionary principle also applies to public health actions. Am J Public Health 91(9):1358–1361. https://doi.org/10.2105/ajph.91.9.1358
12. Afrina N (2019) Covid-19: How pandemic poverty overtakes Malaysians. Retrieved from, https://midas.mod.gov.my/others/22-anything/2-2021/296-covid-19-how-pandemic-poverty-overtakes-malaysians
13. Aziz NA, Othman J, Lugova H, Suleiman A (2020) Malaysia's approach in handling COVID-19 onslaught: report on the Movement Control Order (MCO) and targeted screening to reduce community infection rate and impact on public health and economy. J Infect Public Health 13(12):1823–1829. https://doi.org/10.1016/j.jiph.2020.08.007. ISSN 1876-0341.
14. Bonfanti RC, Oberti B, Ravazzoli E, Rinaldi A, Ruggieri S, Schimmenti A (2023) The role of Trust in Disaster Risk Reduction: a critical review. Int J Environ Res Public Health 21(1):29. https://doi.org/10.3390/ijerph21010029
15. Loganathan T, Rui D, Ng CW, Pocock NS (2019) Breaking down the barriers: understanding migrant workers' access to healthcare in Malaysia. PLoS One 14(7):e0218669. https://doi.org/10.1371/journal.pone.0218669
16. Wahab A (2020) The outbreak of Covid-19 in Malaysia: pushing migrant workers at the margin. Soc Sci Humanit Open 2(1):100073. https://doi.org/10.1016/j.ssaho.2020.100073
17. Amnesty International (2020) Crackdown on migrants and refugees an appalling violation of human rights. Retrieved from, https://www.amnesty.my/2020/05/01/crackdown-on-migrants-and-refugees-an-appalling-violation-of-human-rights/. Accessed 10 July 2025
18. Phua J, Kulkarni AP, Mizota T, Hashemian SMR, Lee WY, Permpikul C, Chittawatanarat K, Nitikaroon P, Arabi YM, Fang WF, Konkayev A, Hashmi M, Palo JE, Faruq MO, Shrestha BR, Kc B, Mat Nor MBB, Sann KK, Ling L, Haniffa R et al (2023) Critical care bed capacity

in Asian countries and regions before and during the COVID-19 pandemic: an observational study. Lancet Reg Health West Pac 44:100982. https://doi.org/10.1016/j.lanwpc.2023.100982
19. Nord-Bronzyk A, Savulescu J, Ballantyne A, Braunack-Mayer A, Krishnaswamy P, Lysaght T, Ong MEH, Liu N, Menikoff J, Mertens M, Dunn M (2025) Assessing risk in implementing new artificial intelligence triage tools-how much risk is reasonable in an already risky world? Asian Bioethics Rev 17(1):187–205. https://doi.org/10.1007/s41649-024-00348-8
20. Gopalan N, Tong WT, Low WY (2025) Navigating health inequities and human rights violations: a narrative review on refugees and asylum seekers in Malaysia. J Public Health Emerg. https://doi.org/10.21037/jphe-24-108
21. Lee C (2023) #KitaJagaKita: (De)legitimising the government during the 2020 movement control order. In: Rajandran K, Lee C (eds) Discursive approaches to politics in Malaysia, Asia in transition, vol 18. Springer, Singapore. https://doi.org/10.1007/978-981-19-5334-7_12
22. Choi K, Choi H, Kahng B (2022) COVID-19 epidemic under the K-quarantine model: network approach. Chaos Solitons, Fractals 157:111904. https://doi.org/10.1016/j.chaos.2022.111904
23. Kaur S, Min MTK, Shu H, Ng CJ (2024) Ethical challenges in clinical practice during the COVID-19 pandemic in an academic healthcare institution in Malaysia: A qualitative study. Clinical Ethics 19(3):243–251. https://doi.org/10.1177/14777509221134499
24. Muuti MZ, Razali HYH, Ooi K, Tan MKM (2020) Ethical considerations for utilizing telemedicine during COVID-19. In: Tan HS, Tan MKM (eds) (2020) Bioethics and COVID-19: guidance for clinicians, 1st edn. Malaysian Bioethics Community. Retrieved from, https://doi.org/10.5281/zenodo.3821126 Bioethics and COVID-10

COVID-19 Practices and Policies in the Philippines: Perspectives from Public Health Ethics

7

Alexander Atrio Lim Lopez and Leander Penaso Marquez

Abstract

During the onslaught of the COVID-19 pandemic, at one point, the Philippines recorded one of the highest numbers of cases and the longest lockdown in the world. This chapter provides a normative analysis of the COVID-19 practices and policies in the Philippines focusing on three issues: the effectiveness of the government response and its proportionality to health outcomes and infringement of liberties, the prioritization of the vulnerable, and relational autonomy and access to information. From this analysis, the chapter provides recommendations that are deemed more ethical with regard to addressing possible future similar instances of pandemics.

Keywords

COVID-19 · Philippines · Public health ethics · Prioritization of the vulnerable · Access to information

Introduction: COVID-19 in the Philippines—A Timeline

This chapter provides a public health ethics analysis of the COVID-19 response in the Philippines. The chapter starts with a general timeline of events regarding the pandemic in the country, and proceeds to detail three thematic issues within a normative analysis: the effectiveness of the government response and its proportionality to health outcomes and infringement of liberties, the prioritization of the

A. A. L. Lopez (✉)
School of Law, Ateneo de Manila University, Makati City, Philippines
e-mail: allopez1@up.edu.ph

L. P. Marquez
College of Education, University of the Philippines Diliman, Makati City, Philippines

L. P. Marquez et al. (eds.), *COVID-19 in the Pacific Rim and Beyond*,
Collaborative Bioethics 6, https://doi.org/10.1007/978-3-032-21218-4_7

vulnerable, and relational autonomy and access to information. From this analysis, recommendations are given to possibly address future pandemics in the country through a more ethical response.

In the application of these principles, distinctive features of the Philippines are taken into account: its archipelagic territory composed of 7641 islands, its national population of around 109 million in the year 2020 [1], having "one of the highest rates of income inequality in South East Asia… [with] an income Gini coefficient of 42.3 percent in 2018" [2], and a community health expenditure of 41.5% being out-of-pocket in 2021 [3].

On January 9, 2020, the World Health Organization (WHO) reported that the outbreak of cases of pneumonia that was first reported in Wuhan, China, in December 2019 was due to a novel coronavirus [4]—later officially named as COVID-19.

A report by Baclig [5] recounts the initial stages and developments of the COVID-19 situation in the Philippines. The Department of Health (DOH) confirmed the first case of COVID-19 in the country on January 30, 2020. The first patient arrived in the country through Hong Kong from Wuhan, China. On February 2, the DOH recorded the first COVID-19 fatality in the Philippines. The first local transmission of the virus was confirmed on March 7.

The next day, then President Rodrigo Duterte declared a state of public health emergency through signing *Proclamation No. 922* as the country recorded its 24th COVID-19 case. In less than 2 weeks, Duterte declared a state of calamity in the entire country by signing *Proclamation No. 929*, which gave local government units (LGUs) the power to utilize funds reserved for emergency situations. On March 25, congress gave Duterte special powers through the *Bayanihan to Heal as One Act*; he was given the authority to reallocate government funds to aid pandemic response.

On April 3, the wearing of face masks was made mandatory in the country. A lockdown was then implemented in Luzon on the same day. On April 12, the local government of Valenzuela City in Metro Manila implemented the first local, targeted virus mass testing. By June 10, the Department of Budget and Management (DBM) had released P355.1 billion for the COVID-19 response. On July 24, the Philippines became a member of the COVID-19 Vaccines Global Access Facility (COVAX). On August 6, the country was reported as having the biggest number of COVID-19 cases in Southeast Asia with a total of 119,460 cases. On September 18, Duterte extended the state of calamity status of the country through *Proclamation No. 1021*.

On October 2, the Philippines was identified to be among the top 20 countries in the world with the most COVID-19 cases, as well as the country that had the longest lockdown. On November 19, Duterte approved "in principle" the granting of Emergency Use Authorization (EUA) by the Food and Drug Administration (FDA) to COVID-19 vaccines. By the end of the year, the Philippines recorded 474,064 COVID-19 cases, wherein 439,796 cases resulted in recoveries and 9244 cases resulted in deaths. The national government continued efforts to secure vaccines from various sources. By December 2020, the government mandated the use of face shields with face masks in public [6, 7].

Abad [8] describes the COVID-19 pandemic situation in 2021 as very similar to that of the previous year in numerous instances. The lockdown was still ongoing for the most part of the year, and the country had a slow vaccination pace. There were also investigations into anomalies with pandemic contracts facilitated by the government. It was late in the year when reported cases decreased that led to ease in restrictions, including the lifting of the mandatory use of face shields. A WHO [9] situation report confirms this decline but also mentions a "sudden and sharp increase" of reported cases as of January 2, 2022. On a positive note, the report also states that 60% of the population who were 12 years old and above were inoculated.

A welcome development in 2022 was that COVID-19 was not included anymore in the top ten leading causes of death in the Philippines, whereas it was identified as the third leading cause of death in the country in 2021 [10]. However, local health authorities recognized that COVID-19 was here to stay [11]. On October 28, 2022, Duterte's successor, President Ferdinand Marcos Jr., signed an executive order making the use of face masks voluntary except in health facilities, medical vehicles, and public transportation [12].

According to the WHO [13], the reported confirmed COVID-19 cases in the Philippines were at 4,110,205 from January 3, 2020, to September 6, 2023; among these cases, there were 66,661 confirmed deaths, while 189,317,158 doses of COVID-19 vaccines have been administered as of March 18, 2023.

In April 2023, the Health Technology Assessment Council (HTAC) of the DOH recommended the use of COVID-19 bivalent vaccines as booster for the general population. The availability of bivalent vaccine doses in the country varied from one municipality or city to the next, with some reporting that there were no doses available for residents [14].

On July 21, 2023, Marcos signed *Proclamation No. 297*, which effectively ended the state of public health emergency in the Philippines [15]. As of May 2024, the DOH revealed that there is no budget allocated for vaccines against new COVID-19 variants and that it is relying on whatever immunity is left from earlier vaccines [16].

From this general timeline of events, specific instances that raised ethical challenges arose during the pandemic response in the Philippines. The analysis will focus on three areas: the effectiveness of the government response to the pandemic and its proportionality to health outcomes and infringement of liberties, prioritizing the vulnerable, and relational autonomy and access to information.

Effectiveness and Proportionality of the Government Response to the Pandemic

It is useful to start the ethical analysis of the Philippine pandemic response with the most powerful entity in the country: the government. After all, the government's policies largely shaped the overall experience of daily life during the pandemic. No one can escape the government, certainly so during the longest lockdown in the world.

But how effective was this superlative length of quarantine? Despite the longest lockdown in the world, the Philippines still experienced one of the worst COVID-19 statistics internationally. But would different contextual specificities lead to a more favorable assessment of the lockdown?

How compliant were the citizens? A study using cellphone-based origin-destination flows from 1387 municipalities and cities in the Philippines covering 20 weeks during the pandemic concluded that "community quarantines were an effective policy for reducing urban mobility in the Philippines during crises, and mobility was highly responsive to different degrees of lockdown stringency" [17].

How strict was the lockdown? A study that used a stringency index concluded that "the Philippines' lockdown measures were excessively stringent, but did manage to minimize the overall cases and mortalities with minimal resources" [18]. This index considered such factors as the cancellation of onsite classes in schools, the prohibition of mass gatherings, the suspension of most modes of public transportation, the limited operations of hotels, and the closure of non-essential businesses.

This study also assessed that while the strictest period of the lockdown was effective in preventing a surge of cases, as a whole, it was "neither effective nor necessary as the nation reached an all-time high unemployment rate of 17.6%... and weekly cases began to increase after the end" of this period [18].

Stringency is also highlighted in another study that analyzed data from the Oxford COVID-19 Government Response Tracker (OxCGRT). This study concluded that "policing and law enforcement" produced "limited outcomes in reducing COVID-19 spread" [19].

Prefacing the ethical analysis with these facts and studies, this chapter now focuses on specific experiences of the lockdown to show that the moral considerations infringed by these experiences were not proportional to the benefits accrued from these experiences.

Proportionality, in this regard, may be differentiated to ask two different questions. One is how proportional are the benefits to the harms produced by a policy? Another is: "Will the probable benefits of the action outweigh the infringed moral considerations?" [14]. The first question relates well to the principles of nonmaleficence and beneficence in biomedical ethics. Nonmaleficence obliges people to prevent doing harm. Harm involves the infliction of a wrong, unjust treatment, or a violation of rights. Rules springing from this principle include the avoidance of both killing and the causing of pain and suffering. While nonmaleficence involves refraining from harm, beneficence involves positive steps undertaken to help others. Beneficence supports rules like protecting others' rights and rescuing people in danger [20]. These two principles are often understood to work together in that harms and benefits are analyzed to come up with the net benefit of an action.

While these two principles are usually analyzed in the interaction between individual patients and healthcare providers, in public health ethics, the objects of analysis become populations. The analysis becomes even more complicated considering the Philippine population of more than 100 million.

Part of the complication is that it can certainly be debated which variables in the first place are to be plugged into a harm-benefit analysis. However, mainstays are present in public health, like mortality and morbidity. These variables make sense considering one of the main goals of public health is the prevention of human sickness [14]. As such, net benefits measured in mortality and morbidity statistics become indicators of effectiveness of a public health intervention. It is in these indicators that the Philippines failed.

While the first-entertained question of proportionality puts to the fore facts and data, the second involves a more qualitative approach. Was the lockdown worth it in terms of how it infringed moral considerations? This question stems from a presumption that a principle of least infringement should be upheld [14].

Specifically, what should not be infringed are rights or liberties. A classic justification of this non-infringement is John Stuart Mill's harm principle: liberties should only be curtailed to prevent harm to society [21]. For the specific harm in this scenario, the discourse circles back to COVID-19 mortality and morbidity statistics. Harm from COVID-19 was not proportionately decreased considering how Filipinos' rights or liberties were curtailed during the lockdown.

This conclusion is ethical: one arrived at by the application of public health ethics principles. This ethical conclusion also cannot be arrived at without the existence of the problems in the implementation of the lockdown. After all, ethical principles have to be applied to facts within a context to generate an ethical conclusion.

The continuing discussion in this section analyzes more specifics of this experience of infringement, especially as disproportional to the intended effects of the government pandemic response: lockdown by the military and police, mandatory face shields with face masks, and alleged government corruption.

Lockdown by the Military and Police

Even before the pandemic, under Duterte, the Philippines already saw an increased visibility of the military and police. One of the most striking examples is their role in the President's war against drugs. With COVID-19, the war was against the virus.

There should be nothing offensive to the notion that the military and police are employed by the state to promote different kinds of public good; for instance, health. However, it is the extent of this employment and the atmosphere produced that may be questioned. Military vehicles were seen in the streets, and soldiers were stationed in shopping malls.

A helicopter aerial inspection was even conducted over the island of Cebu [22]. This inspection met criticism. What exactly did the aerial inspection accomplish? From that altitude, what was visible to give a conclusion of how effective quarantine measures were being observed? Was the inspection part of the government branding or image of military and police control?

To achieve such control by the executive branch of the government, a dramatic apex was when the President, on national television, ordered quarantine violators shot dead [23]. The Philippine National Police (PNP) clarified that the President was just overemphasizing the implementation of order during the pandemic crisis [24].

While killing quarantine violators was not the government policy, violators still suffered varying degrees of punishment that were neither proportional nor effective when it came to reducing morbidity and mortality.

One of the most extreme happened to 28-year-old Darren Peñaredondo. On April 1, 2021, Darren, after violating curfew during the quarantine, was allegedly made to do 300 rounds of squat-like exercises. When he got home, he required assistance in moving and eventually had seizures. On April 3, he died from a stroke [25]. Three policemen were fired from this incident [26]. While administrative and criminal investigations were instituted on the matter, there are no media updates on these investigations, including whether the family of the victim were able to receive compensation.

Further alleged reports of police abuse include confining youth curfew violators in a dog cage and, in another case, commanding curfew violators to sit under the hot midday sun [27]. Invoking the government's emphasis on order, it could be inferred that the point of these drastic measures was to instill fear and compliance with the law. This motivation is consistent with the government's war against drugs: if citizens are scared of harsh punishments, they will more probably obey the law. Transposed to the pandemic, if citizens are scared of harsh punishments, then they will probably obey curfew and other quarantine regulations.

However, even if a case could be made that a quarantine violator may spread the virus, in the timeline of the case of Peñaredondo, there was no possibility of establishing that Peñaredondo indeed contributed to the spread of the virus, and as such, warranted a harsh punishment.

While the few cases of curfew violations mentioned here could be attributed to specific agents of law enforcement rather than the government as a whole, what is argued here is that these isolated cases, seen as harsh punishments, are actually not isolated but stemming from even the pre-pandemic thrust of the government for harsher punishments.

This argument should be read still in light of an ethical analysis. It is ethical to do good (improve health) with the least harm (infringement of liberties). A national atmosphere of fear did not produce the least harm and arguably did not help in increasing the good. Therefore, the infringement of liberties is unethical.

Mandatory Face Shields with Face Masks

Compared to the just-mentioned penalties for curfew violations, the policy of mandatory face shields with face masks seems a rather trivial issue on proportionality. However, the practice produced not only discomfort to wearers but also possible links to government corruption.

At the outset, it should be mentioned that for the prevention of transmission of COVID-19 infections, there is no direct evidence that a face shield with a face mask is better than a face mask alone [28]. Against this set of medical evidence, the

Philippine government mandated the simultaneous use of a face shield with a face mask as the default rule, even outdoors.

The practice is quite uncomfortable. The discomfort should be explained first from the discomforting effects of wearing a face mask. Those who wear eyeglasses are bothered by the fogging effect on their eyewear. Face mask anxiety can occur because of the mask's effects on breathing and other causes [29].

All these sensations and anxieties can be made worse by face shields. The discomfort is considerable in a tropical country like the Philippines, right smack at the equator. The ill effects are experienced longer by those working outdoors for a full day like market and sidewalk vendors.

What is also discomforting is a face shield's handling and storage. Compared to a face mask, a face shield is solid, bulky, and takes up more space. While a face mask can just be easily put in one's pocket, one needs to find a flat planar space to put one's face shield on. One must also be conscious and careful that one's face shield is stored properly or placed somewhere that it would not be damaged for it can be easily bent or misshaped, which may affect clarity of vision. A damaged face shield means one may need to buy another. In 2020, one face shield could cost up to P80 per piece while one face mask could only cost up to P30 per piece [30]. Some lawmakers remarked that the requirement of face shield use was anti-poor [31]. The daily minimum wage was around P537 in 2021 (minimum wage varies per region) [32]. P80 of P537 is around 15%. A minimum wage laborer who goes to the streets faces higher chances of her face shield getting damaged and needing replacement.

To be clear, it is not being advocated that one should not use a face shield or face mask just because doing so can be uncomfortable. Instead, what is being pointed out is how necessary certain practices are in light of medical evidence. The reminder still is that certain practices do have a trade-off, and while physical, mental, or financial discomfort may not be the priority in a scenario where people are literally dying, experiencing these discomforts day after day in the world's longest lockdown magnifies these displeasures. The magnification can lead citizens to irately question the wisdom behind the policies and whether the experiences of those who have to work outdoors daily were even considered. The government should be able to present scientific and other evidence to support its implemented policy. In the speculation of why face shield use was continued to be mandated, allegations of government corruption cropped up.

Alleged Corruption in Government Spending for the Pandemic Response

A government audit found that in 2020, the Procurement Service of the Department of Budget and Management (PS-DBM) bought around 1.3 million face shields at P120 per piece, which was higher than the market price [33]. The audit further noted that around P188,000 worth of face shields were "sitting at regional depots" [34].

Allegations of anomalous spending of pandemic funds would include the DOH poorly spending P67 billion [35, 36]. Investigations would lead to senate hearings uncovering that the government transacted with a company, Pharmally Pharmaceutical Corporation, for around P10 billion [35]. Part of these transactions was that the PS-DBM gave COVID-19-related contracts to Pharmally, which imported PPEs, despite the application of local PPE and surgical mask manufacturers that could have provided the equipment at lower prices [37]. As a result of these investigations, in 2024, government prosecutors filed graft cases against suspected individuals over the alleged misuse of P41 billion for pandemic supply purchases [38].

The misuse of government funds is not only a problem in the implementation of the overall government response against the pandemic but also an ethical issue in public health. In public health, the health of the public should be taken care of; it is ethical to do so. To purposely misuse public funds is to avoid taking care of the health of the public, at least by some reasonable standard. Thus, such misuse is unethical.

If the allegations of corruption turn out to be true, then it makes much more sense why the use of face shields outdoors, even if not backed up by medical data, was insisted on by the government. The Filipino face shield experience becomes now a concrete reminder, quantifiable by public funds spent, of a disproportionate government response to the pandemic. There is disproportion because the funds spent were disproportionately higher than the actual cost of face shields. There is also disproportion because these funds could have been more proportionately spent on actual life-saving equipment backed up by medical data and other useful expenses.

These issues on corruption, tied up with the overbearing use of face shields, experienced in the longest quarantine, coupled with heightened military and police presence speak of the disproportional harms experienced by Filipinos during the COVID-19 pandemic. The harms consisted of the infringement of rights such as harsh punishments for quarantine violators and government losses of funds that could have been put to better use. The harms were not proportional to the benefits gained considering the country's poor statistics on COVID-19 morbidity and mortality. These harms were much harder experienced by the marginalized or vulnerable.

Prioritizing the Vulnerable During the Pandemic

The vulnerable are those with a status that may have an adverse impact on their well-being [39]. Alternatively, "one is vulnerable if one's needs are threatened" [39]. Simply put, one is vulnerable if one has limited access to basic necessities of living, among other things. As such, one may be in a status, state, or condition of having limited access to these necessities. The examples of status that may threaten one's needs or have an adverse impact on one's well-being can cover different classification schemes.

One possible classification is by looking at those who have less enjoyment of the social determinants of health (SDH), or the "non-medical factors that influence health outcomes" [40]. Those with a poorer profile of SDH have more health inequities, "the unfair and avoidable differences in health status seen within and between countries" [40]. Examples of SDH include income, "working life conditions," basic amenities, and education [40]. Those who have less of these determinants or those who experience more health inequities may also be marginalized, those who have systematically experienced oppression through generations.

A discussion on the marginalized or the vulnerable implies that there is a principle of protecting the vulnerable by prioritizing their needs. This principle makes sense given the definition of health inequities being unfair and avoidable. The vulnerable are characterized as such because of the health inequities they experience. A possible explanation of why such inequities are unfair is because these are the products of social conditions that people are born into [41]. Meaning, it is not the fault of individuals that they are born into social conditions that adversely affect their health outcomes. One did not choose in which country or family one was born into. Meanwhile, health inequities can be reduced because these can be at least partly alleviated as evidenced by action from the government or the private sector.

This section focuses on vulnerable groups whose needs needed prioritization during the Philippine pandemic experience: healthcare workers, the incarcerated, and the poor. These vulnerable populations are also contrasted with the experiences of the privileged. The section ends with the private sector's contributions to prioritize the vulnerable.

Lack of Support for Healthcare Workers

Healthcare workers are vulnerable during a pandemic because of their increased exposure to pathogens. This exposure was especially adverse considering how at the pandemic's inception, scientists were still trying to understand how a COVID-19 infection developed, and there were still no clear viable treatment options. Thus, the relevant SDH for health workers is their working life conditions. Health workers' working conditions made them vulnerable.

During the early pandemic, some hospital workers wore garbage bags in lieu of proper protective gear [42]. Recall the discussion on alleged corruption in the government's pandemic transactions. The money that the government could have saved if it procured face masks, face shields, and other equipment at market price could have been spent on more medical equipment, especially protective gear for health workers.

It is crucial to protect healthcare workers not just because of their vulnerability but also because they are part of the frontline response against the virus. Those who take care of the rest of the population must be taken care of. One way to provide this care is by supplying protective gear. Another is to increase their pay to be commensurate with their service.

Regarding compensation, nurses protested the late release of their hazard pay amidst rising hospital admissions and declining medical supplies [43]. In one case, a nurse died from the virus even before getting her hazard pay [44]. Regardless of this nurse's family being compensated, what is wrong in this situation is that the hazard pay that was meant for the nurse was not given to the nurse. Hazard pay is supposed to compensate for a worker's additional efforts, not the worker's surviving family members. After all, it is not their family but the workers who are directly exposed to the hazard. To adopt the line of thinking that because a worker's family may be compensated, then nothing wrong happened, is to be complacent not only with implementing the purpose behind laws, but also giving people what is properly due to them. Late pay, in itself, should also be flagged as wrong. People plan expenses around when they receive money. Interests and penalties may increase because of late payments. One can ask one's own self the seemingly casual but actually grave question: "Is it ok if I receive part of my pay late?"

In 2022, *Republic Act No. 11712* or the *Public Health Emergency Benefits and Allowances for Health Care Workers Act* took effect [45]. This law gave a monthly health emergency allowance to healthcare and non-healthcare workers, with the amount increasing according to risk exposure. Healthcare and non-healthcare workers in health facilities working in low-, medium-, and high-risk areas were to be given at least P3,000.00, P6,000.00, and P9,000.00 monthly, respectively. Low-risk areas are those away from patients. Medium-risk areas are those with patients who are not known or suspected to have COVID-19. High-risk areas are those with suspected or known COVID-19 patients. Furthermore, additional compensation was given to those who contracted COVID-19: P15,000.00 for mild to moderate cases, and P100,000.00 for severe cases. For the workers who died from COVID-19, P1,000,000.00 would be given to their heirs. In 2024, the government finally released funds for around five million claims arising from this law [46]. However, there were still reports of further delay in getting the benefits for some hospitals [47]. These incidents of delayed compensation may be partly addressed with the different government agencies and hospitals being more transparent at different stages with why delays were being incurred.

Persons Deprived of Liberty (PDLs) Suffered Congestion in Facilities

Even before the pandemic, in 2019, the country's detainment facilities were sufficient for only 40,610 PDLs while the actual number of PDLs in these facilities was 215,000 [48]. The congestion rate in 2019 was 438%, and in 2020, 403% [49].

These figures are important because while social distancing was a key preventative measure against contracting COVID-19, there was just no space to accomplish this inside prison cells. As such, by May 2020, 300 PDLs were already infected and 4 died [50].

Measures to combat the pandemic for PDLs faced challenges. The Bureau of Jail Management and Penology (BJMP) locked down more than 400 facilities [50]. Separate isolation facilities were also designated [51]. Those at increased risk of contracting COVID-19, like older PDLs, were put in less cramped facilities [50]. Face masks were distributed to all PDLs, but the discomfort was intense in the summer heat in "spaces that are so cramped that it is impossible to walk two steps without bumping into someone else" [66].

To provide some perspective, the 1955 First United Nations Congress on the Prevention of Crime and the Treatment of Offenders adopted the Standard Minimum Rules for the Treatment of Prisoners, which states guidelines on accommodations for PDLs [52]:

> Where sleeping accommodation is in individual cells or rooms, each prisoner shall occupy by night a cell or room by himself. If for special reasons, such as temporary overcrowding, it becomes necessary for the central prison administration to make an exception to this rule, it is not desirable to have two prisoners in a cell or room.

In addition, from the same guidelines:

> All accommodation provided for the use of prisoners and in particular all sleeping accommodation shall meet all requirements of health, due regard being paid to climatic conditions and particularly to cubic content of air, minimum floor space, lighting, heating and ventilation.

Even the United Nations recognizes the importance of climatic conditions, heating, and ventilation. Discomfort is not such a trivial issue.

"To decongest... overcrowded jails... and further prevent the spread of... Covid-19," the Supreme Court of the Philippines introduced new rules on "electronic filing of informations and transmission of release orders" and videoconference hearings resulting in the release of not less than 9731 PDLs from March to April in 2020. The Supreme Court also reduced bail and ordered recognizance to release more indigent PDLs [53].

By 2022, decongestion was down to 367%, 348% by 2023, and 239% by May 2024 [54]. The BJMP attributed this decongestion to continuously implementing paralegal services that reduce PDLs' prison sentence because of good conduct or studying, mentoring, or teaching [54]. The Department of Justice also mentioned reforms such as disallowing the filing of weak cases and ensuring that evidence is complete before someone can be charged with an offense, which will be tried in court to ultimately determine culpability [55].

The Poor Lacked Access to Various Resources

Congestion was not limited to PDLs. In 2020, the country's population density was the 32nd highest in the world, amounting to about 376 people per sq. km, and rising to around 379 in 2021, still with the same world rank [56]. However, in the National Capital Region (NCR) where the country's capital, Manila, is located, the population density was 21,765 people per sq. km in 2020 [57]. In the urban poor area of Baseco, Manila, with 16,000 households, 10 members of a family could be living in a shack measuring 20 sq. m [58].

The poverty rate in 2018 was 16.7% and increased to 18.1% in 2021 [59]. During the pandemic, 2.2 million Filipinos became poor [59]. In 2023, the poverty rate decreased to 15.5% [59].

In May 2020, the Social Weather Stations (SWS) publicized survey results stating that around 3.9 million families experienced involuntary hunger for three months due to the COVID-19 lockdown [5]. The lockdown was particularly difficult for those who relied on a daily wage, such as market and sidewalk vendors, stevedores, construction workers, and public utility vehicle drivers. These people had little to no savings to begin with, and abrupt announcements on the lockdown worsened their situation [58].

This scenario raises questions on the efficiency of the government to distribute social assistance, while highlighting the gap between those who have the power to provide food for themselves and those who have to rely on government aid to eat.

During the Enhanced Community Quarantine, the transportation of essential goods and services among provinces was exempted from absolute quarantine restrictions. However, some individuals and families engaged in panic buying and hoarding, which led to limited access to food [60]. While those who have money can afford to buy food and hoard essentials, those who do not have the means are faced with the possibility of starvation. The government did not provide a cap on individual purchases.

However, one of the solutions the government instituted to combat the plight of the poor was the giving of cash aid or *ayuda*. Abad [61] recounts the details of this program and its pitfalls [61]. In 2020, for April and May, "'18 million low-income households' were entitled to P5,000 to P8,000 a month." However, the identification of recipients was based on a 2015 listing of poor families, and aid distribution took months. As such, there was confusion on the ground as government lists did not match the actual number of family recipients. Initial distributions also resulted in people crowding over distribution centers. In addition, allegations emerged of local officials favoring certain constituencies.

In March 2021, the government announced it would distribute "P1,000 each to 22.9 million poor Filipino individuals, with a maximum of P4,000 per family" [61]. To address the problems of the first wave of *ayuda*, the money was distributed through mobile banking platforms and banks. The problem with this method was that it favored those who were knowledgeable of these means. The poorest of the poor would not have a cell phone, in the first place, and might not know how to use an ATM. Distribution was also still delayed.

The Vulnerable as Opposed to the Privileged

In stark contrast to the vulnerable, the privileged found ways to avoid the quarantine. There had been reports of quarantine skippers who bribed hotel staff to go out of the hotel where they were quarantined [62]. One such case involved someone who arrived in the Philippines from the United States, and who was in mandatory quarantine in a hotel. She violated protocols by going out of the hotel and attending parties nearby. She later turned out to be COVID-positive; other people from these parties turned out to be COVID-positive as well [63]. Another case of quarantine violation involved a senator who, after already having a sore throat and a 38 °C fever, still went to a hospital to accompany his wife to give birth. He later turned out to be COVID-positive [64].

More government officials were embroiled in controversy. In December 2020, unspecified vaccines from China were administered to the members of the Presidential Security Group (PSG) and the military without local Emergency Authorization Use (EUA) approval [65]. It is worth noting that the Phase 3 trial results for the CoronaVac were published between February and July 2021 [66]. The country's defense minister justified the move of smuggling and administering the then-unapproved vaccine as necessary to protect the country's Head of State [65]. To be clear, there was no justification from any government official involved that these vaccines were used in a clinical trial or experiment. In any event, there was no approval for any such trial or legal exceptions approved for the vaccines to be administered at that time.

In a scenario where the majority of the population, including the unvaccinated poor, were suffering from financial losses and literal hunger from not being able to work during the strictest phases of the lockdown, it was hard to swallow these publicized violations. Government officials, especially, should be expected to set a good example for compliance. These contrasts highlight how social status and conditions were crucial in molding one's experiences of the pandemic. The already existing inequities that afflicted the vulnerable pre-pandemic became even more problematic with the impact of COVID-19.

Efforts from the Private Sector

Prioritizing the vulnerable became a goal not just of the government but also of the private sector. Some of the specific efforts the private sector was responsible for during the early pandemic were donating money that the government used for the manufacture of testing kits and providing free shuttle services to healthcare workers when public transportation was severely restricted [67]. Donations in money and medical equipment from the private sector and civil society continued during the pandemic [68].

A distinct development in voluntary aid was the development of the community pantry. A 26-year-old Ana Patricia Non put out a cart containing groceries with a label, translated to English as, "Give what you can, take what you need" [69]. This gesture went viral over social media, and after a week, 200 other community pantries were organized by different organizations and private citizens. Groceries and funds were donated by individuals from different socio-economic classes and even corporations.

The efforts of the private sector and the development of the community pantry exhibit *bayanihan*. *Bayanihan* is an "indigenous Filipino trait" or "Filipino way of life" from the root word *bayan* ("town or nation") [70]. A derivative of *bayan* is *bayani* or a hero who sacrifices for others [71]. *Bayanihan* refers to how Filipinos mutually help each other, the classic image of which is a group of several men carrying a nipa hut.

Bayanihan can be analyzed as a form of solidarity. Solidarity reflects the "fundamental, relational character of social life": communities can face common problems, and collective decisions can produce collective solutions [14]. The mutual aid of *bayanihan* reflects the collective efforts of community members to help each other implement a solution to a problem.

Applied to community pantries, *bayanihan* exhibits how Filipinos came together to help each other to address hunger and the loss of income during the pandemic. The community pantry mantra of "Give what you can, take what you need" shows the mutuality of offering help and at the same time getting help from others. Solidarity is also exhibited as the community pantry is a solution implemented by the community to solve a community problem.

Community pantries, exhibiting *bayanihan*, also show the resilience of Filipinos. The classic image of Filipino resilience is that of smiling Filipinos being interviewed by a newscaster, right after a typhoon, with flooding and destroyed houses in the vicinity. Zosa and Ombao (2025) [72] argue that the *bayanihan* of community pantries showed how Filipinos had to come up with their own solutions because of the inability of the government to address problems during the pandemic such as the loss of income from the lockdown. These problems were much worse for the vulnerable who had little income in the first place.

Bayanihan and solidarity can help in prioritizing the vulnerable. This section of the chapter highlights healthcare workers, PDLs, and the poor as vulnerable populations. An important seeming contrast to solidarity is autonomy, which is the topic of the next section.

Relational Autonomy

The principle of autonomy in biomedical ethics refers to one's capacity to freely act according to one's own plan [20]. While it can be theorized that an individual is autonomous, in a practical sense, one is not completely free to do whatever one wants. Government laws are imposed for the interest of all individuals, more so in a pandemic. "Relational autonomy emphasizes that individual actions occur in the context of other people whom these actions will affect" [14]. With this important disclaimer, the impact of the pandemic measures with regard to relational autonomy is specifically examined relating to three instances: first, how the Philippine government would announce confusing quarantine measures abruptly; second, how the non-uniform presence of internet connectivity in an archipelagic country like the Philippines affected access to these announcements and information about COVID-19; and third, how data privacy was possibly violated in the handling of information for contact tracing.

Confusion from Unclear and Abrupt Government Rules

The government's response to the pandemic was at times relayed in a manner that brought about immense confusion. For instance, in reaction to Duterte's announcement of the initial lockdown covering the period of March 15 to April 14, Assistant Professor Gene Nisperos of the University of the Philippines Manila College of Medicine said: "Every agency has its own interpretation of what happened [Thursday] night" [73]. This announcement involved the implementation of the *Inter-Agency Task Force for the Management of Emerging Infectious Disease Resolution No. 11*, stating the imposition of a "community quarantine" [74]. However, Duterte also said that there is a hesitation to use the term "lockdown" because the term "intimidates" people, "but it's a lockdown" [75]. Duterte's Cabinet Secretary would clarify that "It's not lockdown, we are under community quarantine" [76]. Thus, there was confusion of what really was being implemented. Was it a lockdown? A mere quarantine? What would make a quarantine not a lockdown?

Nomenclature aside, the Resolution mandated LGUs and government agencies to implement the said community quarantine without providing specifics. To illustrate, LGUs were told to abide by the guidelines in the Resolution, but the guidelines merely mentioned the minimum number of COVID-19 cases to trigger the quarantine, and to avoid mass gatherings. Then what?

Adjustment in work arrangements shall be in accordance with guidelines to be issued by the Department of Labor and Employment and the Department of Trade and Industry. Social distancing in public railways shall be in accordance with guidelines to be issued by the Department of Transportation.

The problem is that the Resolution already took effect, but the Resolution also stated that the concerned agencies shall issue guidelines. The guidelines were already needed at the start of the community quarantine.

As such, "less than 24 hours before community quarantine... [began], most people still didn't understand what it meant," resulting in panic buying in malls. Another effect was the surge of temporary residents of the country's capital Metro Manila booking trips to return to their provinces before the lockdown began.

Even a year after the pandemic, frustration was still present due to government announcements regarding quarantine measures. In September 2021, the transition to a new and less strict lockdown system was suddenly postponed leading to Metro Manila mayors being blindsided, and business establishments and churches cancelling their preparations to resume in-person operations [77].

While the government should be expected to adjust to ongoing developments in the pandemic, this should not be used as an excuse for lack of coordination among different government units. According to one mayor: "*We just found out about news of... [quarantine] extension from the media*" [73].

The clarity of government instructions and the timeliness of this delivery are important factors to respect the autonomy of individuals' decision-making. Understanding is one of the essential requisites of autonomy, and abrupt and confusing announcements and guidelines undermine this requisite.

In a pandemic with quarantine rules and other regulations, there is already an infringement in the exercise of one's autonomy, and it is crucial to make people understand why these infringements are enforced and how. As such, access to information is important individually and institutionally. The government does not help this access with insufficient guidelines.

This access to information that promotes autonomy is also compromised because of the lack of and unequal internet access in the Philippines, an archipelagic country composed of 7641 islands.

Non-uniform Internet Connectivity

Communication and coordination via the internet are crucial in an archipelago with the Philippines's more than 110 million inhabitants. However, according to the Department of Information and Communication Technology (DICT), "Sixty-five percent of Filipinos are still not connected to the internet" [78]. Only around half of households have access to the internet [79].

Internet access remains Manila-centric; the NCR has the highest proportion of its households with internet access at 74.6% [79]. From a 2019 government poll conducted prior to the onset of the COVID-19 pandemic, the number of telecom towers in urban areas was thrice as much as those in rural areas, and 95% of these telecom towers were private-owned [80]. Thus, internet access is largely provided by businesses, and not everyone has the money for this additional expense, and even if some of the *ayuda* can be used to buy prepaid internet access for mobile phones, the problem in far-flung areas, as pointed out, is the lack of telecom towers in the first place.

The internet connectivity problem is important especially because government agencies have active official social media pages where they post their updated announcements, and it is these pages that Filipinos usually visit first. Thus, access to information essential for understanding that supports autonomous choices is compromised.

Information received may affect health outcomes. An illustration of the relationship between access to information and health outcomes is the availability of telehealth. The use of telehealth for accessing healthcare services during the pandemic is beneficial for those residing in places with reliable internet connectivity, but it may not be suitable for everyone or advantageous all of the time [81]. For instance, populations experiencing homelessness or residing in economically disadvantaged regions may have difficulty accessing telehealth services. Private consultations are only afforded by those who can shell out funds, while others primarily depend on LGU support in public health centers. This internet situation in the country persisted despite a law enacted in 2017 to provide free internet to the public [82]. However, the free internet can only be accessed in certain public areas like government offices, schools, hospitals, plazas, libraries, and transportation terminals for the public. But then, the lockdown limited people's mobility and mass gatherings. To maximize telehealth, internet access was needed even outside these public areas.

Violations Against Privacy

Aside from the importance of correct information to understanding, and thus, autonomy, rules that respect autonomy include respecting privacy and protecting confidential information [20].

By May 2020, the National Privacy Commission was investigating 22 complaints involving the breach of privacy of at least 150 COVID-19 patients and probable and suspected cases [83]. Government entities were sometimes involved in privacy issues. For instance, a province's official Facebook page, for a time, disclosed the names of COVID-19 patients for "faster contact-tracing" [83]. In addition, the government's official tracing app StaySafe had its share of issues. Researchers from the University of Toronto wrote:

> Through a vulnerability in the backend database used by this app, we were able to access the geolocation data of hundreds of thousands of users. We are concerned, but did not confirm, that it would be possible to use users' movement patterns to deanonymize them and to discover their health status.
>
> We disclosed this vulnerability to Multisys, the developer of StaySafe PH, who released updates for the Android and iOS versions of the app and secured the platform's backend database in response [84].

Aside from government contact tracing, private establishments also had their own. A congressman suspected these contact tracing systems to be the source of personal data used in mobile phone text scams [85]. Examples of these scams included text messages that would lead mobile phone users to click links to sites that asked for user passwords or other sensitive personal information.

These privacy issues, together with the previously discussed problems with internet access and abrupt and unclear guidelines, violate relational autonomy. The violation of autonomy via diminishing people's understanding and compromising their privacy may lead to decision-making resulting in harmful consequences. As a central illustration, the section on unclear government guidelines showed harmful consequences such as confusion in government administration, panic buying, and halted business operations. That these consequences involve not just an individual's limited interactions, but how individuals decide and act in a community points to how relational and not just personal autonomy is important in a pandemic.

Lessons and Recommendations

First, the COVID-19 pandemic highlighted numerous issues in the Philippine healthcare system, even those existing pre-pandemic, and the capacity of the national government to effectively respond to national health threats. While the pandemic caught the world unawares, nations that had reliable healthcare systems fared better than those that did not. The government should not wait for another pandemic before trying to drastically improve its healthcare system.

Second, the national government should ensure better coordination among its different agencies and with LGUs. While urgency is important in releasing guidelines, guidelines with insufficient coordination cause a gamut of issues on the ground.

Third, it should be clear by now that while the fear of imposition of penalties is a deterrent to infractions, relying significantly on such a strategy may lead to abuses in implementation, which further incite unrest among the citizenry. Policies should be backed up by appropriate evidence, scientific and otherwise, and these should be communicated to citizens in a manner that is comprehensible and accessible.

Fourth, the government should construct more infrastructure to provide internet access to more people, especially in far-flung areas. The government should be more cognizant of the different factors that it has to rely on in the delivery of its services. Reliance on apps should entail well-thought-out medium- and long-term solutions to improve free internet access.

Fifth, there should be better accountability for leaders. Initial investigations on violations committed by public officers should be followed through. If officials are keen on imposing punishment on citizens, it should not be forgotten that officers are citizens, too.

These lessons and recommendations accentuate the interaction between government actions, attainment of health outcomes, and the normative ethical evaluation of such actions. Problems in planning and implementation are linked to ethical critique. Ethical principles are applied to a set of facts, and government actions partly constitute such facts. As such, failures in planning and implementation reasonably link to failures in ethical obligations to the public.

References

1. Philippine Statistics Authority (2020) 2020 Census of population and housing report no. 3 Population density. https://library.psa.gov.ph/cgi-bin/koha/opac-detail.pl?biblionumber=27822. Accessed 2 Feb 2025
2. Llorito D (2022) Philippines: reducing inequality key to becoming a middle-class society free of poverty. World Bank Group. https://www.worldbank.org/en/news/press-release/2022/11/24/ph-reducing-inequality-key-to-becoming-a-middle-class-society-free-of-poverty. Accessed 02 Feb 2025
3. Ordinario C (2022) Health expenses remain a big burden for Pinoys despite UHC. Philippine Institute for Development Studies. https://www.pids.gov.ph/details/news/in-the-news/health-expenses-remain-a-big-burden-for-pinoys-despite-uhc. Accessed 2 Feb 2025
4. World Health Organization (2020) Listings of WHO's response to COVID-19. World Health Organization. https://www.who.int/news/item/29-06-2020-covidtimeline. Accessed 2 Feb 2025
5. Baclig CE (2021) Timeline: one year of Covid-19 in The Philippines. Inquirer. https://newsinfo.inquirer.net/1406004/timeline-one-year-of-covid-19-in-the-philippines. Accessed 2 Feb 2025
6. Esguerra DJ (2020) Wearing of face shield now required in public places. Inquirer https://newsinfo.inquirer.net/1371933/wearing-of-face-shield-now-required-in-public-places. Accessed 2 Feb 2025
7. Presidential Communications Office (2020) Palace: Government now requires wearing of both face shield and mask in public places. Presidential Communications Office. https://pco.gov.

ph/news_releases/palace-government-now-requires-wearing-of-both-face-shield-and-mask-in-public-places/. Accessed 2 Feb 2025
8. Abad M (2021) Pandemic in 2021: how The Philippines responded to COVID-19. Rappler. https://www.rappler.com/newsbreak/iq/list-review-2021-covid-19-pandemic-philippines/. Accessed 2 Feb 2025
9. World Health Organization (2022) Coronavirus Disease 2019 (COVID-19) Situation Report #92. https://www.who.int/docs/default-source/wpro%2D%2D-documents/countries/philippines/emergencies/covid-19/who_phl_sitrep_92_covid-19.pdf?sfvrsn=bbbff1d2_5. Accessed 2 Feb 2025
10. Cabico GK (2023) COVID-19 not among top 10 causes of death in 2022. Philstar. https://www.philstar.com/headlines/2023/01/25/2240134/covid-19-not-among-top-10-causes-death-2022. Accessed 2 Feb 2025
11. Nazario D (2022) DOH says it can see the end of Covid-19, but stresses virus here to stay. Manila Bulletin. https://mb.com.ph/2022/09/16/end-of-the-covid-19-pandemic-doh-says-it-can-see-it-but-virus-here-to-stay/. Accessed 2 Feb 2025
12. Marcos F (2022) Executive order no. 7. https://pco.gov.ph/wp-content/uploads/2022/10/20221028-EO-7-FRM.pdf. Accessed 2 Feb 2025
13. World Health Organization (2023) Philippines: WHO coronavirus disease (COVID-19) dashboard with vaccination data. WHO Data. https://covid19.who.int/region/wpro/country/ph. Accessed 2 Feb 2025
14. Ortmann LW, Barrett DH, Saenz C et al (2016) Public health ethics: global cases, practice, and context. In: Barrett DH, Ortmann LW, Dawson A et al (eds) Public health ethics: cases spanning the globe. Springer Open, pp 3–35
15. Yu LS (2023) Marcos lifts COVID-19 public health emergency. Rappler. https://www.rappler.com/nation/marcos-lifts-covid-19-public-health-emergency-july-22-2023/. Accessed 2 Feb 2025
16. Cabalza D (2024) DOH: no budget for new vaccines vs 'FLiRT' variants of COVID-19. Inquirer. https://newsinfo.inquirer.net/1945940/doh-no-budget-for-new-vaccines-vs-flirt-variant. Accessed 2 Feb 2025
17. Jiang Y, Laranjo JR, Thomas M (2022) COVID-19 lockdown policy and heterogeneous responses of urban mobility: evidence from The Philippines. PLoS One 17(6). https://doi.org/10.1371/journal.pone.0270555
18. Park S (2022) Gauging the efficacy of The Philippines lockdown policies in response to COVID-19. GSC Adv Res Rev. https://doi.org/10.30574/gscarr.2022.12.1.0172
19. Vallejo BM Jr, Ong RAC, Rye RS (2023) An assessment of The Philippines government's response to the COVID-19 pandemic based on policy response indices. SciEnggJ 16(2):431–441
20. Beauchamp TL, Childress JF (2019) Principles of biomedical ethics, 8th edn. Oxford University Press, New York
21. Mill JS (2015) On liberty. In: Philp M, Rosen F (eds) On liberty, utilitarianism, and other essays. Oxford University Press, Oxford, pp 1–112
22. Malasig J (2020) 'Virus is invisible': Cebu City aerial survey triggers concerns on COVID-19 response. Interaksyon. https://interaksyon.philstar.com/politics-issues/2020/06/29/171691/virus-invisible-cebu-city-aerial-survey-triggers-concerns-covid-19-response/. Accessed 2 Feb 2025
23. Amnesty International (2020) Philippines: President Duterte gives "shoot to kill" order amid pandemic response. Amnesty International. https://www.amnesty.org/en/latest/news/2020/04/philippines-president-duterte-shoot-to-kill-order-pandemic/. Accessed 2 Feb 2025
24. Tupas E (2020) PNP: we won't shoot them dead. Philstar. https://www.philstar.com/headlines/2020/04/03/2005146/pnp-we-wont-shoot-them-dead. Accessed 2 Feb 2025
25. Bolledo J (2021) Quarantine violator dies after being forced to do 300 rounds of exercise – family. Rappler. https://www.rappler.com/philippines/quarantine-violator-cavite-dies-forced-do-rounds-exercise/?utm_medium=Social&utm_campaign=Echobox&utm_source=Facebook#Echobox=1617617820. Accessed 2 Feb 2025

26. Mangosing F (2021) General Trias police chief, 2 other cops sacked for curfew violator's death. Inquirer. https://newsinfo.inquirer.net/1415866/pnp-sacks-cavite-police-exec-for-allegedly-punishing-harshly-curfew-violators. Accessed 2 Feb 2025
27. Human Rights Watch (2020) Philippines: curfew violators abused. Human Rights Watch. https://www.hrw.org/news/2020/03/26/philippines-curfew-violators-abused. Accessed 2 Feb 2025
28. Gregorio GEV, Sanchez-Tolosa MT, San Jose MCZ et al (2022) Is face mask with face shield more effective than face mask alone in reducing SARS-CoV-2 transmission? Acta Med Philipp 56(9):67–75
29. Saunders GH, Jackson IR, Visram AS (2020) Impacts of face coverings on communication: an indirect impact of COVID-19. Int J Audiol 60(7):495–506
30. Cordero T (2021) Lopez: high prices of face mask, shield in 2020 due to short supply. GMA News. https://www.gmanetwork.com/news/topstories/nation/800256/lopez-high-prices-of-face-mask-shield-in-2020-due-to-short-supply/story/. Accessed 2 Feb 2025
31. Magsambol B (2021) Review sought for mandatory wearing of face shields in PH. Rappler. https://www.rappler.com/philippines/review-sought-mandatory-wearing-face-shields-philippines/. Accessed 2 Feb 2025
32. Trading Economics (2025) Philippines daily minimum wages. Trading Economics. https://tradingeconomics.com/philippines/minimum-wages. Accessed 11 May 2025
33. Buan L (2021) Tale of pandemic face shields: costly, overstocked, unauthorized gov't buys. Rappler. https://www.rappler.com/philippines/tale-covid-19-pandemic-face-shields-costly-overstocked-unauthorized-government-buys/. Accessed 2 Feb 2025
34. Rey (2021) P95 million worth of face masks, face shields sitting in government depots. Rappler. https://www.rappler.com/philippines/millions-worth-face-masks-shield-sitting-government-depots-as-of-august-2021/. Accessed 2 Feb 2025
35. Cepeda M (2021) LIST: everything you need to know about the Pharmally pandemic deals scandal. Rappler. https://www.rappler.com/newsbreak/iq/list-everything-need-to-know-pharmally-covid-19-pandemic-deals-scandal/. Accessed 2 Feb 2025
36. Commission on Audit (2021) COA clarification on audit findings of DOH COVID-19 funds. Commission on Audit. https://www.coa.gov.ph/coa-clarification-on-audit-findings-of-doh-covid-19-funds/. Accessed 2 Feb 2025
37. Tomacruz S (2021) How the Duterte Gov't shut out local PPE producers during a pandemic. Rappler. https://www.rappler.com/newsbreak/in-depth/how-duterte-government-shut-out-filipino-ppe-producers-during-pandemic/. Accessed 2 Feb 2025
38. Bolledo J (2024) PNP arrests ex-DBM official tagged in multibillion-peso COVID-19 supplies mess. Rappler. https://www.rappler.com/philippines/pnp-arrests-christopher-lao-covid-19-supplies-mess-september-18-2024/. Accessed 2 Feb 2025
39. Wrigley A, Dawson A (2016) Vulnerability and marginalized populations. In: Barrett DH, Ortmann LW, Dawson A et al (eds) Public health ethics: cases spanning the globe. Springer Open, pp 203–241
40. World Health Organization (n.d.) Social determinants of health. World Health Organization. https://www.who.int/health-topics/social-determinants-of-health#tab=tab_1. Accessed 2 Feb 2025
41. Rawls JT (2001) Justice as fairness: a restatement. Harvard University Press, Cambridge
42. Garcia MA (2020) Laguna hospital staff use plastic, trash bags as improvised PPEs. GMA News. https://www.cnnphilippines.com/news/2021/4/5/PPE-shortage-COVID-19-Philippines-hospitals.html. Accessed 2 Feb 2025
43. Portugal A (2021) Philippines health workers protest neglect as COVID-19 strains hospitals. Reuters https://www.reuters.com/world/asia-pacific/philippines-health-workers-protest-neglect-covid-19-strains-hospitals-2021-09-01/. Accessed 2 Feb 2025
44. Ranada (2020) Cainta nurse gets P60 daily hazard pay and dies of COVID-19 before receiving it. Rappler. https://www.rappler.com/newsbreak/in-depth/cainta-nurse-maria-theresa-cruz-dies-covid-19-before-getting-hazard-pay/. Accessed 2 Feb 2025

45. Supreme Court E-Library. Republic Act No. 11712, April 27, 2022. https://elibrary.judiciary.gov.ph/thebookshelf/showdocs/2/94929. Accessed 2 Feb 2025
46. Valmonte K (2024) DBM to release P27-B overdue health allowance for pandemic workers on July 5. Rappler. https://www.rappler.com/philippines/dbm-release-overdue-health-allowance-covid-19-pandemic-workers-july-2024/. Accessed 2 Feb 2025
47. Delilan A (2024) COVID-19 heroes forgotten? Negros health workers still waiting for promised allowances. Rappler. https://www.rappler.com/philippines/visayas/covid-heroes-forgotten-negros-occidental-health-workers-waiting-promised-allowances/. Accessed 2 Feb 2025
48. World Prison Brief (n.d.) Philippines. World Prison Brief. https://www.prisonstudies.org/country/philippines. Accessed 25 Feb 2025
49. Commission on Human Rights (2022) Statement of CHR Executive Director, Atty Jacqueline Ann de Guia, on the decongestion efforts of the Bureau of Jail Management and Penology. Commission on Human Rights. https://chr.gov.ph/statement-of-chr-executive-director-atty-jacqueline-ann-de-guia-on-the-decongestion-efforts-of-the-bureau-of-jail-management-and-penology/. Accessed 2 Feb 2025
50. Santos AP (2020) 'Waiting to die': coronavirus enters congested Philippine jails. Aljazeera https://www.aljazeera.com/news/2020/5/4/waiting-to-die-coronavirus-enters-congested-philippine-jails. Accessed 2 Feb 2025
51. Batnag D (2020) Coronavirus: Philippines scrambling to head off public health crisis in prisons. The Straitstimes. https://www.straitstimes.com/asia/se-asia/coronavirus-philippines-scrambling-to-head-off-public-health-crisis-in-prisons. Accessed 2 Feb 2025
52. United Nations Office on Drugs and Crime (1955) Standard minimum rules for the treatment of prisoners. https://www.unodc.org/pdf/criminal_justice/UN_Standard_Minimum_Rules_for_the_Treatment_of_Prisoners.pdf. Accessed 11 May 2025
53. Peralta DM (2020) Administrative Circular no. 38 – 2020. https://sc.judiciary.gov.ph/wp-content/uploads/2023/02/AC-38-2020.pdf. Accessed 2 Feb 2025
54. Presidential Communications Office (2024) Jail congestion rate dropped last year in implementing BJMP paralegal programs – DILG. Presidential Communications Office. https://pco.gov.ph/news_releases/jail-congestion-rate-dropped-last-year-in-implementing-bjmp-paralegal-programs-dilg/. Accessed 2 Feb 2025
55. Pulta B (2024) DOJ: almost 12K PDLs freed, jail congestion rate down to 239%. Philippine News Agency. https://www.pna.gov.ph/articles/1228294. Accessed 2 Feb 2025
56. World Bank Group (n.d.) Population density (people per sq. km of land area). https://data.worldbank.org/indicator/EN.POP.DNST. Accessed 2 Feb 2025
57. Philippine Institute for Development Studies (2024) Population density. PIDS Economic and Social Database. https://seis.pids.gov.ph/index.php?r=databank%2Ffrontend%2Fdataset%2Fview-table-2&id=296. Accessed 2 Feb 2025
58. Tenolete J, Racelis M (2020) [OPINION] coronavirus in Baseco: a community leader's assessment. Rappler. https://www.rappler.com/voices/ispeak/256081-opinion-coronavirus-baseco-community-leader-assessment/. Accessed 2 Feb 2025
59. Sousa LP (2024) Poverty & equity brief October 2024 Philippines. https://pip.worldbank.org/country-profiles/PHL. Accessed 2 Feb 2025
60. Garcia MA (2020) Panic buying at Metro Manila stores over fears of COVID-19 lockdown. GMA News. https://www.gmanetwork.com/news/money/economy/729309/panic-buying-at-metro-manila-stores-over-fears-of-covid-19-lockdown/story/. Accessed 2 Feb 2025
61. Abad M (2021) What went wrong in 2020 COVID-19 'ayuda,' lessons learned for 2021. Rappler. https://www.rappler.com/newsbreak/explainers/coronavirus-ayuda-government-aid-what-went-wrong-2020-lessons-learned-2021/. Accessed 2 Feb 2025
62. Caliwan CL (2022) Quarantine skippers bribe to escape protocols: DILG chief. Philippine News Agency. https://www.pna.gov.ph/articles/1164335. Accessed 2 Feb 2025
63. Cos W (2021) Authorities probe reports female from US escaped quarantine, attended parties. ABS-CBN. https://news.abs-cbn.com/news/12/30/21/authorities-probe-female-from-us-who-escaped-quarantine-attended-parties. Accessed 2 Feb 2025

64. Cepeda M (2020) Makati Med denounces Pimentel's 'irresponsible, reckless' protocol breach. Rappler. https://www.rappler.com/nation/255874-makati-medical-center-denounces-koko-pimentel-protocol-breach-coronavirus-quarantine/. Accessed 2 Feb 2025
65. Lema K, Petty M, Cameron-Moore S (2020) Philippine president's guards used 'smuggled' COVID-19 vaccines: minister. Reuters. https://www.reuters.com/article/us-health-coronavirus-philippines-vaccin-idUSKBN2940OC. Accessed 2 Feb 2025
66. Huzar T (2021) CoronaVac vaccine shows 83.5% protection in phase 3 trial. Medical News Today. https://www.medicalnewstoday.com/articles/coronavac-vaccine-shows-83-5-protection-in-phase-3-trial. Accessed 2 Feb 2025
67. Gita-Carlos RA (2020) Private sector's Covid-19 response overwhelms palace. Philippine News Agency. https://www.pna.gov.ph/articles/1097133. Accessed 2 Feb 2025
68. Philstar (2021) Private sector, civil society efforts vs COVID-19 in Philippines. Philstar. https://www.philstar.com/happens/859?fbclid=IwAR1tppD93fB64s9RrtHwiNeZtOTpwF7cRtx7NIwXEnnMGRgigAYTjf8ACM4. Accessed 2 Feb 2025
69. Kusuma N (2021) Maginhawa community pantry in The Philippines: a new hope. Green Network. https://greennetwork.asia/news/maginhawa-community-pantry-in-the-philippines-a-new-hope/. Accessed 2 Feb 2025
70. Ang GR (1979) The bayanihan spirit: dead or alive? Philippine Q Cult Soc 7(1):91–93
71. Bankoff G (2020) Old ways and new fears: Bayanihan and Covid-19. Philippine Stud Hist Ethnogr Viewp 68(3–4):467–475
72. Zosa MAH, Ombao RP (2025) Bayanihan and community pantries: redefining Filipino resilience in the COVID-19 pandemic. In: Llanera T (ed) Resilience and the brown babe's burden: writings by Filipina philosophers. Routledge, London, pp 114–131
73. Asplnwall N (2020) Coronavirus lockdown launches Manila into pandemonium. Foreign Policy. https://foreignpolicy.com/2020/03/14/duterte-quarantine-philippines-coronavirus-lockdown-launches-manila-into-pandemonium/. Accessed 2 Feb 2025
74. Inter-Agency Task Force for the Management of Emerging Infectious Disease (2011) Resolution no. 11 Series of 2020 Recommendations for the management of the coronavirus disease 2019 (covid-19) situation. https://doh.gov.ph/wp-content/uploads/2023/08/IATF-RESO-11.pdf. Accessed 11 May 2025
75. Rappler (2020) Metro Manila to be placed on lockdown due to coronavirus outbreak [Youtube video]. https://www.youtube.com/watch?v=DWxIvQlpJK8. Accessed 11 May 2025
76. Cudis C (2020) No lockdown, just 'community quarantine': Nograles. Philippine News Agency. https://www.pna.gov.ph/articles/1096510. Accessed 11 May 2025
77. De Leon (2021) Frustration mounts after gov't abruptly defers new lockdown system in Metro Manila. Rappler. https://www.rappler.com/nation/frustration-after-government-defers-new-lockdown-system-metro-manila/. Accessed 2 Feb 2025
78. Clepano JR (2023) DICT: 65% of Philippines not connected to internet. Philstar. https://www.philstar.com/headlines/2023/06/14/2273740/dict-65-philippines-not-connected-internet. Accessed 2 Feb 2025
79. Ordinario CU (2023) Internet access in PHL still Manila-centric–PSA. Business Mirror https://businessmirror.com.ph/2023/07/06/internet-access-in-phl-still-manila-centric-psa/. Accessed 2 Feb 2025
80. Romero (2021) PH gov't still 'very far' from achieving wider internet access: dev't study agency. ABS-CBN. https://news.abs-cbn.com/spotlight/11/11/21/wider-internet-access-in-ph-still-a-problem-devt-study-agency-says. Accessed 2 Feb 2025
81. Cordero DA (2023) Telehealth during the COVID-19 pandemic in The Philippines. Fam Pract 40(1):207–208
82. Supreme Court E-Library. Republic Act No. 10929, August 2, 2017. https://elibrary.judiciary.gov.ph/assets/dtSearch/dtSearch_system_files/dtisapi6.dll?cmd=getdoc&DocId=21081&Index=%2a2ceb34621b4a207c3a52af138f1e7bbf&HitCount=8&hits=1+2+3+4+15+16+17+18+&SearchForm=C%3a%5celibrev2%5csearch%5csearch%5fform. Accessed 11 May 2025
83. Dizon N (2020) Unauthorized disclosure of COVID-19 patients' identities continues. Rappler. https://www.rappler.com/newsbreak/

in-depth/264851-unauthorized-disclosure-covid-19-patients-identities-continues-npc/. Accessed 2 Feb 2025
84. Lin P, Knockel J, Poetranto I, et al (2020) Unmasked II an analysis of Indonesia and The Philippines' government-launched COVID-19 apps. The Citizen Lab. https://citizenlab.ca/2020/12/unmasked-ii-an-analysis-of-indonesia-and-the-philippines-government-launched-covid-19-apps/. Accessed 2 Feb 2025
85. Aurelio JM, Piad TJC (2022) Salceda suspects contact tracing as source of data leak. Inquirer. https://newsinfo.inquirer.net/1659610/salceda-suspects-contact-tracing-as-source-of-data-leak. Accessed 2 Feb 2025

COVID-19 Policies and Practices in Thailand

8

Pramon Viwattanakulvanid and Yuru Wang

Abstract

Thailand has entered a new normal of living with the COVID-19 pandemic after more than three years of battling it since 2020. As of October 2022, the Thailand Ministry of Public Health has reclassified COVID-19 as a "Disease under surveillance" instead of a "Dangerous infectious disease." Despite Thailand's recognized success in reopening its borders to tourists as the first country in Southeast Asia, there is still room for reflection and advancement. A comprehensive reflection on government policies and practices, and involvement from various sectors, is necessary to identify areas of enhancement and share valuable insights for other countries. This report synthesizes the latest reports, scientific publications, and official government guidelines, and incorporates a cultural perspective to facilitate this analysis. The literature used for data synthesis in this report was collected from January 2020 to December 2024. By examining the government's implemented measures and the efforts of multi-sectoral entities, identifying areas for enhancement, it is hoped that the lessons learned from this report will not only benefit Thailand but also provide valuable insights for other countries facing similar contexts. Consequently, countries can enhance their preparedness and response strategies to effectively combat future pandemics.

Keywords

COVID-19 pandemic · Border control policy · Contracting unit for primary care (CUP) · Social protection programs · Thailand

---

P. Viwattanakulvanid (✉) · Y. Wang
College of Public Health Sciences, Chulalongkorn University, Bangkok, Thailand
e-mail: pramon.v@chula.ac.th

L. P. Marquez et al. (eds.), *COVID-19 in the Pacific Rim and Beyond*,
Collaborative Bioethics 6, https://doi.org/10.1007/978-3-032-21218-4_8

Introduction

Updated COVID-19 Situation in Thailand

In response to the initial case in Wuhan, China, at the end of 2019 [1], Thailand took immediate action to combat COVID-19. Shortly after the first case, on January 4, 2020, Thailand activated its Emergency Operation Centre (EOC) under the Department of Disease Control, Thai Ministry of Public Health (MOPH) [2], showcasing their proactive measures in dealing with the outbreak.

On January 13, 2020, Thailand recorded its first imported case of COVID-19, a Chinese person from Wuhan City. Not long after, on January 31, 2020, the first domestic case was confirmed in a taxi driver who had close contact with a foreign passenger [3]. In response to the gradually rising number of domestic cases, the MOPH declared COVID-19 a dangerous communicable disease under the Communicable Disease Act on February 29, 2020 [2]. Remarkably, it was also the same day when the first death attributed to the virus was recorded [4]. Despite having the situation under control, the government determined it was crucial to take proactive measures in order to curb the further spread of the coronavirus. On March 26, 2020, a state of emergency was declared, which involved closing borders, together with the formation of the Center for COVID-19 Situation Administration (CCSA) [5, 6]. Subsequently, a nationwide curfew was imposed from 10 pm to 4 am the following week on April 3, 2020, with the purpose of limiting how frequent individuals left their residences and effectively containing the virus spread. Additionally, around the same time, the government enforced mandatory state quarantine for all incoming international flights [5]. In 2020, Thailand demonstrated its success in containing the outbreak with 6884 reported cases and 61 fatalities, with a mortality of 0.89% [7].

However, the situation took a turn for the worse in 2021 as Thailand witnessed a significant surge in infected cases, largely due to the Delta variant, and later in 2021, the Omicron variant. In response, the government initiated a series of lockdowns in various regions of the country, starting from July 20, 2021, to mitigate the spread of the virus [8]. While dealing with the challenges of the pandemic, Thailand made a strategic move to become one of the first countries in Asia to reopen its borders for international tourism, beginning in early July 2021 [9], with initiatives such as the Phuket Sandbox, Samui Plus, and 7 + 7 Extension. Subsequently, various destinations within the country progressively welcomed back tourists since October 2021.

After a year, on July 1, 2022, the CCSA made the decision to lift the Thailand Pass registration program and the mandatory US$10,000 health insurance requirement for all travelers. Moreover, all provinces were classified as either surveillance areas or "Green Zones," in combination with the reopening of entertainment venues [10].

It has now been more than 3 years since Thailand faced the enduring consequences of COVID-19. As of July 4, 2023, the country has reported almost 4.8 million confirmed cases, along with a total of 34,371 recorded deaths [11]. Bangkok, the capital city and a major tourist destination, continues to have the highest proportion of COVID-19 cases in the country [11]. By June 2024, the vaccination rate in Thailand reached a significant milestone, with 77% of the population having completed the COVID-19 vaccine primary series [12].

Navigating COVID-19: An Overview of Thailand's Healthcare System, Primary Care, and Emergency Response

Thailand, situated in Southeast Asia, is renowned for its rich culture, thriving tourism industry, and public health system. The delivery care system is currently characterized by a dominant public sector, supplemented by the private sector since the 1970s. Thailand's public healthcare system, overseen by the MOPH as the national health authority, is responsible for health policy and ensuring accessible and affordable healthcare. This is achieved through a network of primary centers, district hospitals, and provincial hospitals that deliver comprehensive healthcare services to the population. In addition, several autonomous health agencies, established by legislation, work in collaboration with the MOPH to support its goals and initiatives [13]. However, in urban areas such as Bangkok, the private healthcare sector often plays a prominent role and fills the gap left by public healthcare facilities [14, 15]. Alternatively, healthcare facilities catering to the population are classified into three-level service systems: primary, secondary, and tertiary care facilities [16].

Since achieving Universal Health Coverage (UHC) in 2002, Thailand has witnessed a significant rise in public health expenditures [13], resulting in a remarkable reduction in the number of uninsured individuals and notable improvements in healthcare access [17]. Under the Universal Coverage Scheme, primary healthcare in Thailand is delivered through contracting units for primary care, which consist of networks that typically include a primary healthcare center and a hospital. In rural areas, where qualified staff are primarily concentrated in hospitals, health centers collaborate with district hospitals to establish a Contracting Unit for Primary Care (CUP), with each CUP representing one district. The scale of a CUP is determined by the specific healthcare service capabilities in real contexts. In urban settings, there could be several hospitals in the same area and doctors in health centers. Each CUP can consist of several health centers plus one hospital, a group of health centers, or even private clinics to meet the human resources criteria. While for private clinics, each facility connects a CUP with only one primary health care (PHC) unit. This contracted PHC unit is known as a "warm community clinic." In 2010, the public sector had 937 CUPs and 11,051 contracted PHC units, while the private sector had 218 CUPs and 224 contracted PHC units. Secondary and tertiary care services are predominantly provided in hospitals, following a referral system from primary healthcare facilities to district and provincial/regional levels [18]. Figure 8.1 [18] illustrates the levels of health services in Thailand.

With the system relying on primary health care (PHC), the success of healthcare in Thailand has been attributed to the political endorsement of PHC as the fundamental pillar of healthcare delivery [19]. Furthermore, the Prime Minister's commitment to ensuring that Thai citizens can access healthcare services from medical

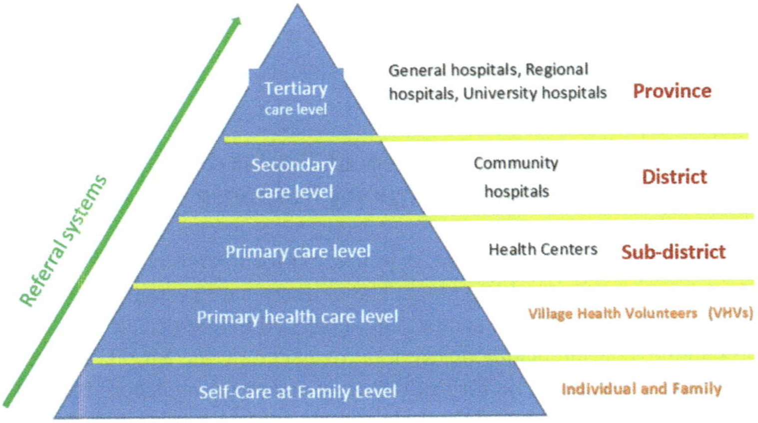

Fig. 8.1 Levels of health services in Thailand

practitioners at the primary healthcare level has further reinforced this approach [20]. The effective functioning of PHC can be attributed to a well-equipped and committed health workforce, complemented by the invaluable contributions of village health volunteers (VHVs). Together, they form a strong foundation for the success of PHC in Thailand. Therefore, Thailand's public health system's strength lies in its network of VHVs and community-based initiatives [21], which effectively respond to disasters and address health issues, and help promote health and prevent illnesses within their communities. Among many factors that contributed to Thailand's success in coping with the COVID-19 pandemic, the VHVs' network has gained attention, recognition, and acceptance from global communities [22, 23]. In particular, the role of VHVs was highlighted by the World Health Organization (WHO) [23] as an essential element in the effective response to public health challenges. With committed leadership, backed by early screening efforts and scientific evidence, VHVs have showcased a remarkable ability to assist local public health workers in tracing and isolating high-risk groups.

Regarding public health emergency management, in 2007, Thailand initiated the implementation of a health security program and public health emergency management, which was further revamped in 2015 to enhance its effectiveness [24]. In addition, the country has proactively strengthened its capacities in line with the International Health Regulations 2005 [25] and became a member of the Global Health Security Agenda in 2014. The country's experiences in effectively managing various epidemics and pandemics, such as SARS, avian influenza H5N1, H1N1, etc., have contributed to the enhancement of its public health system's preparedness against emerging threats. Additionally, with support from the US Centers for Disease Control (US-CDC) and the WHO, Thailand's National Field Epidemiology Training Programme (FETP) is the oldest program of its kind outside the United States since 1980 [26]. It has trained a large number of experts in the investigation and control of disease outbreaks, many of whom have been instrumental in managing the COVID-19 epidemic at the provincial and district levels.

Thailand's robust health system encompasses strong laboratory and research capacities. Notably, the Thai Red Cross Emerging Infectious Diseases-Health Science Centre, located in Chulalongkorn University and serving as a WHO Collaborating Centre, contributes to the country's expertise in combating emerging infectious diseases [27].

Together, all these components of the system have made significant contributions to the successful fight against COVID-19.

COVID-19 Policies from the Government and Relevant Sectors and Their Implementation Status

Thailand's effective response to the pandemic has contributed to its rise in global health security rankings. The country has been ranked fifth in the world, notably, being the only non-high-income country among the top ten [28]. This highlights Thailand's efforts in safeguarding public health and ensuring a strong healthcare system. During the pandemic, Thailand has showcased numerous successful elements throughout the various phases of its preparedness and response framework, which are presented in this section.

Testing and Surveillance

In early 2020, as the pandemic surged with increasing domestic cases, the Thai government swiftly implemented strong containment measures. Alongside the establishment of CCSA in March 2020, other government and private agencies, as well as the general public, were also called upon to collaborate in preventing and containing the virus. The focus involved two key measures: preventing the entry of new cases into the country and reducing their transmission within the country [29]. Furthermore, the measures were heightened by declaring a state of emergency and enforcing a curfew that prohibited individuals from leaving their homes between 10:00 pm and 4:00 am [5]. Isolating, providing treatment, and tracing and quarantining their contacts were also implemented. All individuals diagnosed with COVID-19 were isolated in dedicated facilities rather than their homes. To enhance diagnostic capabilities, a laboratory network for RT-PCR testing was established. By the end of July, 78% of Thailand's 77 provinces had the capacity to independently diagnose COVID-19 [27].

As for surveillance, three levels were implemented: points of entry, healthcare facilities, and community case finding. Health workers were instructed to expand screening to areas with confirmed cases, unknown infections, risk factors, as well as communities with continuous confirmed cases in the past 4 weeks [30]. These initiatives were intended to enhance the early detection of uncommon situations and cases. A district-level surveillance, monitoring, and inspection committee was established to oversee the situation [31]. In addition, high-risk areas were promptly placed under lockdown to prevent the spread of outbreaks, with a particular focus on districts, as the primary healthcare level, playing a vital role in effectively managing the situation [32]. At the subdistrict level, the VHVs network and health center staff supported the response team, assessing infection risks and liaising with health centers.

To enhance the efficiency of testing and surveillance, laboratories play a vital role in the prevention, detection, and response to virus outbreaks. The Thailand National Laboratory System, which comprises private diagnostic laboratories, public health laboratories, regional public health laboratories, and one public health reference laboratory [30], is an essential key stakeholder for virus detection. Also, laboratories at the sub-national level are situated across 14 regional medical science facilities [30]. However, the significant challenge during that time was the absence of laboratory licensing and a legislative framework [30] to support the system, ensure the quality of service, and monitor the cost-effectiveness.

Healthcare Infrastructure

During a seminar in late 2021, Dr. Jos Vandelaer, WHO Representative to Thailand, highlighted Thailand's Universal Health Coverage as a crucial system in combating COVID-19 and facilitating post-pandemic recovery. This system encompasses access to services, affordable prices, and quality healthcare, addressing the challenges faced during the ongoing pandemic [33].

The private sector also played a role in supporting the government's action plans [27]. Siam Biosciences, a domestic company, developed COVID-19 diagnostic kits, and the web application Thai Chana aided in contact tracing. Hotels were transformed into isolation facilities for mild cases and quarantine purposes. Additionally, private manufacturers in Thailand ramped up production or repurposed operations to meet the demand for masks and PPE [27]. Alongside the existing system, Thailand initiated a comprehensive project, guided and supported by WHO, to enhance healthcare facilities and personnel's response and improve the post-pandemic resilience of the healthcare system. The project encompasses all healthcare facilities in Bangkok and other Regional Health Offices, intending to establish a "new normal" model for the delivery of health services [34].

Travel Restrictions and Border Control

Thailand implemented a comprehensive strategy to control travel and secure its borders during the COVID-19 pandemic. During the first phase, as early as April 4, 2020, all international flights were suspended, allowing only emergency or authorized flights to operate [35]. Entry Measures into Thailand were constantly updated on the official website of the Ministry of Foreign Affairs, Kingdom of Thailand

Fig. 8.2 Thailand Pass QR code

[36]. For instance, inbound passengers were required to cooperate with customs inspection and present the QR code of Thailand Pass (shown in Fig. 8.2).

Finally, travelers to Thailand have been exempt from providing COVID-19 documents, including vaccination certificates and test results, upon arrival since October 1, 2022, according to the Tourism Authority of Thailand [37].

Public Health Guidelines

During the early stage of the pandemic, when comprehensive global recommendations on effective social measures were not yet established, the government implemented strategies to minimize droplet transmission. These included promoting face masks, emphasizing hand hygiene, maintaining physical distance, and discouraging social gatherings. To enforce these measures, a State of Emergency Decree was declared on March 26, 2020, and curfews were imposed between 22:00 and 04:00 on April 2, 2020.

In early April 2020, the government decided to close all kindergartens, schools, and universities, suspending in-person learning. Nonessential employees were also advised to work from home, and national holidays were canceled to prevent large social gatherings and domestic travel [35]. Both Thais and foreign residents were advised to wear face masks, maintain a distance of over 2 meters, and practice good hand hygiene. Throughout the pandemic period, the CCSA monitored the population's compliance with social measures through regular surveys [35].

Social Support

According to a survey by the World Bank, approximately 70% of households experienced income decline, particularly affecting a majority of households in rural areas, the Southern region, and lower income groups [38].

Various social protection programs were implemented in Thailand to provide financial support to informal workers, farmers, low-income individuals, vulnerable populations, older adults, individuals without access to smartphones, and those residing in health facilities [39]. These initiatives included the reduction of electricity and water bills for all citizens in early 2020, as well as the distribution of electronic vouchers to eligible individuals [40]. Such e-vouchers contain the State Welfare Card (SWC) cash benefits that can be used in registered stores. Formal workers in Thailand are covered by the social security scheme, while all workers, including informal sector workers, receive cash assistance to mitigate the impact of poverty. However, the cash handouts may not fully address the income loss, and certain vulnerable groups may still be excluded based on eligibility criteria [40].

Also in 2020, the Thai government introduced emergency assistance programs (e.g., SWC cash benefits) that benefited over 80% of households, especially low-income households and those facing income shocks, with proportions nearing 90%. The number of social assistance beneficiaries was reported to have nearly doubled compared to the previous year, 2019 [38].

Communication, Public Awareness, Risk Communication System, Technology

Amidst the pandemic, a communication plan was prioritized as one of the key strategies for preventing and responding to emergency public health crises. Clear and transparent communication was mentioned as essential to enable effective contributions from all individuals and personnel, as seen in Fig. 8.3 [29].

Risk communication preparedness involved continually assessing risks, monitoring people's behaviors, and developing suitable communication strategies for diverse populations in Thailand, utilizing various channels. The MOPH appointed a

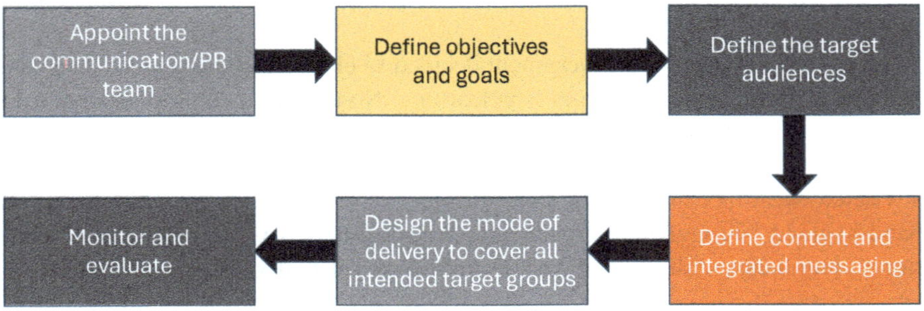

Fig. 8.3 Framework for implementation of the communication plan [29]

spokesperson to provide daily updates and COVID-19 knowledge through live broadcasts. The Department of Disease Control Hotline at 1422 served as a platform to disseminate public health and COVID-19 information in Thai, English, and neighboring languages. Interestingly, in Thailand, social media and celebrities were crucial in raising awareness among the general population about the risk of COVID-19 [30].

The pandemic accelerated and fostered innovation in Thailand's digital infrastructure, building upon the country's prior investments in this area. The MOPH encouraged the use of telemedicine and postal delivery of pharmaceuticals. Digital technology was piloted to develop new models of health service delivery, ensuring protection against COVID-19 and enhancing efficiency [27]. The DDC Care app and Thai Chana web application, developed by the Department of Disease Control and the Ministry of Digital Economy and Society, respectively [30], were developed to monitor patients and track population movement for contact tracing purposes.

Making use of the existing technology, "Aor Sor Mor Online," a mobile application used to provide health news to VHVs since 2015, was utilized by VHVs to update the situation, record data, educate the community, and monitor quarantine in their respective areas. The application also aided VHVs in preparing for their tasks by providing information on high-risk countries, COVID-19 knowledge, prevention measures, and surveillance mechanisms [41].

Specifically, in Bangkok, the Thai government developed the BKK HI/CI Care platform, allowing care providers to remotely monitor COVID-19 patients in Bangkok. The platform enables monitoring of crucial factors such as food intake, temperature, and treatments, providing a convenient and efficient method of healthcare delivery [42]. For privacy consideration, only nurses and doctors can access this platform via Windows, iOS and Android mobile phones, tablets, and computers. Additionally, expert volunteers from various sectors collaborated to enhance interaction through health technologies, including the Sabaidee Bot and Covid19Bot, which provided valuable information and assistance during the COVID-19 pandemic [43, 44]. For example, Sabaidee Bot has partnered with the Ministry of Public Health and affiliated public hospitals to develop a system for registering suspected COVID-19 patients (PUI—Patient Under Investigation) by completing intake information and preliminary symptom data automatically via chatbot. Since the data are collected automatically without the need for medical staff, the risk of transmission is reduced and the safety of nurses and doctors is protected. The medical staff can verify the accuracy of the intake information and forward the data to the relevant agencies electronically. Hospitals will also notify the patient of their test results and next steps via Sabaidee Bot. Moreover, not only patients will receive proper healthcare tips, but their feedback will also help the Department of Disease Control to analyze the situation more precisely.

Furthermore, the government leveraged digital platforms, including social media, to disseminate health news, updates, and educate the public. The adoption of digital technology played a crucial role not only in enhancing surveillance response but also in enhancing care recipients' satisfaction, reducing the number of readmission cases, and maximizing health resources [45] during the pandemic in Thailand.

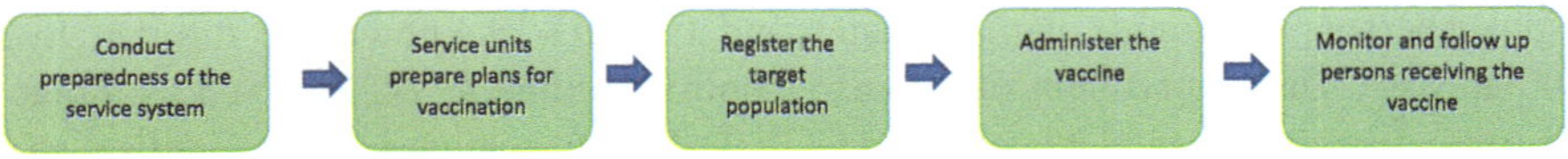

Fig. 8.4 Steps of COVID-19 vaccine services implementation in Thailand [46]

Vaccination Strategy

At the beginning of 2021, Thailand prepared to manage the number of COVID-19 vaccines to deliver vaccines for all citizens and residents in Thailand.

To ensure widespread vaccine coverage in Thailand, every province had to prepare for vaccine administration, side effects monitoring, and service record systems. Five key strategies for managing COVID-19 vaccines include [46]: *(1) Information Distribution*: Disseminate accurate and up-to-date COVID-19 information through appropriate channels. *(2) Quality Services*: Train personnel, establish a cold chain system, distribute vaccines for full population coverage, and foster cooperation between the government and the private sector. *(3) Vaccine Quality Assurance*: Validate vaccine quality according to standards, create information packages for reporting and monitoring side effects, and manage cases with adverse effects. *(4) Vaccine Management Information System*: Develop an effective system for vaccine registration, follow-up, and control using technology, and establish a database linkage between the public and private sectors. *(5) Knowledge Management*: Prepare technical information packages for healthcare workers and the public, design plans for vaccine and service quality research and evaluation, and develop policy recommendations.

Additionally, MOPH established the guidelines to help the preparation of the obtainment, storage, delivery, and administration of COVID-19 vaccines at every health system level. The guidelines were also flexible for being adjusted according to the change of nature of the strains of COVID-19, the supply of vaccines, and side effects. The Government Pharmaceutical Organization (GPO) was the Central COVID-19 stock under the MOPH. The Vendor Managed Inventory (VMI) system was used for the distribution of vaccines to MOPH-regional/Provincial/District hospitals. Meanwhile, GPO distributed the COVID-19 vaccines to health centers in Bangkok, non-MOPH government hospitals, medical schools, as well as private hospitals. The steps of the COVID-19 vaccine service were organized as the flow in Fig. 8.4.

Implications of Thailand's COVID-19 Policies and Public Health Measures

Thailand implemented early and strict policies, including border closures and curfews, which significantly helped control the initial spread of the virus. However, these measures also had long-term effects on the economy, particularly in the tourism sector. The tourism industry plays a vital role in Thailand's economy, contributing a substantial portion to the country's gross domestic product (GDP). While social assistance programs provided various support and helped to relieve some financial burden, many households experienced income decline, especially low-income groups in rural areas [38]. The reclassification of COVID-19 as a "Disease under surveillance" indicates a shift toward a more sustainable approach to managing the virus, but it also raises questions about the potential impact on public health and healthcare resource allocation. A more detailed economic impact assessment is needed to understand the comprehensive economic impact and to develop effective recovery strategies. In the future, policymakers need to consider more the impact of the COVID-19 pandemic crisis on various sectors, including manufacturing, agriculture, and services, as well as the distributional effects of available healthcare services.

In comparison to countries that depended more on centralized, top-down disease control measures, Thailand's network of Village Health Volunteers (VHVs) played a pivotal role in responding to the COVID-19 pandemic. With over one million volunteers nationwide, VHVs served as a vital link between local communities and the formal healthcare system. The VHVs' success can be attributed to the following factors: (i) Thailand's well-established primary healthcare system, with VHVs as a key component, providing a solid foundation for pandemic response; (ii) sustained political endorsement reinforced the importance of VHVs in the healthcare delivery system; (iii) VHVs received training and support from health centers and other healthcare professionals, equipping them with the knowledge and skills needed to perform their roles effectively; and (iv) community trust facilitated VHVs to build rapport with residents, effective communication, and encourage them in COVID-19 prevention efforts.

Although Thailand prepared for the vaccination in terms of policies and services before and during the pandemic, the nation still faced a lot of challenges as follows [47].

Sustained Investment in Research and Development (R&D): Thailand relies heavily on imported vaccines. Along with other factors, such as global vaccine shortages, logistical bottlenecks, and public hesitancy, these issues have contributed to delays in vaccine access. To address these, the government must consistently invest in domestic pharmaceutical R&D and production, emphasizing technology-based research.

Public-Private Collaboration in Vaccine Development: Private organizations can play various roles in vaccine development, and their support during the COVID-19 pandemic was crucial. The public sector should incentivize advanced research, innovation, medical devices, tests, vaccines, and therapeutic development through grants and support proposed by the private sector.

Streamlining Vaccine Introduction: Thailand, as a vaccine self-procuring nation with limited resources, should carefully consider the cost-effectiveness of vaccines during the introduction process. Prioritization in decision-making should focus on preventive vaccine effects, not just treatment effects like medicines. Revision of procurement regulations is needed to ensure vaccine availability.

Reforming Vaccine Procurement Regulations: Stringent procurement regulations, involving epidemiological needs assessments, budget allocations, supplier negotiations and agreements, vaccine registrations and approvals, vaccine importation and distribution, and pharmacovigilance, have hindered quick decision-making to a certain degree, leading to delayed vaccine access during emergencies. Revising procurement regulations to align with the limited vaccine market competition is essential for effective vaccine procurement.

In summary, as the global vaccine market evolves, the Thai government should prepare for these changes, manage public awareness related to health promotion and pandemic prevention, and revise vaccine policies. A long-term plan to support domestic R&D is necessary to enhance the vaccine value chain and elevate pharmaceutical manufacturing standards.

Conclusion

Thailand's response to the challenges posed by the COVID-19 pandemic has been a lesson for its future readiness and provides a roadmap for other similar settings. While the achievements are significant, it is important to acknowledge the challenges and limitations that were faced. The pandemic highlighted the necessity of investing in domestic vaccine research, adaptability in policies, and a robust healthcare system. These experiences may serve as a learning opportunity, which helps Thailand prepare for potential future pandemics.

In summary, using a multi-sectoral approach, including primary healthcare as one of the remarkable strengths, combined with technological innovation, effective communication, and collaborative strategies, enabled Thailand to establish a strong foundation for handling the pandemic. Looking ahead, the insights gleaned from COVID-19 will significantly influence Thailand's pandemic readiness plans. The lessons learned highlight the importance of a balanced approach between public health measures and socio-economic considerations, the power of community and primary care level involvement, and the potential of technology-driven solutions.

References

1. Zhu N, Zhang D, Wang W, Li X, Yang B, Song J, Zhao X, Huang B, Shi W, Lu R, Niu P, Zhan F, Ma X, Wang D, Xu W, Wu G, Gao GF, Tan W, China Novel Coronavirus Investigating and Research Team (2020) A novel coronavirus from patients with pneumonia in China, 2019. New Engl J Med 382(8):727–733. https://doi.org/10.1056/NEJMoa2001017
2. Namwat C, Suphanchaimat R, Nittayasoot N, Iamsirithaworn S (2020) Thailand's response against coronavirus disease 2019: challenges and lessons learned. OSIR 13(1):33–37
3. Pongpirul WA, Pongpirul K, Ratnarathon AC, Prasithsirikul W (2020) Journey of a Thai taxi driver and novel coronavirus. New Engl J Med 382(11):1067–1068. https://doi.org/10.1056/NEJMc2001621

4. World Health Organization (2020) Coronavirus disease 2019 (COVID-19) WHO Thailand Situation Report – 1 March 2020. World Health Organization, Geneva. Available from: https://www.who.int/thailand/emergencies/novel-coronavirus-2019/situation-reports-(archive). Accessed June 19, 2023
5. Yorsaeng R, Suntronwong N, Thongpan I, Chuchaona W, Lestari FB, Pasittungkul S, Puenpa J, Atsawawaranunt K, Sharma C, Sudhinaraset N, Mungaomklang A, Kitphati R, Wanlapakorn N, Poovorawan Y (2022) The impact of COVID-19 and control measures on public health in Thailand, 2020. PeerJ 10:e12960. https://doi.org/10.7717/peerj.12960
6. Prime Minister's Office (2020) Declaration of an Emergency Situation in all areas of the Kingdom of Thailand. Retrieved June 19, 2023, from http://www.mfa.go.th/main/contents/files/news3-20200326-161207-994002.pdf
7. World Health Organization (2020) Coronavirus disease 2019 (COVID-19) WHO Thailand Situation Report – 31 December 2020. World Health Organization, Geneva. Available from: https://cdn.who.int/media/docs/default-source/searo/thailand/2020_12_31_tha-sitrep-111-covid19.pdf?sfvrsn=e6b73214_3. Accessed June 19, 2023
8. Thai Government Extends Curfew and Tight Restrictions in Dark Red Provinces until 2 August. National News Bureau (2021). https://thainews.prd.go.th/en/news/detail/TCATG210719151735814
9. Thaicharoen S, Meunrat S, Leng-Ee W, Koyadun S, Ronnasiri N, Iamsirithaworn S, Chaifoo W, Tulalamba W, Viprakasit V (2022) How Thailand's tourism industry coped with the COVID-19 pandemics: a lesson from the pilot Phuket tourism sandbox project. J Travel Med:taac151. Advance online publication. https://doi.org/10.1093/jtm/taac151
10. Tourism Authority of Thailand (2022) Thailand's entry requirements from 1 July 2022: Thailand Pass removed. https://www.tatnews.org/2022/06/thailands-entry-requirements/
11. World Health Organization (2023) Coronavirus disease 2019 (COVID-19) WHO Thailand Situation Report – 4 July 2023. World Health Organization, Geneva. Available from: https://cdn.who.int/media/docs/default-source/searo/thailand/2023_07_04_tha-sitrep-266-covid-19.pdf?sfvrsn=623fdf24_1. Accessed August 23, 2023
12. WHO. COVID-19 Dashboard. Thailand. Available at: https://covid19.who.int/region/searo/country/th/. Accessed June 15, 2023
13. Jongudomsuk P, Srithamrongsawat S, Patcharanarumol W, Limwattananon S, Pannarunothai S, Vapatanavong P et al (2015) The kingdom of Thailand health system review, vol 5(5). World Health Organization, Regional Office for the Western Pacific, Manila
14. Marshall AI, Witthayapipopsakul W, Chotchoungchatchai S, Wangbanjongkun W, Tangcharoensathien V (2023) Contracting the private health sector in Thailand's Universal Health Coverage. PLOS Glob Public Health 3(4):e0000799. https://doi.org/10.1371/journal.pgph.0000799
15. Aljunid SM, Srithamrongsawat S, Chen W, Bae SJ, Pwu RF, Ikeda S, Xu L (2012) Health-care data collecting, sharing, and using in Thailand, China mainland, South Korea, Taiwan, Japan, and Malaysia. Value Health 15(1 Suppl):S132–S138. https://doi.org/10.1016/j.jval.2011.11.004
16. Ministry of Public Health (MoPH)- Health Administration Division (2012) Health service system development plan. Ministry of Public Health, Nonthaburi
17. Thaiprayoon V, Wibulpolprasert S (2017) Political and policy lessons from Thailand's UHC experience. ORF Issue Brief, p 174
18. Viwattanakulvanid P, Chen IPJ (2021) Primary Health Care – Thailand, Taiwan and Singapore. College of Public Health Sciences. Chulalongkorn University
19. Tejativaddhana P, Briggs D, Singhadej O, Hinoguin R (2018) Developing primary health care in Thailand: innovation in the use of socio-economic determinants, Sustainable Development Goals and the district health strategy. Public Admin Pol 21(1):36–49. https://doi.org/10.1108/PAP-06-2018-005
20. Ministry of Public Health (2016) Guideline on the operation of primary care cluster for health providers (in Thai language). Ministry Of Public Health, Nondhaburi. Available at: http://bps.moph.go.th/new_bps/sites/default/files/Guidelines%20PCC.pdf. Accessed June 21, 2023

21. United Nations (2022) Village health volunteers: unsung heroes for Thailand's health crisis. https://sdgs.un.org/partnerships/village-health-volunteers-unsung-heroes-thailands-health-crisis. Accessed June 21, 2023
22. World Health Organization. Coronavirus Disease 2019 (Covid-19) WHO Thailand Situation Report-12 April 2020 (2020). Retrieved 21 June 2023 from https://www.who.int/docs/default-source/searo/thailand/2020-04-12-tha-sitrep-50-covid19-finala0161210534642308b23a2965a57a44c.pdf?sfvrsn=5150506c_0
23. World Health Organization. Thailand's 1 million village health volunteers - "unsung heroes" - are helping guard communities nationwide from COVID-19, 2020. Available: https://www.who.int/thailand/news/feature-stories/detail/thailands-1-million-village-health-volunteers-unsung-heroes-are-helping-guard-communities-nationwide-from-covid-19. Accessed June 21, 2023
24. Plipat T, Panlar P, Muangdang C (2015) The framework for the development of the emergency operation center and the public health emergency command system, Department of Disease Control B.E. 2016-2021. Bureau of Epidemiology, Department of Disease Control, Nonthaburi. (In Thai)
25. Department of Disease Control (2015) Strategic plan for the development of international sanitation law B.E. 2005 for B.E. 2017-2021. Department of Disease Control, Nonthaburi. (In Thai)
26. Thailand Field Epidemiology Training Program. http://164.115.45.89/th/
27. World Health Organization (WHO) (2020) Thailand: how a strong health system fights a pandemic. WHO, Geneva. COVID-19: WHO's Action in Countries. https://www.who.int/publications/m/item/thailand-how-a-strong-health-system-fights-a-pandemic. Accessed June 21, 2023
28. Homepage - GHS Index. Available: https://www.ghsindex.org/. Accessed June 21, 2023.
29. Thai Ministry of Public Health (2021) Strategic plan: Covid-19 strategy: managing the new wave of the Covid-19 Epidemic Ministry of Public Health, January 2021. Available at: https://ddc.moph.go.th/viralpneumonia/eng/file/main/en_Thailand%20Covid-19%20plan_MOPH_2021.pdf. Accessed June 22, 2023
30. Ministry of Public Health Thailand. Thailand's experience in the COVID-19 response. Available: https://ddc.moph.go.th/viralpneumonia/eng/file/pub_doc/LDoc9.pdf. Accessed June 22, 2023
31. Disease Prevention and Control (2020). COVID-19 Situation in Office of Disease Prevention and Control. https://ddc.moph.go.th/odpc11/. Accessed June 22, 2023
32. Kaewchandee C, Hnuthong U, Thinkan S, Rahman MS, Sangpoom S, Suwanbamrung C (2023) The experiences of district public health officers during the COVID-19 crisis and its management in the upper southern region of Thailand: a mixed methods approach. Heliyon 9(1):e12558. https://doi.org/10.1016/j.heliyon.2022.e12558
33. Thailand National Health Security Office. Description News (2021). UHC has well prepared Thailand for coping with COVID-19. Available at: https://eng.nhso.go.th/view/1/DescriptionNews/UHC-has-well-prepared-Thailand-for-coping-with-COVID-19/400/EN-US
34. World Health Organization (2020) Thailand launches "New Normal" healthcare system to build back better after COVID-19 [Internet]. Available from: https://www.who.int/thailand/news/feature-stories/detail/thailand-launches-new-normal-healthcare-system-to-build-back-better-after-covid-19
35. Rajatanavin N, Tuangratananon T, Suphanchaimat R, Tangcharoensathien V (2021) Responding to the COVID-19 second wave in Thailand by diversifying and adapting lessons from the first wave. BMJ Glob Health 6(7):e006178. https://doi.org/10.1136/bmjgh-2021-006178
36. Ministry of Foreign Affairs, Kingdom of Thailand. News. COVID-19. https://www.mfa.go.th/en/content/entry-measures-into-thailand-by-air-travel-0105202?cate=5f069ee372a783584326eb24
37. Tourism Authority of Thailand. Thailand fully reopens from 1 October 2022. https://www.tatnews.org/2022/10/thailand-fully-reopens-from-1-october-2022/

38. The World Bank (2021) Impact of COVID-19 on Thailand's households: insights from a rapid phone survey. World Bank Group, Washington, DC. https://documents1.worldbank.org/curated/en/390821633313679563/pdf/Impact-of-COVID-19-on-Thailand-s-Households-Insights-from-a-Rapid-Phone-Survey.pdf. Accessed January 2, 2023
39. DFDL. Thai Finance Ministry approves additional THB 3.04 billion for the Rao Chana ('We Win') stimulus program for at-risk communities affected by COVID-19. Published April 28, 2021. https://www.dfdl.com/resources/legal-and-tax-updates/thai-finance-ministry-approves-additional-thb-3-04-billion-for-the-rao-chana-we-win-stimulus-program-for-at-risk-communities-affected-by-covid-19/. Accessed June 22, 2023
40. Oxford Policy Management (OPM) (2020) United Nations Thailand. Social impact assessment of COVID-19 in Thailand. OPM, Oxford. https://www.unicef.org/thailand/media/5071/file/SocialImpactAssessmentofCOVID-19inThailand.pdf. Accessed June 22, 2023
41. Viwattanakulvanid P (2021) Ten commonly asked questions about Covid-19 and lessons learned from Thailand. J Health Res 35(4):329–344. https://doi.org/10.1108/JHR-08-2020-0363
42. Ang A (2021) Thailand closely monitors COVID-19 patients via telemedicine platform. Healthcare IT News. https://www.healthcareitnews.com/news/asia/thailand-closely-monitors-covid-19-patients-telemedicine-platform
43. Opendream. Sabaidee Bot. https://www.opendream.co.th/en/project/sabaidee-chatbot-en. Accessed June 22, 2023
44. Sea Thailand. Garena partners with Sabaidee Bot, encouraging Thai gamer to join forces in COVID-19 battling. Published April 3, 2020. https://www.seathailand.com/content/212. Accessed January 2, 2023
45. Intawong K, Olson D, Chariyalertsak S (2021) Application technology to fight the COVID-19 pandemic: lessons learned in Thailand. Biochem Biophys Res Commun 538:231–237. https://doi.org/10.1016/j.bbrc.2021.01.093
46. Health MoP (2021) Strategy: managing the new wave of the Covid-19 Epidemic: Division of Strategy and Planning, Office of the Permanent Secretary, Ministry of Public Health; [cited 16 August 2023]. Available from: https://ddc.moph.go.th/viralpneumonia/eng/file/main/en_Thailand%20Covid-19%20plan_MOPH_2021.pdf
47. Kategeaw W, Youngkong S, Taychakhoonavudh S, Techathawat S, Chaiyakunapruk N (2022) Potential changes in vaccine access and policy landscape in Thailand post COVID-19 pandemic. Hum Vaccin Immunother 18(6):2095823

Vietnam: The Demands for Candid, Democratic, and Civilised Dialogue

9

Hai Thanh Doan

Abstract

The first section of this chapter demonstrates that the precautionary principle is the overarching principle governing epidemic prevention and control. The second section identifies, and is confined to, important paradigms and shifts in paradigms of Vietnam's COVID-19 responses and argues that Vietnam applied the precautionary principle in responding to COVID-19, which accounts for certain successes, especially in the initial periods. The third section demonstrates that Vietnam's responses, especially in and as manifested by the later periods, are not quite successful and that deficiencies in the application of the precautionary principle, namely, disproportionality, outdated and inappropriate approach, and corruption, undermine the legitimacy, merit, and efficacy of Vietnam's responses. The fourth section argues that the more fundamental cause of the shortcomings is the insufficiency of candid, democratic, and civilised dialogue, which stems from long-established practices across social, political, and legal institutions. Consequently, while future responses to risks and dangers necessarily implement the precautionary principle as envisaged in the Law on Prevention and Control of Contagious Diseases (LPCBTN) and as has been practised in Vietnam with laudable results, two key aspects are demanded for more refined application of the precautionary principle and more justified responses: (i) the assertion of the imperatives of bioethics and public health ethics, and (ii) the installation of jurisprudence embracing duties to protect, to give due regard, and to give reasons.

Keywords

Precautionary principle · Candid, democratic, and civilised dialogue · Duty to protect · Duty to give due regard · Duty to give reasons

H. T. Doan (✉)
University of Otago, Dunedin, New Zealand

L. P. Marquez et al. (eds.), *COVID-19 in the Pacific Rim and Beyond*,
Collaborative Bioethics 6, https://doi.org/10.1007/978-3-032-21218-4_9

The Precautionary Principle as the Overarching Principle Governing Epidemic Prevention and Control

Article 4 of Luật phòng Chống bệnh Truyền nhiễm or the Law on Prevention and Control of Contagious Diseases (LPCBTN), which sets out core principles of epidemic prevention and control, can be conceptualised and abstracted into *the precautionary principle*[1] with use of and emphasis paid on *mass mobilisation,* rooted in Vietnamese and communist political constructions and maxims [1]. Specifically:

(i) Prevention, which includes, inter alia, information, education and communication, surveillance and the combination of technical medical measures with social and administrative measures;
(ii) Coordination, cooperation, and social mobilisation, and the integration of activities of prevention and control of infectious diseases into socio-economic development programmes;
(iii) Ensuring the prompt publication of accurate information concerning epidemics; and
(iv) Anti-epidemic activities must be proactive, timely, and comprehensive.

This principle is further elaborated and implemented via the State's policies concerning epidemic prevention and control (Article 5), addressing diseases not just on a one-off basis but progressively with arrangements for planning, investment, preparation, and anticipation.[2]

[1] An analogy can be drawn between the language and structure of Vietnamese instruments concerning disease prevention and control and EU instruments, e.g., the Treaty on the Functioning of the European Union, Charter of Fundamental Rights of the European Union, and the European Commission's Communication concerning the precautionary principle in 2000 as well as instruments of the countries concerning disease prevention and control. However, it is not (and not necessary) to see that the precautionary principle within the Vietnamese context has a Western origin [105]

[2] Various areas are covered:

- The prioritising and encouragement of and support for investment, training, research concerning preventive medicine, the capacity for surveillance personnel, and systems to detect infectious diseases.
- The production of vaccines and bio-medical products; exchange and training of specialists, and transfer of technology.
- The support for medical attendance and care for, for example, persons suffering from infectious diseases due to occupational risks.
- The support for compensation for the cull of cattle and poultry carrying agents of infectious disease.
- The mobilisation of contributions in terms of finance, technique, and labour from the entire society.
- The expansion of cooperation with international organisations and other countries in the region and the world.

In light of the precautionary principle, Article 3 provides for the classification of infectious diseases, allowing for the differentiation of measures, thus facilitating appropriate measures:

- Class A consists of extremely dangerous infectious diseases that can transmit very rapidly and spread widely, with high mortality rates or *with unknown agents*.
- Class B consists of dangerous infectious diseases that can rapidly transmit and be fatal.
- Class C consists of less dangerous infectious diseases that are not rapidly transmissible.

The specific categorisation is adjusted and updated by the Ministry of Health, indicating a duty to monitor circumstances and adjust measures accordingly.

COVID-19 Pandemic Phases and the Implementation of the Precautionary Principle as the Basis of the Successes

Phase Zero: Preparations

Soon after the airing of the news about the at-the-time "Wuhan pneumonia cases", Vietnam prepared for and took precautionary measures, including working with international organisations and planning for various scenarios to handle the epidemic [2–4], calibrated to the risk scenarios assessed at the time.[3]

Phase One: "Fighting-the-Disease-Like-Fighting-the-Enemy"

Following the first COVID-19 cases recorded on 23 January 2020, the Prime Minister issued Directive 05/CT-TTg, coining the *"fighting-the-disease-like-fighting-the-enemy" ["Chống dịch như chống giặc"]* approach. While Directive 05/CT-TTg read that the target would be to minimise deaths, VCP's[4] newspapers suggested that the objective was *zero deaths* [5]. The Vietnamese Politburo and the Prime Minister emphasised that citizens' life and health should be treated as a priority and that the State "would *accept to incur some economic losses to protect people's lives and health*" [6]. This orientation reflects an attempt to balance public-health protection with broader socio-economic considerations. Then, the Standing Member of the Secretariat of VCP issued Official Dispatch No. 79, importantly, asserting that the prevention and control of the epidemic caused by the new strain of

[3] For example, on 3 January 2020, Vietnam tightened the Vietnam-China border control (Official Dispatch 02/DP-DT). From 7 January, Viet Nam started health checks of all passengers departing from Wuhan [137]. On 21 January 2020, Decision No. 181/QD-BYT was issued, which drew different scenarios and determined surveillance measures relative to the scenarios.

[4] VCP stands for the Vietnamese Communist Party.

coronavirus would be *a key and urgent task of the entire political system*. It required, inter alia, agencies and units to ensure economic and social development, social order and safety, maintenance of people's lives, and to prepare plans to respond to the disease. It also allowed restrictive measures. These overarching guidelines, which informed Vietnam's COVID-19 response during the height of the pandemic, may be understood as reflecting the precautionary principle.

Phase 1.1: Initial Set of Stringent Measures and the First Localised "Social Distancing" Measures

Following political-policy determinations, Decision 219/QD-BYT added nCoV to the list of class A infectious diseases and Directive 06/CT-TTg provided for, inter alia,

- The limitation of gatherings and travelling
- The wearing of masks in public places
- The establishment of communication channels between localities with borders with China and the respective Chinese local authorities to grasp daily updates
- The discouragement of trade and exchange with China
- The suspension of tourist activities with China
- The suspension of all flights from and to China

However, because the discouragement of trade and exchange soon proved to be disproportional and associated with materially adverse effects [7], the relevant measures were permanently withdrawn [8, 9] and were never reimplemented.

On 12 February 2020, the entire Sơn Lôi Commune was subject to social distancing, marking the first commune-level social distancing concerning COVID-19, after a group of labourers returned from Wuhan had tested positive.[5] The social distancing was lifted after 21 days [10]. On 7 March 2020, case number 17 was recorded in HN[6]—Vietnam's capital city; only a section of Trúc Bạch Street was cordoned as a result [11].

[5] The reason for this measure was that, according to the Deputy-Chairman of Vĩnh Phúc People's Committee, "[o]ut of 10 cases of COVID-19 infection in Vĩnh Phúc Province, Sơn Lôi Commune alone had 5 cases" and "all the initial outbreak was in Sơn Lôi" [138].

[6] HN stands for Hà Nội.

Phase 1.2: From Bạch Mai Case to Sweeping Public-Health-Societal Measures and National Social Distancing

Bạch Mai Case: Testing to Trace Source(s), Measures Applied to Crucial Healthcare Settings, and Experimental Sweeping Public-Health-Societal Measures

On 20 March 2020, two cases related to Bạch Mai Hospital were confirmed, but were considered not the index ('source') cases. Bạch Mai Hospital is a central tertiary hospital and a major referral destination for severe cases, reportedly receiving approximately 10,000 patients per day (including around 6000 inpatients). Initially, HN People's Committee proposed closing and restricting access to certain departments of the hospital. The Ministry of Health (BYT) proposed restricting measures to certain floors and departments associated with confirmed cases [12]. After the confirmation of 12 more cases, the Chairman of the HN People's Committee authorised the Chairman of the Đống Đa District People's Committee to issue a decision to place the hospital under a cordon measure in accordance with the maxim "no-one-can-go-in-or-out" [nội bất xuất, ngoại bất nhập] [12].[7] All staff were tested, and both the facilities and people were disinfected [13]. Simultaneously, local district authorities closed surrounding eateries and provided takeaway food to support hospital operations (Vũ, 2020b). *The people's committees of districts in HN decided to test and quarantine people who had symptoms related to the outbreak* [12] *to trace back the source(s).* It was ultimately found out that the infection originated from Trường Sinh Ltd., a contractor of Bạch Mai [14].

Notwithstanding, subsequently, the authorities recognised that any blockage of the hospital could have detrimental consequences because the most serious cases would not receive clinical support, and the transfer or channelling of these cases to other hospitals would be infeasible [15]. Consequently, it was decided that

- Patients not in critical condition could be transferred to some central medical facilities and military hospitals.
- While on-demand examination and re-examination at the hospital were suspended [16], special procedures were to be established for receiving severe and critical cases.
- It was recognised that it would be infeasible and undesirable to quarantine the medical staff at Bạch Mai Hospital for 14 consecutive days, as the general rule applicable to medical staff of other healthcare facilities.
- The authorities admitted that if most of the medical staff of Bạch Mai Hospital were concentrated for 14 days, the quarantine requirements could not be guaranteed. The authorities suggested that it would be necessary, inter alia, to urgently arrange a separate quarantine area as part of the Hospital, as well as some accommodation facilities and hotels for doctors to rest. Separate transportation would be provided.

[7] However, the authorities denied the saying of "blockage" [15, 16].

- Doctors and nurses who had tested negative (even for the first time) were considered not the source of infection and would participate in the treatment of patients [15].

Although the isolation of the Hospital was lifted after 14 days, certain restrictions lingered for months. Until the middle of May 2020, only 30% of hospital beds were expected to be in use [17].

Directives 15 and 16

On 27 March 2020, believing that "the next 2 weeks would be decisive in defeating the epidemic", following "experimental" measures implemented by HN's People Committee [18], Prime Minister Nguyễn Xuân Phúc issued Directive No 15/CT-TTg. In addition to reiterating previous maxims, this Directive also termed the maxim "*[e]ach person is a soldier*" and required chairmen of provincial people's committees to direct the application of measures to restrict mass gatherings from 00:00 on March 28 to the end of 15 April 2020, including:

- The cancellation of meetings and events with more than 20 people in a room and 10 or more people outside of offices, schools, hospitals (religious activities shall *not be considered* for exceptions).
- The suspension of *all* businesses, except for those providing essential goods and services—the chairmen shall decide what would be considered essential goods and services.
- The restriction of the movement of people: All passenger transportation activities in HN, HCMC, and areas with active transmission or from those areas to other places were to be minimised. Public transportation, except for the transportation of goods, would also be suspended. It was also requested, inter alia, that *unless absolutely necessary*, people should not leave their houses or visit *medical facilities*.
- Strict disciplinary measures for heads of local authorities of locations where large gatherings occurred.

On 30 March 2020, VCP's General Secretary and President, Nguyễn Phú Trọng, appealed to the nation for solidarity "to ensure victory over the COVID-19 pandemic". He also suggested that [the State must] "*[p]repare for the worst-case scenarios to have the best plans*" [19]. Invoking the call of VCP's General Secretary, the Prime Minister signed Directive No 16/CT-TTg, prescribing stricter measures, including:

- Nationwide social distancing for 15 days *from 0:00 am 1 April 2020*, according to the principle "*households isolate from households, villages isolate from villages, communes isolate from communes, districts isolate from districts, provinces isolate from provinces*".
- The closure of all businesses, except for those providing essential goods and services.

- The requirement of a two-metre distance in minimum in producing facilities—if such facilities could not comply with the rule, they would be closed (Official Dispatch 2601/VPCP-KGVX).
- The prohibition of two-person gatherings.
- Work-from-home for cadres, civil servants and public employees, except for strictly necessary cases.
- Suspension of all public passenger transportation and all movements in areas with active transmission and from those areas to other localities, except for the provision of food and essential necessities, vehicles to pick up workers, experts of enterprises, and transport raw materials for production.
- The minimisation of all other movements.

These measures impacting the entire nation are *highly restrictive*; however, attempts to differentiate measures corresponding to risk levels are *negligible*.

Phase 1.3. The Dual-Objective Strategy and the Differentiation of Measures Corresponding to Risk Levels

Around late April 2020, noting that cases were significantly reduced, the State adopted the phrase "new normal" [20] and issued Directive 19, ending the nation-wide social distancing and signalling economic recovery attempts. Although the State has never declared that it would neglect economic considerations, this Directive clarifies or redefines the Government's approach as *"balancing the dual goals of fighting the pandemic and ensuring economic growth while placing safety in a high position"* [21, 22].

The most important point of this Directive is that it provides for the differentiation of measures based on *"the risk level of disease transmission"*, making a conceptual shift in the approach with a view to localising and narrowing down measures. However, Directive 19 only made a rudimentary binary distinction between high-risk areas where strict adherence to Directive 16 is required and low-risk areas where promoting awareness would be conducted as the primary—if not only–measure. At the same time, Directive 19, while only addressing the differentiation at the provincial level, did not set out specific criteria for determining risk levels.[8]

Another important point is that while Directive 19 still maintains the distinction between essential and unessential activities, the method of categorising has been shifted. Whereas Directive 16 listed some essential activities, goods, and services, Directive 19 listed non-essential ones. However, Directive 19 also allowed provincial People's Committees to determine which activities, goods, services, and business premises were non-essential, blurring the line between the three directives.

[8] The possibility that the provincial authorities might make the risk differentiation for subordinate administrative units can only be found as a kind of implication. Notwithstanding, subsequently, the differentiation was extended to lower units [139].

Phase 1.4. Infections with Untraceable Source(s) and Initial Mass-Testing Activities

On 25 July 2020, the 416th case was confirmed, linked to ĐN[9] Hospital; however, the source(s) of the infections could not be traced. Three big hospitals in ĐN were blocked with medical staff and patients were isolated inside [23, 24]. It was initially planned for an early reopening of three hospitals [25]; however, after the confirmation of many new cases, social distancing was broadened, covering the entire provincial city, and extended [26]. On 28 July 2020, the Prime Minister allowed ĐN to apply Directive 16 at epidemic hotspots and to apply *certain measures* provided under Directive 16 at the provincial city level without invoking Directive 16 [27].

Because the source(s) of cases could not be traced, epidemiological data was used to develop hypotheses concerning the source(s) [28]. According to Deputy Prime Minister Vũ Đức Đam, if the source(s) was/were in ĐN, measures shall be applied to ĐN primarily; otherwise, measures at the national scale could be needed [28]. The prevailing view was that the initial case appeared in ĐN [28, 29], so measures were focused on and limited to ĐN without the need for direct, definite/beyond-reasonable-doubt proof that the source(s) of infections were in ĐN, though some nearby provinces also adopted social distancing.[10] On 2 August 2020, the Prime Minister articulated the maxim, *"[e]ach household shall be a fortress, and each citizen shall be a soldier in the fight against the pandemic"* [30].

On 16 August 2020, for the first time, ĐN launched citywide testing at three-day intervals to tease out all cases in preparation for reopening [31]. This measure was likely informed by the experience associated with an earlier unsuccessful attempt to ease social distancing. In early September 2020, social distancing measures were eased in most places, with some reportedly lingering until the middle of September [31].

Phase 1.5: Mass Testing aimed at Rapidly Stamping Out Cases in the Community

On 27 January 2021, a Vietnamese worker tested positive upon entering Japan [31], signalling the possibility that there were community cases which had escaped surveillance. Following this, various cases were detected in different provinces, and many of them were linked to industrial settings. On 18 May 2021, BG[11] decided to lock down four industrial parks to keep 67,000 workers from moving or returning to other locations and apply Directive 16/CT-TTg to various districts. On 26 May 2021, Prime Minister Phạm Minh Chính requested rapid and overnight mass testing to be implemented in provinces experiencing active transmission, presumably to

---

[9] ĐN stands for Đà Nẵng.

[10] Some provinces invoked Directive 16 [140], while some invoked Directive 15 but quoted the maxims and applied certain measures provided under Directive 16 [141].

[11] BG stands for Bắc Giang.

rapidly localise cases and stamp them out of the community, thereby *ending the distancing as soon as practicable*. This measure was, thus, different from that applied in ĐN. In ĐN, the chains of infection were cut primarily by social distancing and mass testing was applied before reopening to ensure that cases were not left undetected in the community.[12] On 26 July 2021, it was announced that the outbreaks in BG had been controlled [31].

Phase Two: The Fall of "Fighting-the-Disease-Like-Fighting-the-Enemy" Approach

On 31 May 2021, Decision 2686/QĐ-BCĐQG was issued, differentiating risks into four levels with criteria for the determination by and for specific administrative units. This reflected renewed efforts to differentiate measures corresponding to risk levels, allowing highly restrictive measures to be applied to the highest risk-level ("red") areas, while allowing "green"—new normal—areas to "do business as usual". However, the political-institutional landscape—including officials' concerns about being disciplined when outbreaks happen [32]—together with subsequent situational and epidemiological developments frustrated Decision 2686/QĐ-BCĐQG's purpose.

On 19 May 2021, HCMC's[13] first Delta-variant case was confirmed [31]. From 31 May to 8 July, HCMC implemented social distancing according to Directive 15 with stricter measures ["Chỉ thị 15 Tăng Cường"]. Gò Vấp District and Thạnh Lộc Ward (District 12) applied Directive 16. However, outbreaks could not be contained. On 9 July 2021, HCMC applied Directive 16 [31]. From 17 July to 23 July 2021, various instruments were issued, notably Resolution 78/NQ-CP of the Government, Official Dispatch 969/TTg-KGVX of the Prime Minister. Accordingly, importantly, Directive 16 and then "*Chỉ thị 16 tăng cường*" [the enhanced version(s) of Directive 16] purporting to permit any additional measures beyond those expressly provided for under Directive 16, were to be applied to 19 southern provinces. The measures ultimately included the suspension of all transportation activities. Despite not being covered by those instruments, some other provinces also implemented Directive 16 and the enhanced version of Directive 16 on their own initiative.[14] These signalled the difficulty of maintaining firm control of the situation.

On 31 July 2021, the Prime Minister required *"[a]i ở đâu thì ở đó"*—"[w]herever you are, stay there" and requested local authorities "[a]bsolutely do not allow people to move out of their places until the end of the social distancing" [33]. However, due to the hardship, people massively self-repatriated [34–36], de facto

[12] This approach could be more time-consuming than the approach of rapidly stamping out infections in the community via testing and could provide more room for transmissions to spread, compared to the latter. The latter approach, i.e., rapidly stamping infections out of the community via testing, however, was not advised by the WHO [142].

[13] HCMC stands for Hồ Chí Minh City.

[14] See, for example, Directive 17/CT-UBND of HN People's Committee.

lapsing Vietnam's protracted commitment to zero [community] infection [before the later official promulgation of Resolution 128/NQ-CP for this purpose].

Phase Three: Living with COVID-19

Necessary Preparations

Vaccination

Though following the lifting of social distancing in ĐN in October 2020, Deputy Prime Minister Vũ Đức Đam acknowledged that *"[t]his war will last until there is a vaccine and a specific treatment. We have only won each battle and each campaign, but not the whole war"* [29]. This vision, however, was delayed in being effectuated. Just on 1 February 2021, AstraZeneca was the first COVID-19 vaccine conditionally authorised (Decision 983/QĐ-BYT). The first batch of 204,000 AstraZeneca doses was permitted to be imported on 17 February 2021 (Official Letter 1215/QLD-KD). On 26 February 2021, the government issued Resolution 21/NQ-CP, providing the framework and principles for the purchase and use of COVID-19 vaccines. On 8 March 2021, the first vaccination campaign was carried out with the objective of returning life to normal [37]. On 1 April 2021, Vietnam started asking for foreign support in procuring COVID-19 vaccines [38]. Around June and July 2021, Vietnamese authorities started discussions and negotiations about vaccination supply and transfer of vaccine technology from foreign parties, later termed "*vaccination diplomacy*" ["Ngoại giao vắc xin"] [39]. On 10 July 2021, the Prime Minister launched the COVID-19 vaccination campaign with the goal of injecting 150 million doses for about 75 million people [31]. On 10 September 2021, HCMC authorities organised a mass vaccination campaign with the goal of basically completing the first dose of COVID-19 vaccination for people aged 18 and over and inoculating the second dose on time; rapid and mass vaccination campaigns were also implemented in various locations (Official Dispatch 6202/BYT-DP).

At-Home Treatment

On 28 July 2021, HCMC piloted the treatment of positive cases at home and the controlled use of antiviral medicines for people with mild symptoms at home [31]. On 16 August 2021, BYT permitted the controlled use of molnupiravir for mild COVID-19 cases at home [31]. On 20–21 August 2021, HCMC deployed mobile medical stations[15] and permitted at-home treatment, signalling a major shift in the authorities' response orientation against the backdrop of the overloaded healthcare system and availability of more COVID-19 countermeasures.

[15] Each mobile medical station was expected to manage and care for 50–100 F0 cases.

Political Orientational Decisions

From 25–29 August 2021, the Prime Minister articulated several orientational decisions, significantly shifting the paradigm of COVID-19 responses. He acknowledged that "[w]e must determine that this war is long-term, we have to live with the epidemic for a long time, we cannot control it absolutely, but must adapt..." [40]. He also suggested that

- "Screening, rapid testing, and classifying cases *to both prevent the spread and care for and treat infected cases promptly and effectively*".
- "Safely and flexibly adapting to and effectively controlling the COVID-19 epidemic with the spirit of economic recovery and development, corresponding to the developments and experiences of epidemic prevention and control and proportional to the capacity of the health system. Meanwhile, people's health should always be prioritised first".

The plan was to *fully* return to normal around 30 September 2021 [41].

Heading towards Normalcy

With about 23 million people receiving the first dose of vaccine (accounting for 43.9% of the population aged 18 and over) and 8.6 million people receiving two doses (around 27 September 2021) [42], on 13 October 2021, the government issued Resolution 128/NQ-CP *officially* terminating the implementation of Directives 15, 16, and 19. However, some provinces still maintained highly restrictive measures [43, 44] to the extent that the Prime Minister had to urge them to comply with Resolution 128 [45–47]. On 3 January 2022, more than 90% of the population aged 18 and over had received two doses of the COVID-19 vaccine [48]. On 19 October 2023, COVID-19 was relegated to class B (Decision 3896/QĐ-BYT).

The Implementation of the Precautionary Principle as the Cause of the Successes

It can and should be asserted *unequivocally* that Vietnam had various successes in handling COVID-19, especially in earlier periods. In the first 7 months of the pandemic, the number of deaths was zero, and the first recorded death involved a 70-year-old with multiple severe underlying conditions, including chronic kidney disease, hypertension, and heart failure [49][16]; by the end of 2020, Vietnam had reported only 35 deaths [50, 51]. These outcomes may be attributed, in significant part, to the implementation of the precautionary principle.

The Party-State asserted authoritative leadership in responses to the pandemic and attempted to *balance public health protection and economic considerations,*

[16] By mentioning this, it is to suggest that the Party-State tried their best to save lives up to the point that the endeavour is practically unfeasible in light of the respective surrounding factors and constraints.

with due consideration and emphasis paid to public health protection. In formulating its responses, Vietnam cooperated closely with international organisations and entities. It can also be observed that, in many instances, before implementing certain measures for the broader society, the Party-State first piloted them in limited localities. Measures have been informed by experiences and lessons learnt and updated in light of field situations and the progression of the disease. Many of the decisions and measures are well-reasoned, reasonable, and suit the corresponding scenarios.[17] The Party-State also took measures and initiatives to relieve the adverse impacts of COVID-19[18] and restrictive measures, aligned with the precautionary principle and the rooted Vietnamese ethical values [52]. The Bạch Mai Hospital case is just one example, showing a flexible and reasonable approach of the government in light of the accrued experience and various medical necessities.

To gather the public support and involve the public in the prevention and control of COVID-19 while minimising backlash, the Party-State leaders delivered various addresses to the public with clear, authoritative messages. For this purpose, authorities also conducted plausible public health education attempts [53, 54].

Consequently, the populace reportedly had high levels of confidence in the responses of the Party-State [55]. Various private actors also participated in COVID-19 prevention and control affairs through, inter alia, providing financial support for the purchase of and donating medical facilities, devices, and vaccines [56], assisting in the evacuation and repatriation of people [57], and organising and participating in charitable activities to aid the worst-off population [58, 59].

Shortcomings and Fallouts

Notwithstanding the foregoing, Vietnam's responses, especially in and as manifested by the later periods, are not generally quite successful. The number of deaths increased sharply to around 19,000 in September 2021 [60]. Ultimately, the total recorded deaths were 43.206 [61]. Although the data sets are indicative only, according to Our World in Data, the excessive mortality related to COVID-19 per one hundred thousand people in Vietnam is higher than or proximate to many liberal European countries [62]. Meanwhile, Vietnam was not among the earliest countries returning to normal [63]. The legitimacy, merit, and efficacy of Vietnam's responses are undermined by deficiencies in the application of the precautionary principle, namely, disproportion, outdated and inappropriate approach, and corruption [64].

[17] It is not the case that more or highly restrictive measures are more desirable. Measures with the highest degree of restrictiveness and intensity can be not just unnecessary but even harmful.

[18] For example, during the distancing of Sơn Lôi commune, each person in the commune was reimbursed an amount of 40.000 VND (around 2 dollars) per day, and each person who was isolated in government-administered facilities was reimbursed an amount of 60.000 VND (around 3 dollars) per day [138]. Later, Directive 19 provided for the implementation of social security measures to support people affected by the pandemic.

Disproportionality

There can be hard cases and grey issues which are subject to sustained contestation; however, there are also instances in which unlawfulness or wrongfulness may be assessed with relative clarity [65, 66]. While several issues during the pandemic remain controversial, certain measures may be *arguably* characterised as legally and/or ethically wrongful on the grounds of disproportionality.[19] For example, measures discouraging trade with China appear difficult to reconcile with provisions of the Agreement on the Application of Sanitary and Phytosanitary Measures and the International Health Regulations because it is difficult to establish a rational connection between such measures and their asserted aim of public health protection; and even if some connection could be posited, the measures appear manifestly excessive relative to that aim [67]. Such measures risk significant adverse effects on the national and individual economy, supply chains, and people's well-being. Likewise, many measures implemented by local authorities during the height of the pandemic, such as confining persons to or locking persons inside their dwellings, or entering dwellings and compelling COVID-19 testing through the use of force, raise serious concerns under universal human rights norms [64].

With respect to certain other measures, where illegality may be more difficult to establish, ethical concerns nonetheless arise:

Against the backdrop of a healthcare system that has been described as overburdened and resource-constrained, the blockage of hospitals and the imposition of quotas on hospital beds may have further restricted healthcare access. The advice to limit visits to medical facilities (even without enforcement) or the suspension of new medical appointments and re-examinations could have adverse implications for health. The authorities could assume that new medical appointments and re-examinations are generally not urgent in the context of the pandemic emergency, having regard to competing disease-prevention priorities. Nevertheless, certain illnesses or injuries may have mild initial symptoms or may be easily misdiagnosed without close monitoring, yet deteriorate rapidly; however, such outcomes may not be salient and attended to. Meanwhile, because the measures were often sine die, they also had psychological impacts, including anxieties associated with uncertainty as to access to treatment. The difference in treatment between Bạch Mai Hospital and other hospitals, especially private ones, may raise concerns about equal treatment and proportionality.

Meanwhile, the suspension of all transportation activities is problematic. It may constrain access to essential goods and services, disrupt basic social functioning, and undermine the livelihoods of workers and persons in vulnerable circumstances. Although deploying military forces from other provinces to undertake procurement and shipping functions in substitution for civil transportation activities is a commendably adaptive measure, it has reportedly been ineffective in part due to the forces' unfamiliarity with the activities and with the locations [68]. Many restrictions, such as the requirements of mandatory testing every time entering a province

[19] However, as will be shown, it is unlikely that the Vietnamese courts will make findings of wrongfulness in relation to these measures.

or district [44] or of changing drivers (to a local resident driver) whenever a vehicle enters a province or district (and only local residents were allowed to drive within the fortress of the province or district) [69], are unreasonable.

In light of numerous disproportionate measures, the inquiry extends beyond assessing the proportionality of any particular measure to deliberating about the underlying drivers of recurrent disproportionality and how proportional measures can be designed and implemented.

Lags

Although Vietnam adjusted its measures and policy conceptual paradigms during the pandemic, it fell short in certain important aspects. Once it became clear that COVID-19 could no longer be eradicated, the core objectives of an effective strategy were *simultaneously* minimising deaths and preserving the healthcare system's capacity. The prevention of infections is instrumental to those objectives and should not be treated as an end in itself—still less a standalone overarching objective. Measures, thus, should enable the population to acquire immunity *as safely as practicable*, in *a controlled manner, thereby contributing to maintaining healthcare-system capacity within manageable limits*. Vietnam did not adopt this orientation in a timely manner and instead maintained the 'fighting-the-disease-like-fighting-the-enemy' approach longer than was warranted. The healthcare system was overstretched and medical care in some places reportedly disrupted [70, 71].

Delayed Vaccination

Vietnam was slow in vaccinating, which contributed to lags in adjusting its approach. Globally, vaccination campaigns began as early as November–December 2020, and many countries launched their campaigns in January–February 2021 [72, 73]. However, as of 17 February 2021, doses of AstraZeneca imported (Official Letter 1215/QLD-KD) would cover only around 0.389% of the population aged 18 and over with one dose, and initial vaccination campaigns began 3 weeks later (Decision 1467/QĐ-BYT). At that time, according to the authorities, the purchased and donated doses of AstraZeneca were 60 million, and those of Moderna were 31 million [37]. However, according to BYT, the plan for vaccine imports was to deliver 1.3 million doses in the first quarter of 2021, 9.5 million in the second quarter, 25.9 million in the third quarter, and 51.1 million in the fourth quarter [74]. In fact, as of 13 May 2021, when the first campaign ended, only 959182 doses had been administered, much slower than planned and far below the target. *The Straits Times* estimated that, at that pace, it would take Vietnam 10 years to vaccinate 75% of its population [75]. Although inequity in vaccine access might *partially* account for the delay [76], inadequate planning for the early vaccination rollout may have been the more decisive factor. *First,* although Vietnam is a lower-middle-income country [77], it has been among the world's 40 largest economies since 2020 [78, 79], suggesting that economic and political-strategic factors are not the primary constraints to more timely vaccination efforts. *Second,* once Vietnam had committed to a vaccination rollout, it was able to complete it within three months and became one of the countries with the highest vaccination rates (although it is worth acknowledging that authorities and medical professionals worked very hard towards that end) [80]. *Third,* several facts attest to the slow start of vaccination and to some officials' confidence in containment measures: Resolution 21/NQ-CP was adopted on 26 February 2021; while the importation plan envisaged importing 58.2% of total doses only in the final quarter of 2021,[20] the 'made-in-Vietnam' vaccine was expected to be available for use only after June 2021 [81].

Inappropriate Prioritisation

The provision concerning vaccination prioritisation was also problematic. Accordingly, people prioritised for being vaccinated include[21]:

(i) Healthcare workers, those who join COVID-19 prevention and control, and military and public security forces
(ii) Vietnamese diplomats to be sent abroad, customs and immigration officers
(iii) Essential service workers in sectors such as aviation, transport, tourism, electricity and water supply, among others.
(iv) Teachers and those who work at education and training facilities, those who work at administrative agencies that have regular contact with various people
(v) People with chronic illnesses, people aged above 65
(vi) Residents in pandemic-hit areas
(vii) Poor people, policy beneficiaries
(viii) Those who will be sent abroad for business, learning, or working by competent regulatory agencies
(ix) Other people subject to BYT's decision depending on the COVID-19 situation

Given the scarcity of resources, the long list's lack of further criteria for prioritising certain categories over others undermines its operational prioritising function. These suggest a lack of clarity about the purpose and function of prioritisation, and/or deficiencies in the articulation of the prioritisation framework. The fact that the list prioritised workers and teachers may indicate the State's consideration of maintaining essential economic activities and minimising transmission. The policy may have been shaped by the State's observation that, during that period, outbreaks could occur with some frequency in industrial settings—settings regarded as the backbone of the economy [82]. The State might proceed on the assumption that the purpose of vaccination is infection prevention and that, to the greatest extent

[20] It was planned that in the first quarter, only 1.48% of total doses would be imported and in the first two quarters, only 12.3% of total doses. However, in light of the circumstances, the State could secure a significantly higher number of doses after the second quarter, allowing it to *basically* finish the vaccination campaigns in September 2021.

[21] Article 2.(1) of Resolution 21/NQ-CP

possible, this might be treated as equivalent to or even be regarded as a more valuable objective than minimising deaths. However, the State's assumptions and objectives would not be practical and reasonable, especially in the given context.

First, as the State had been advised [83], limited vaccination coverage cannot offer community immunity and, thus, cannot reasonably be expected to deliver comprehensive infection prevention.

Second, minimising deaths is not coterminous with minimising infections. Many of the groups prioritised for vaccination (e.g., workers, teachers, and related categories) are relatively young (in their 20s–30s) and do not have underlying health conditions. Accordingly, in a context of vaccine scarcity, prioritising those groups might have displaced the priority for older persons and clinically vulnerable individuals, thereby potentially increasing the latter groups' exposure to severe outcomes.

Meanwhile, it is open to question whether certain prioritisations were rationally connected to the policy purposes they purported to pursue. By way of example, while Directives 15 and 16 were in force, in-person teaching was suspended and classes were conducted online, thereby calling into question the justification for prioritising younger, clinically low-risk teachers for vaccination.

Delayed Departing from the Zero-Community-Infection Objective and Shifting to the Strategy of Care

As shown, even after the Delta Variant had become widespread, the State continued to pursue a "fighting the disease as fighting the enemy" approach. In the circumstances, however, a number of considerations suggest that this approach may have been ill-suited and, in several respects, counterproductive:

First, material differences between Ho Chi Minh City (HCMC) and Bac Giang (BG) meant that transposing the BG model and measures to HCMC without adaptation (or treating HCMC as generally analogous to an industrial setting) was liable to error. It also follows that contact tracing and preventing infection in HCMC were likely to have been more difficult but less effective than in BG.[22]

(i) HCMC is the most populous and densely populated centrally governed city in Vietnam and is 8–10 times more populous and dense than BG.[23]
(ii) Though HCMC is the city with the highest GDP in Vietnam [84], a portion of its residents may experience financial hardship and live in substandard sanitary conditions.[24] By contrast, industrial settings in BG may involve more stan-

[22] The Party-State may have been aware of this insofar as, for example, Directives 15 and 16 set measures applied to HN and HCMC separate from measures applied to other areas.

[23] HCMC's official population is around 9,389,720 people and a population density of 4481 people/km [143]. In fact, the de facto population was estimated to be around 13–14 million people [144], with many people just not having legal ID documents [145].

[24] This can be drawn from the preceding footnote.

dardised living and working arrangements, which can be further standardised or modified to reduce risk factors, particularly in light of governmental initiatives to prevent and control COVID-19.[25]

(iii) The BG outbreak was associated with relatively isolated industrial settings involving a partially vaccinated, younger population. HCMC, by contrast, is a megacity characterised by high levels of commercial activity and a highly mobile population, with frequent movement within the city and across southern provinces. Most people in HCMC and the southern provinces had not been vaccinated at the relevant time.

Second, the personnel and equipment required for mass testing and associated follow-up activities across 20 provinces, including the most populous,[26] were substantially greater than those required in a single province. This had several implications, inter alia, (i) some medical workers might have been diverted from other healthcare functions and/or required to assume heavier workloads, with attendant risks of physical and psychological strain; (ii) medical equipment may have been in short supply and therefore reused. However, changing gloves after every sample would likely have been impracticable, thereby potentially increasing the risk of transmission.

Third, mass testing conducted at publicly administered testing sites frequently entailed the congregation of large numbers of persons, which sat uneasily with rules prohibiting gatherings. It was reported, in many instances, that the minimum distance requirement was also not complied with [85].

Fourth, as suspected and confirmed cases increased sharply, state-managed facilities were reportedly unable to consistently comply with distancing, density, and sanitation requirements [70, 71, 86]. This, in turn, could create or increase infection risks within such facilities [70, 71, 86].

Fifth, the stamping-out strategy and isolation measures became increasingly ineffective once community transmission persisted, while cases also emerged and circulated within state-managed facilities.[27]

Sixth, although the Party-State envisaged field hospitals and demanded that they be adequately supplied with medicines and equipment, due to shortages of workers, medicines, and equipment [70, 87], it was reported that some patients, once isolated, may not have received timely clinical care [71]. This, in turn, may have increased the risk of avoidable morbidity and mortality—outcomes that appropriate interventions might have mitigated, and that may not have materialised absent those constraints. In this context, the State ultimately permitted home-based treatment and established mobile medical stations.

---

[25] For example, mô hình ba tại chỗ [the three-on-the-spot model], i.e., work on the spot, eat on the spot, sleep on the spot.

[26] Out of the 10 most populous provinces in Vietnam, 4 are southern provinces [146].

[27] The measures may have contributed to a reduction in bed availability for patients with non-COVID conditions, as beds were reserved for COVID-19 caseloads and infection-control requirements.

Seventh, follow-up activities and transportation of close contacts and suspected, asymptomatic, or mild cases may have occurred at the expense of the timely transfer of patients with severe illness to healthcare facilities [88].

Eighth, BYT and provincial departments of health were slow to provide guidance for the use, distribution, and dispensing of anti-inflammatory and anticoagulant medicines for treatment; and from May to August 2021, some patients were reportedly neither supplied with, nor able to access, these medicines [89]. Illustratively, a district-level VCP secretary was *reportedly commended for acting outside the scope of existing guidance* in delivering medicines to persons in isolation [89]. While that intervention may be understood as motivated by humanitarian concern, it nonetheless raises questions from the standpoint of medical law and ethics, particularly regarding professionalism and the proper allocation of competence and responsibility.

Ninth, social distancing and blockages impeded the transportation of medical equipment, medicines, and essential goods, thereby undermining health and basic subsistence. It was reported that oxygen tanks, sanitary napkins, and diapers were treated as "non-essential" and blocked at cordon sanitaires [71, 90–92].

Tenth, the State reportedly spent 541 million USD on importing KITs in the year 2021 alone [93], implying that overall expenditure on mass testing and isolation was substantial, which may have been inflated by corrupt procurement practices. By comparison, 541 million USD alone could, as of April 2021, have been exchanged for 135.25 million doses of AstraZeneca [94] or redirected to more urgent life-saving activities. This indicates that, had resources been allocated more effectively, the overall pandemic response could plausibly have been improved.

Corruption

During the pandemic, corruption appears to have been widespread across multiple levels of government and diverse aspects of COVID-19 responses. Many provincial-level officials have been imprisoned for receiving bribes in connection with authorisations for isolation facilities [95]. Some senior officials of the Ministry of Foreign Affairs have been imprisoned for receiving bribes in relation to repatriation (evacuation) flights [96]. The Minister of Health, the Minister of Science and Technology, and some other senior officials across governmental departments have been imprisoned in connection with the Việt Á JSC test-kit scandal, including allegations of bribery and price-fixing at approximately 20 USD per kit [97], among many other charges [98, 99]. Although any specific causal link between improper motivations and particular policy choices may be difficult to establish to the criminal-law standard of proof beyond a reasonable doubt, insofar as policymakers and implementing officials are able to obtain illicit gains, speculations have arisen as regards the possibility that at least some policy errors were influenced by undue interests [100].

The Insufficiency of Candid, Democratic, and Civilised Dialogue as a Fundamental Cause of the Shortcomings

It is often argued that candid, democratic, and civilised dialogue is a fundamental principle of governance [101–103] and is essential for the proper application of the precautionary principle [104]. This proposition is borne out by the experience of COVID-19, particularly in Vietnam.

The elimination and prevention of all risks and dangers are; rather, governance aims at and must proceed through risk management, which necessarily entails precaution. Addressing a risk or danger inevitably entails burdens and trade-offs, including inconvenience and the creation or redistribution of countervailing risks, even where overall risk is reduced [105]. By way of example, addressing a risk or danger entails resource costs that can impose burdens on individuals, sectors, and the national economy, with distributive consequences. This, in turn, indirectly hollows individuals' liberty in the sense that it reduces the de facto [pecuniary] capacity to act as they want or raises barriers for the realisation of one's wishes [106, 107]. These effects are associated with material determinants of well-being, including standards of living and life expectancy [108–111]. Moreover, resources devoted to one activity cannot be allocated to others; accordingly, other activities may need to be curtailed, including, in some circumstances, the provision of life-saving treatments and social welfare programmes. In certain contexts—including COVID-19—risk-control measures can impose direct and intensive constraints on liberty and rights and can materially disrupt the functions of public and private institutions and entities and the society, thereby further limiting individuals' capacity to pursue well-being. The scale of repatriation during COVID-19, notwithstanding the existence and perception of potential penalties, including imprisonment, underscores the practical limits of the "zero-risk" approach and the welfare costs that may be associated with extended confinement and severe economic insecurity.

Risk management—which necessarily entails the toleration of some risks and the timing of risk-toleration—depends not only on technical assessments of risk characteristics and available countermeasures, but also, and often decisively, on relevant social factors and preferences and *normative value judgments*. In other words, what counts as a justified measure depends *substantively* on the balancing of risks and interests, as constrained by the *polity's established commitments* and *jurisprudence*. The intrinsic values of people, which are now recognised by *worldwide polities*, mean, inter alia,

(i) There are interests and rights that are so valuable and, thus, cannot be abrogated and cannot be limited without legitimate justifications of grave and urgent risks and dangers to society and individuals, if not at all.
(ii) People and parties, especially those at stake or vulnerable, should have meaningful opportunities to raise concerns and provide input for decisions and measures.

Meanwhile, candid, democratic, and civilised dialogue can, inter alia, enhance the quality of assessments and subsequent measures by pooling knowledge, experiences, and viewpoints, bringing matters under scrutiny, requiring decision-makers to give reasons—thereby inducing careful consideration of alternatives, foreseeable consequences, and remedies, including for vulnerable groups.

Consequently, candid, democratic, and civilised dialogue can be understood as a precautionary arrangement that should be put in place to enhance the quality of decisions and measures and to protect human beings' sacred interests and rights. Conversely, it should, in turn, be protected by other precautionary arrangements. On this view, deficiencies in public deliberation may arguably be implicated in Vietnam's policy shortcomings.

At the political level, the VCP-State acknowledged that the corruption scandals were due to violations of the principle of democratic centralism insofar as many decisions had not been discussed within and by the relevant authorities [98, 99]. However, the capacity to engage in candid, democratic, and civilised dialogue should not be confined to intra-party institutions and processes. Nevertheless, certain policy statements and practices may have had a chilling effect on public dialogue. For example, the Vietnamese National Television, in response to calls for the State to reconsider the approaches and measures, claimed that "[l]iving with COVID-19 is the discourse of those seeking to oppose [the State]" ["[s]ống Chung với dịch là luận điệu của những kẻ rắp tăm chống phá"]; the statement may signal low tolerance for certain policy alternatives. In such circumstances, participants in public discussion may perceive non-trivial reputational or institutional costs in advancing views regarded as unacceptable to be articulate, which may in turn reduce the visibility of concerns that may later prove important for policy adjustment. Further, for reasons addressed in part below, there were reported currents of heightened nationalist sentiment or chauvinism [112] in the Vietnamese public political life, which may have influenced the tone and boundaries of public discussion. Such dynamics may render particular concepts socially or politically sensitive, and discourage and sanction frank expression of conscience, opinion, and concern, and thereby weaken the brainstorming, information-pooling, problem-solving, and error-correction functions of public dialogue. They may also hinder adaptive learning from other jurisdictions' experiences and from the evolving epidemiological situation insofar as "fighting-the-disease-like-fighting-the-enemy" and zero community infection were hailed with celebratory terms. In some contexts, discussions about hardship or suffering may be perceived as politically fraught or socially unwelcome, which may reduce the visibility of implementation harms and impede timely policy adjustment. Situations in other countries and their measures were treated dismissively by some prominent social-media figures [113–115]. Critical commentary or comparative assessments between Vietnam's response and those of other countries may, in some instances, attract social sanctions and—depending on framing and context—disciplinary or legal consequences.[28] In such an environment, where costs of public disagreement are perceived as non-trivial, even well-founded concerns (including in relation to Việt Á) may have limited incentives to be openly expressed. Moreover, heightened sentiment may also influence governmental and

[28] For example, a lecturer at a university was reportedly dismissed after criticising the social security system in her English class about [147]. It may be argued, on professional grounds, that personal policy views should not be introduced into teaching where they are not germane to the subject matter. Nevertheless, the dismissal may be regarded as a disproportionate response.

private actors [116] through risk-avoidant or appeasing responses.[29] For example, a Deputy Director of the Department of Natural Resources and Environment was disciplined for "causing public confusion and fear" after requesting some subordinate agencies to report on contingency cremation plans in response to COVID-19 [117]. This may illustrate institutional sensitivity to public reaction against transparent representation of potential (worst-case) scenarios and, thus, to communications concerning worst-case planning. However, it was later reported that Bình Hưng Hòa Crematorium operated beyond capacity [118, 119]. In these circumstances, authorities may wish to clarify that good-faith contingency planning and evidence-based comparison are compatible with responsible governance and do not necessarily undermine public health efforts.

In adjudication settings, the case concerning physically compelled testing, in which the claimant's submissions[30] were rejected, also illustrates potential risks associated with voicing sincere safety-related concerns. On appeal, in rejecting the claimant's submission that she had taken the state-administered test which should mean that the subsequent fine for not taking the state-administered test was therefore unjustified, the High Court of HCMC held that, notwithstanding the fact that the claimant had taken the state-administered test, the fine was justified because her concerns that public testing did not comply with safety principles[31], together with her stated intention to take the self-test, indicated that, *absent the fine, she would not comply*; accordingly, an administrative violation was constituted (Judgment No.: 988/2023/HC-PT, pp. 16–17). On one reading, the judgment treated the expression of sincere safety-related concerns and the proposal of an alternative measure as probative of anticipated non-compliance and therefore supportive of a finding of an administrative violation.[32] Although the alleged conduct of officials, such as breaking the lock, entering private dwellings, and the use of physical coercion, was not squarely adjudicated, it may be inferred that the court's reasoning left limited space for reasonableness- and proportionality-based engagement with such conduct, insofar as the justification for the fine rested on the view that enforcement was necessary to secure compliance.[33] This approach may be contrasted with practices in some

[29] This practice might be called populism [148] by some people or mass mobilisation by others [124].

[30] The woman asserted that she had already taken a COVID test by herself but was still subjected to physical force by public officials to coerce her into taking the state-administered COVID test (apparently to ensure that she was not left unchecked by the authorities) [64].

[31] She argued, "While the disease is transmitted through the respiratory tract, people were required to remove their masks in a public place; nasal samples were collected in a crowded setting".

[32] Clearly, the woman, because of her promise to undertake a self-test, was subjected to two coercive sanctions: a physically forced testing and a fine. In effect, she was not afforded a meaningful choice as to the measures by which she might protect her own safety.

[33] In response to the applicant's contention that she was not a subject under infectious disease surveillance, the administrative authorities and courts relied primarily on the administrative plans designating the respective ward as a "very high risk" locality. They further recorded that, before and during the mass testing campaign, the ward disseminated notices via its public loudspeaker system and instructed the building manager to notify the residents living of the campaign. The judgments did not address the interpretation of the relevant provisions of LPCBTN or examine the applicable legal sources. The reasoning appears to proceed on the premise that persons present

other jurisdictions, where—even where claims do not succeed on the merits—judicial engagement with the claimant's reasons and evidence can nonetheless provide a forum for deliberation, reason-giving, expressions of humanitarian considerations, and for the interests at stake to be acknowledged within the court's reasoning.[34]

Causes for the Insufficiency of Candid, Democratic, and Civilised Dialogue

The insufficiency of candid, democratic, and civilised dialogue may be attributed to established legal structures and political, legal practices.

The Legal Model of Pandemic Responses

There were reportedly more than 2000 COVID-19 instruments [64], including a significant number of non-legislative instruments that, despite having legal effects and, in some instances, operating in ways that can even override or supersede or derogate from primary legislation, do not follow ordinary legislative procedures. Many of these instruments were reportedly overlapping, inconsistent and mutually contradictory [120]. Although these directive-styled instruments may facilitate rapid response, the resulting volume and fragmentation may complicate or impede public understanding and, in turn, meaningful deliberation [64]. For example, although Directives 15, 16, 19 and various official dispatches are formally ad hoc non-legislative instruments, they were applied repeatedly, sometimes, without clearly articulation of the criteria for their activation or their relationship with higher-order legal instruments. This pattern is consistent with certain long-established features of Vietnam's legal system, including the prominent role of directive-style instruments—a feature that some comparative accounts have linked to the influence of Soviet-derived models of governance and administration [121].

A Potentially Over-Binary Distinction Between Accurate and Inaccurate Information and Associated Penalties

Vietnam maintains legal provisions addressing misinformation and disinformation, notably, Article 8.(1). (d) of the 2018 Cybersecurity Law, which are very much in line with the declaration of the WHO and leaders of several countries [122]. However, drawing a bright line between accurate and inaccurate information is often difficult, especially in relation to scientific and rapidly evolving matters [123]. Accordingly, debates and discussions on freedom of expression have occupied extensive jurisprudential and scholarly attention and analysis in several jurisdictions. By contrast, the Vietnamese legal and jurisprudential treatment appears comparatively under-elaborated and tends to operate through relatively binary

---

within the designated area were, by that fact alone, automatically subject to social and public health measures of unspecified scope, without any discernible assessment of less restrictive means to achieve the stated objective. The decisions likewise do not engage with constitutional guarantees or the State's human rights obligations.

[34] There are various cases where the courts offered condolences during COVID-19. See, for example, Taylor v. Newfoundland and Labrador Taylor v. Newfoundland and Labrador, [2020] N.J. No. 1912020 NLSC 125; [2020] N.J. No. 191.

classifications, with potentially disproportionate consequences. Within the context of Vietnam, statements that depart in phrasing or emphasis from official communications may be treated as "inaccurate" and may attract sanctions.[35] Likewise, policy and political expression or concerns that diverge from prevailing official narratives may be subject to sanction where they are characterised as "inaccurate" or as causing public alarm.

Militarised rhetoric, Heightened Nationalist Sentiment, and Chauvinism

Mobilisation has been recognised as a political principle of communist governance [1, 124]. In fact, mobilisation tactics may also be employed by a range of political actors and parties to secure public support. It has been reported that the Party-State supported organised communications efforts—including units commonly referred to as "Force 47" and, at times, external public-relations contractors—aimed at monitoring and shaping public opinion in support of governmental policies, countering dissenting narratives, and encouraging particular behaviours [123, 125, 126]. It is well recognised that informal channels (including entertainment-oriented platforms) and the messaging of influential opinion leaders, particularly when framed in accessible, "trending" and "catchy" language, can be more readily received and influential than formal communications [127–129]. Repeated messaging may also reinforce certain narratives and frames over time [130].[36] In turn, public-opinion dynamics can exert discourage dissenting views and reduce incentives for independent critical assessment, even in the absence of formal sanctions, particularly when deviation from prevailing norms attracts social disapproval [131–133]. During the pandemic, war-related rhetoric and appeals to nationalism were used extensively. Given Vietnam's historical experience, references to war and national resilience may function as potent symbols that promote solidarity [134]. However, war-framing may narrow the space for open dialogue insofar as it may emphasise urgency, unity, and—at times— confidentiality and militarist hierarchical compliance, and reduce tolerance for contestation.[37] Meanwhile, enforcement and moderation practices may operate unevenly: statements alleged to be inaccurate or defamatory may attract comparatively less official attention when framed or received as supportive of the Party-State than when comparable claims are framed

[35] By way of illustration, the authorities may have tended to employ the word "giãn cách xã hội"—"social distancing" when referring to measures applied on a large territorial scale, such as at the provincial or municipal level, while reserving "phong tỏa" for measures imposed on a small- or medium-scale basis, including specific localities or establishments [149]. In this usage, "phong tỏa" in domestic context was commonly understood to denote a "blockage" or "cordon" measure, and was treated as distinct from the term "phong tỏa" as used by the authorities and the press when describing measures adopted in other countries, where it was typically intended to convey lockdown [150]. Individuals may be exposed to sanctions where they have not been given adequate notice of these distinctions. Moreover, in many instances, these distinctions may not have been defined or clarified in advance in a sufficiently clear and unambiguous manner [151, 152].

[36] A Vietnamese saying—"mưa dầm thấm lâu" (literally, "a steady drizzle soaks deeply")—may be used to describe the cumulative effect of sustained messaging on public attitudes [153, 154].

[37] This can result in catastrophic ramifications. See Korematsu v United States, 323 US 214, 65 S Ct 193, 89 L Ed 194, 1944 US LEXIS 1341 and related issues.

as oppositional, provided the former do not trigger substantial public controversy. To the extent that such asymmetries occur, they may constrain reasoned public deliberation by narrowing the range of positions and options perceived as acceptable to articulate in public.

The Mandatory Missions for the Future

While future responses to risks and dangers necessarily implement the precautionary principle as envisaged in LPCBTN and as has been practised in Vietnam with laudable results, two key aspects are demanded for more refined application of the precautionary principle and more justified responses.

Asserting the Imperatives of Bioethics and Public Health Ethics

The management of risks and dangers is *inherently* controversial, and such controversy cannot, as a practical matter, be eliminated. Accordingly, risk and danger management should include structured and constructive engagement with relevant controversies and debates. Attempts to avoid or withdraw from contested ideas are likely to unduly narrow the range of available options and to constrain the development, calibration, and quality of policy and measures. However, facilitating and engaging with dialogue of or with controversial ideas should not be construed as authorising discriminatory, defamatory, derogatory, harmful discourses and behaviours, harassment, or uncontroversial crimes ("undesirable conduct"). For example, acknowledging excessive deaths and undertaking transparent contingency planning for excess mortality should not be equated with normalising such outcomes. Although some undesirable conduct is intentional and can be readily identified, some may arise inadvertently. *Thus,* humane, empathetic, and proportionate responses to risks and dangers *demand careful, evidence-based, and critical deliberation, with and by various parties, especially those whose interests are at stake, with due regard to ethical and legal principles.* It follows that critical and ethical thinking, concerned with how to think rather than what to think, should be systematically developed and applied by decision-makers. These considerations are very much within and consistent with the visions, purposes, and objectives of bioethics, in particular, public health ethics, which seek to *cultivate knowledge and skills to protect the public while maintaining appropriate care and respect for individuals, especially with regard to their vulnerabilities and needs,* via *dialogue* [135, 136]. That means bioethics and public health ethics should be recognised and operationalised in Vietnam as resources for policy deliberation, and contributions from these should be welcomed, consulted, and commissioned in and for the management of risks and dangers.

Installing the State's Duties to Protect, to Give Due Regard, and to Give Reasons

Freedom of conscience, speech, expression, bioethical inquiry, and the capacity to engage in dialogue can only exist and thrive in an environment where individuality and the plurality of perspectives and opinions are *respected, guaranteed, and protected*.[38] The negative right of being protected from public powers' arbitrary interference is necessary, but it is not, by itself, sufficient to secure the conditions under which those freedoms can be effectively exercised. There must also be *strong positive precautionary arrangements and institutions for the protection of the dignity and security* of people and entities against interference from *both public and private subjects* and for lodging voices and concerns to be addressed in democratic and civilised manners.

Meanwhile, for dialogue and discussions to have a practical effect, related parties must engage in them seriously and in good faith. This, then, implies that authorities should give due regard to discussions and dialogue wherever they are relevant to the matters they decide and should demonstrate that they have been considered. It follows that relevant administrative and judicial institutions and arrangements must be designed in such a way as to afford and facilitate discussions and dialogue, to take parties' concerns seriously, and safeguard their rights and interests to the extent reasonably practicable. This, then, indicates that authorities should also bear the duty to give reasons: decisions should be accompanied by intelligible explanations, particularly how they pursue a legitimate aim, how they are proportionate to that aim and to the interests engaged, and how they minimise—so far as reasonably available—the impairment of fundamental rights. By way of illustration, where differential restrictions are imposed between Bạch Mai Hospital and other (including private) hospitals, additional assessment and justification may be required. COVID-19 threats, uncertainties, healthcare needs, and mitigation strategies may justify differentiated measures; however, where a high-throughput facility can operate under less restrictive conditions without a material increase in risk, the imposition of more restrictive measures on less crowded settings—particularly where precautionary capacity is stronger—requires specific and publicly intelligible reasons. Such reasons should address the relevant factors typically engaged in risk governance, including the magnitude and likelihood of risk, the feasibility and effectiveness of precautions, system-wide capacity constraints, and the competing demands of access to healthcare. These considerations are articulated in what is known as "jurisprudence," as principles of "reasonableness" and "due process," which translate and integrate ethical principles into the legal system as legal standards through which law operates as an instrument directed to the end of justice.

[38] A "duty to protect" may be identified across legal systems, notwithstanding variation in, inter alia, the modalities through which protection is delivered and the historical pathways through which the duty has been articulated and developed. For example, in Germany, Grundgesetz für die Bundesrepublik Deutschland—the German Constitution—asserted the Schutzpflicht—the duty of protection, which German jurisprudence has reaffirmed. Accordingly, the State may be required—under specified conditions—to adopt measures affecting private relationships to secure the effective protection of constitutional rights and legally protected interests. By contrast, in the United States, the assertion of the state's duty to protect is traditionally met with scepticism, including concerns that such claims can serve as an agenda to expand state power beyond accepted constitutional limits or to promote authoritarianism. Notwithstanding, protection duties can be found in and guaranteed by the US Constitution, judicial doctrine in specific contexts, and the design and commitment of institutions dedicated to freedom from censorship, procedural fairness, and the security of persons.

Dialogue and reason-giving are therefore not merely optional; they are the ethical and legal requirements of good governance practices and, as such, warrant recognition and strengthening in Vietnam.

Conclusion

This chapter demonstrates that Vietnam endorses and applies the precautionary principle in epidemic prevention and control, which accounts for the successes in handling COVID-19, especially during the initial periods. However, Vietnam's responses, especially in and as manifested by the later periods, are not entirely successful. The legitimacy, merit, and efficacy of the responses are undermined by deficiencies in the application of the precautionary principle, namely, disproportionality, outdated and inappropriate approach, and corruption. The insufficiency of candid, democratic, and civilised dialogue, rooted in long-established practices, is the more fundamental cause. Consequently, while future responses to risks and dangers necessarily implement the precautionary principle as envisaged in LPCBTN and as has been practised in Vietnam with laudable results, two key aspects are demanded for more refined application of the precautionary principle and more justified responses: (i) the assertion of the imperatives of bioethics and public health ethics, and (ii) the installation of jurisprudence embracing duties to protect, to give due regard, and to give reasons.

References

1. Minh HC (1949) Dân vận [Mass mobilization]. Sự Thật [Truth] 120
2. Ban biên tập trang thông tin điện tử Cục Y tế dự phòng [Editorial Board of the website of the Department of Preventive Medicine] (2020) Tại sao Việt Nam chiến thắng dịch COVID-19? [Why did Vietnam win the COVID-19 epidemic?]. In: Dep. Prev. Med. Portal. https://vncdc.gov.vn/tai-sao-viet-nam-chien-thang-dich-covid-19-nd15787.html. Accessed 16 Apr 2023
3. Thanh V (2020) Không để Việt Nam rơi vào vòng xoáy của viêm đường hô hấp cấp do chủng mới của virus Corona [Vietnam must not fall into the spiral of acute respiratory inflammation caused by the new strain of corona virus]. In: Communisty Party Sect. Ho CHi Minh City. https://www.hcmcpv.org.vn/tin-tuc/khong-de-viet-nam-roi-vao-vong-xoay-cua-viem-duong-ho-hap-cap-do-chung-moi-cua-virus-corona-1491861869. Accessed 12 Jul 2022
4. Ly (Tracy) TTH (2020) How Vietnam is winning its war against COVID-19. Asia Pacific Found
5. Thoa Đ (2020) Việt Nam lập kỷ tích một tháng không có ca lây nhiễm COVID-19 trong cộng đồng [Vietnam sets a record of one month without community COVID-19 infections]. Báo Điện tử Đảng Cộng Sản Việt Nam [Vietnamese Communist Party's News]
6. (2020) Thủ tướng: Chấp nhận một số thiệt hại về kinh tế để bảo vệ tính mạng, sức khỏe người dân [Prime Minister: Willing to take some economic losses to protect people's lives and health]. Công lý
7. V.Tuấn BN (2020) Ngàn tấn thanh long ùn ứ ở biên giới, tiểu thương 'chết đứng' giữa đại dịch corona ["Thousands of tons of dragon fruit stuck at the border, small traders "frozen in shock" amid the coronavirus pandemic."]. Tuổi Trẻ Online
8. Nghĩa H (2020) Lạng Sơn: Cửa khẩu phụ, cặp chợ biên giới mở lại vào ngày 10/2 tới [Lạng Sơn: Auxiliary border gates and border markets to reopen on February 10]. Tài nguyên và Môi trường

9. Hải T (2020) Thông quan hàng hóa tại cửa khẩu Tân Thanh từ hôm nay (20/2) [Customs clearance at Tân Thanh border gate resumes today (February 20)]. Đầu tư
10. (2020) Chính thức dỡ phong toả xã Sơn Lôi sau 21 ngày cách ly phòng dịch COVID-19 [Officially lifting the blockade of Son Loi commune after 21 days of isolation to prevent COVID-19]. Lao Động
11. Yến N, Sơn P (2020) Hà Nội cách ly toàn bộ các hộ dân từ số nhà 125–139 phố Trúc Bạch [Hanoi quarantines all households from house number 125 to 139 Trúc Bạch Street]. Báo Công an Nhân dân điện tử
12. Long X (2020) "Giải cứu" 3 bệnh viện ở Đà Nẵng: kinh nghiệm từ Bệnh viện Bạch Mai ["Rescuing" 3 hospitals in Đà Nẵng: experience from Bạch Mai Hospital]. Tuổi Trẻ Online
13. Hưởng N, Thanh BH (2020) Đề xuất khử khuẩn Bệnh viện Bạch Mai, xét nghiệm toàn bộ nhân viên y tế, người bệnh [Propose to disinfect Bạch Mai Hospital and test all medical staff and patients]. Người Lao động [The Employee]
14. An N (2020) Nguồn lây nhiễm chính ở Bệnh viện Bạch Mai là từ Công ty Trường Sinh [The source of infection at Bạch Mai Hospital is from Trường Sinh Company]. Tuổi Trẻ Online
15. (2020) Bệnh viện Bạch Mai không thể dừng tiếp nhận, cứu người [Bạch Mai Hospital cannot stop receiving and saving people]. Cổng thông tin Bộ Y tế [Ministry Heal. Portal]
16. Thanh H, Dung N, Thanh H (2020) Bác bỏ việc phong toả Bệnh viện Bạch Mai [It is refuted that Bạch Mai Hospital is being sealed off]. Người Lao động [The Employee]
17. Thu D (2020) Sau gỡ phong toả, Bệnh viện Bạch Mai vẫn chưa nhận khám thông thường [After lifting the blockade, Bạch Mai Hospital still does not accept regular examinations]. In: Người Lao động [The Employee]. https://nld.com.vn/suc-khoe/sau-go-phong-toa-benh-vien-bach-mai-van-chua-nhan-kham-thong-thuong-20200424062746958.htm. Accessed 14 Nov 2024
18. Chính N (2020) Hà Nội đang đối mặt với 4 nguồn lây nhiễm Covid-19 [Hanoi is facing 4 sources of Covid-19 infection]. Báo Điện tử Đảng Cộng Sản Việt Nam
19. Huyền TC (2020) Việt Nam chiến thắng COVID-19 bằng bản lĩnh, trí tuệ, kỷ luật, đoàn kết [Vietnam won against COVID-19 with bravery, wisdom, discipline and solidarity]. Xây dựng Đảng
20. (2020) Vừa thực hiện giãn cách xã hội, vừa tạo điều kiện phát triển sản xuất kinh doanh [Concurrently implementing social distancing and creating conditions for production and business] development]. In: VNCDC. https://web.archive.org/web/20220718102943/https://vncdc.gov.vn/vua-thuc-hien-gian-cach-xa-hoi-vua-tao-dieu-kien-phat-trien-san-xuat-kinh-doanh-nd15505.html. Accessed 20 Nov 2024
21. (2020) Phó Thủ tướng Vũ Đức Đam: "Mục tiêu kép" nhưng trên hết phải an toàn [Deputy Prime Minister Vũ Đức Đam: 'Pursuing dual objectives, but safety remains the top priority]. Sức khỏe đời sống
22. Mạnh T, Nam Đ (2020) Thực hiện mục tiêu kép nhưng phải tuyệt đối an toàn [Dual goals pursued, but safety above all]. Báo Chính phủ [Government News]
23. Trang T (2020) Tạm thời phong toả Khoa Hồi sức của Bệnh viện Đà Nẵng [Temporarily block the Resuscitation Department of Đà Nẵng Hospital]. Lao Động [Labor Newspaper]
24. Bình T (2020) Thêm 11 ca mắc COVID-19 liên quan đến Bệnh viện Đà Nẵng, trong đó có 4 nhân viên y tế, Việt Nam có 431 ca bệnh [Eleven more COVID-19 cases linked to Đà Nẵng Hospital, among them there are four medical staff; Vietnam now has 431 confirmed cases]. Sở Y tế TPHCM
25. Dũng TB (2020) 'Sớm mở cửa lại 3 bệnh viện đang bị phong tỏa tại Đà Nẵng' [Reopening 3 locked down hospitals in Đà Nẵng soon]. In: Tuổi Trẻ Online. https://tuoitre.vn/som-mo-cua-lai-ba-benh-vien-dang-bi-phong-toa-tai-da-nang-20200804211945404.htm. Accessed 19 Nov 2024
26. Bình T (2020) Tiếp tục áp dụng biện pháp cách ly y tế tại Bệnh viện Đà Nẵng, từ 0 giờ 00 ngày 11-8-2020 [Medical quarantine measures will remain in effect at Đà Nẵng Hospital as of midnight, August 11, 2020]. In: Cổng Thông tin điện tử thành phố Đà Nẵng. https://danang.gov.vn/chinh-quyen/chi-tiet?id=40492&_c=100000155. Accessed 19 Nov 2024
27. An N (2020) Thực hiện giãn cách xã hội toàn Đà Nẵng từ 0h ngày 28-7 [Social distancing measures to be enforced across Đà Nẵng from 12:00 a.m. on July 28]. Tuổi Trẻ online

28. (2020) Nhận định tình hình dịch COVID-19 ở Đà Nẵng qua truy vết, phân tích dữ liệu [Evaluation of the COVID-19 situation in Đà Nẵng based on contact tracing and data analysis]. VNCDC
29. Khang M (2020) Đã truy được nguồn gốc dịch Covid-19 bùng phát ở Đà Nẵng [Source behind Đà Nẵng's COVID-19 outbreak successfully traced]. In: Sở Y tế TP. Hồ Chí Minh. https://medinet.hochiminhcity.gov.vn/tin-tuc-su-kien/da-truy-duoc-nguon-goc-dich-covid-19-bung-phat-o-da-nang-cmobile1780-31375.aspx. Accessed 19 Nov 2024
30. Hà T (2020) Mỗi gia đình là một pháo đài, mỗi người dân là một chiến sĩ phòng, chống dịch [Every home is a stronghold, every citizen a frontline warrior in the battle against the pandemic]. In: Báo Ấp Bắc. https://baoapbac.vn/suc-khoe-y-te/202008/thu-tuong-chinh-phu-nguyen-xuan-phuc-moi-gia-dinh-la-mot-phao-dai-moi-nguoi-dan-la-mot-chien-si-phong-chong-dich-905634/. Accessed 19 Nov 2024
31. Minh H, Anh V, Lê T, et al Dấu mốc quan trọng 2 năm đại dịch Covid-19 tấn công Việt Nam [Major milestones in Vietnam's two-year battle with COVID-19]. In: Nhân dân. https://special.nhandan.vn/daumoc-Covid-19/index.html. Accessed 10 Nov 2024
32. Doan H (2025) Expectations and Vietnam's responses during COVID-19: potential human rights violations and related propositions. Wellcome Open Res. https://doi.org/10.12688/wellcomeopenres.18972.3
33. Anh L (2021) "Ai ở đâu, ở yên đấy" góp phần phòng chống dịch bệnh ["Whoever is where, stay there" contributes to pandemic prevention]. In: Báo Điện tử Đảng Cộng Sản Việt Nam. https://dangcongsan.vn/cung-ban-luan/ai-o-dau-o-yen-day-gop-phan-phong-chong-dich-benh-586797.html. Accessed 21 Nov 2024
34. Minh L, Huỳnh T (2021) Người dân miền Tây rời TP.HCM, đi bộ hàng trăm cây số để về quê [Southwest people leave Ho Chi Minh City, walk hundreds of kilometers to repartriate]. In: VOV https://vov.vn/xa-hoi/tin-24h/nguoi-dan-mien-tay-roi-tphcm-di-bo-hang-tram-cay-so-de-ve-que-895563.vov. Accessed 7 Jan 2023
35. Biên CA (2021) Người dân chạy xe máy về quê giữa lúc TP.HCM giãn cách: Chốt chặn mướt mồ hôi thuyết phục [People ride motorbikes to repatriate in the middle of Ho Chi Minh City distancing: Checkpoints sweaty dissuade]. In: Thanh Niên. https://thanhnien.vn/nguoi-dan-chay-xe-may-ve-que-giua-luc-tphcm-gian-cach-chot-chan-muot-mo-hoi-thuyet-phuc-post1101380.html. Accessed 7 Jan 2023
36. Nhân T (2021) Công an Ninh Thuận lý giải việc xử phạt người dân từ vùng dịch về quê [Ninh Thuan police explain the fines against people who are repatriated from epidemic areas]. In: Thanh Niên. https://thanhnien.vn/cong-an-ninh-thuan-ly-giai-viec-xu-phat-nguoi-dan-tu-vung-dich-ve-que-post1389790.html. Accessed 7 Jan 2023
37. Anh L (2021) Hôm nay Việt Nam bắt đầu tiêm vắc xin ngừa COVID-19 [Vietnam rolls out COVID-19 vaccine today]. Tuổi Trẻ Online
38. (2021) Vietnam asks for foreign support in procuring Covid-19 vaccines. The Straits Times
39. Phương T (2021) Ngoại giao vaccine-Sứ mệnh đưa nguồn vaccine quý giá về với nhân dân Việt Nam [Vaccine Diplomacy—The mission to bring precious vaccines home to the Vietnamese people]. Báo Tin tức
40. Dũng T (2021) Thủ tướng: Xác định sống chung lâu dài với dịch, không thể khống chế tuyệt đối [Prime Minister: Determining to Live Long-termed With the Epidemic, Cannot Control Absolutely]. VnEconomy
41. Tiếp P (2021) Thủ tướng: Phấn đấu đến ngày 30/9 trở lại trạng thái bình thường mới [Prime Minister: Strive to return to the new normal by September 30]. Báo Tin tức
42. An Q (2021) 10 tỉnh, thành phố có tỷ lệ tiêm vắc xin phòng COVID-19 cao và thấp nhất tính đến 29/9 [10 provinces and provincial-cities with the highest and lowest COVID-19 vaccination rates as of September 29]. SOHA
43. (2021) Nhiều địa phương vẫn yêu cầu phải có giấy xét nghiệm COVID-19 khi qua chốt [COVID-19 test certificates still required at checkpoints in many areas]. Sức khỏe và Đời sống
44. Huyền K (2021) Thực hiện Nghị quyết 128: Vì sao có sự thiếu đồng bộ giữa các địa phương? [Implementation of Resolution 128: Why do discrepancies persist between provinces?]. QĐND

45. B.T.V (2021) Thủ tướng: Các ngành, địa phương phải thực hiện nghiêm Nghị quyết 128 [Prime Minister: Sectors and localities must strictly implement Resolution 12]. Người Lao động [The Employee]
46. Dũng T (2021) Thủ tướng: Vẫn có nơi thực hiện chưa đúng Nghị quyết 128, gây ách tắc, phiền hà cho dân [Prime Minister: Certain localities have not properly implemented Resolution 128, resulting in disruptions and difficulties for citizens]. In: Người Lao động [The Employee]. https://nld.com.vn/chinh-tri/thu-tuong-van-co-noi-thuc-hien-chua-dung-nghi-quyet-128-gay-ach-tac-phien-ha-cho-dan-20211017094914578.htm. Accessed 23 Nov 2024
47. Dũng T (2021) Thủ tướng: Triển khai Nghị quyết 128, cấp dưới phải phục tùng cấp trên [Prime Minister: The implementation of Resolution 128 requires lower-level authorities to follow the directives of higher-level authorities]. In: Người Lao động [The Employee]. https://nld.com.vn/chinh-tri/thu-tuong-trien-khai-nghi-quyet-128-cap-duoi-phai-phuc-tung-cap-tren-20211017162630577.htm. Accessed 23 Nov 2024
48. (2022) Việt Nam: hơn 90% dân số 18 tuổi trở lên đã tiêm 2 mũi vaccine ngừa COVID-19 [Vietnam: More than 90% of population aged 18 and over have received 2 doses of COVID-19 vaccine]. In: VOV. https://vov.gov.vn/viet-nam-hon-90-dan-so-18-tuoi-tro-len-da-tiem-2-mui-vaccine-ngua-covid-19-dtnew-341151. Accessed 29 Nov 2024
49. (2020) Việt Nam có ca COVID-19 tử vong đầu tiên do nhồi máu cơ tim [Vietnam has first COVID-19 death due to myocardial infarction]. CAND
50. Anh L (2020) Bệnh nhân COVID-19 số 761 tử vong, ca tử vong thứ 35 [COVID-19 patient No. 761 passes away, marking the 35th death]. Tuổi Trẻ Online
51. Chi L (2021) Việt Nam ghi nhận ca tử vong do COVID-19 thứ 36 [Vietnam records the 36th COVID-19 fatality]. In: VTV. https://vtv.vn/suc-khoe/viet-nam-ghi-nhan-ca-tu-vong-do-covid-19-thu-36-20210515150530246.htm. Accessed 23 Nov 2024
52. Doan HT (2025) "Đồng Bào" ('from the same fetus'): from implications for transplantation and ethics in crises to implications for Global Health. J Med Ethics:1–2. https://doi.org/10.1136/jme-2024-110599
53. Sunder G (2020) Vietnamese song 'Jealous Coronavirus' created to spread awareness on COVID-19 goes viral. The Hindu
54. O'Kane C (2020) WORLD Catchy PSA about coronavirus turns into viral TikTok challenge about washing your hands. CBS News
55. (2020) Vietnamese confident in Government's response to COVID-19: international survey. Viet Nam News
56. Trường X (2022) Tỉ phú, Doanh nhân Việt Nam và những đóng góp cho công tác phòng chống dịch Covid - 19 [Contributions of Vietnamese Billionaires and Business Leaders to COVID-19 Prevention and Control]. Pháp lý
57. Thịnh G (2024) Phương Trang cùng lúc đưa dân 8 tỉnh về quê từ TP.HCM [Phương Trang helps residents from 8 provinces repatriate from HCMC at the same time]. Tuổi Trẻ Online
58. Tiến N (2021) Những tấm lòng nhân ái giữa mùa dịch Covid-19 [Hearts of Compassion in the Midst of the COVID-19 Crisis]. Trang Thông tin Điện tử Đảng bộ TPHCM
59. CM (2020) Lan tỏa mô hình "ATM gạo" cho người nghèo [The "Rice ATM" model gains traction in supporting underprivileged communities]. Báo Điện tử Đảng Cộng Sản Việt Nam (Vietnamese Communist Party's News)
60. (2021) Bản tin tình hình chống dịch ngày 23/9/2021 tại Việt Nam [News on the situation of epidemic control on September 23, 2021 in Vietnam]. In: Cổng thông tin Bộ Y tế. https://moh.gov.vn/tin-tong-hop/-/asset_publisher/k206Q9qkZOqn/content/ban-tin-tinh-hinh-chong-dich-ngay-23-9-2021-tai-viet-nam. Accessed 29 Nov 2024
61. Phương N (2024) Gần 100.000 ca mắc, 20 người tử vong do Covid-19 trong năm 2023 [Nearly 100,000 cases, 20 deaths from Covid-19 in 2023]. In: Dân trí. https://dantri.com.vn/suc-khoe/gan-100000-ca-mac-20-nguoi-tu-vong-do-covid-19-trong-nam-2023-20240110072932030.htm. Accessed 23 Nov 2024
62. Vietnam: Coronavirus Pandemic Country Profile. In: Our World Data. https://ourworldindata.org/coronavirus/country/vietnam?fbclid=IwY2xjawGvnZlleHRuA2FlbQIxMAABHSrOVMJUKOs8P7M5IWwff9bHxZuo2CmwqW3tK5lI9HkcHNvLNxmw-8LVjA_aem_YFznIFUr%2D%2D-bPW3GJqYcKQ. Accessed 23 Nov 2024

63. Lane I, Trình Đ (2021) Sau khi mở cửa, tình hình đại dịch COVID-19 ở các nước như thế nào? [How has the COVID-19 situation been like in other countries following their reopening?]. In: SBS Tiếng Việt. https://www.sbs.com.au/language/vietnamese/vi/article/sau-khi-mo-cua-tinh-hinh-ai-dich-covid-19-o-cac-nuoc-nhu-the-nao/o02k9rqzu?fbclid=IwY2xjawGvnZRleHRuA2FlbQIxMAABHQz8g_cwm610EIxoVA9A9evceYKDWpXZe-BT1cR5M4gh3d91R4hLW7uHDtA_aem_cdsLWJUEzxzkUeHNxqwXXQ. Accessed 23 Nov 2024
64. Doan HT (2023) The public health response to COVID-19 in Vietnam: decentralization and human rights. Asian Bioeth Rev 15:103–123. https://doi.org/10.1007/s41649-022-00226-1
65. Hart HL (1961) The concept of law. Oxford University Press UK, Oxford, New York
66. Hart HLA (2013) Discretion. Harv Law Rev 127:652–665
67. Doan T-H (2025) The precautionary principle as a general principle of international law and Global Health law. Asian J WTO Int Heal Law Policy 20
68. (2021) Đi chợ hộ tại TP HCM: Quân đội "chào thua", shipper trở lại [HCMC's proxy shopping system: as the military withdraws, delivery drivers return to service]. https://www.bbc.com/vietnamese/vietnam-58381909. Accessed 21 Nov 2024
69. Liêm T (2021) Quy định "sang xe, đổi tài xế" ở Cần Thơ đang gây khó khăn cho lưu thông hàng hóa [Cần Thơ's Vehicle and Driver Swap Rule Hampers Freight Transport]. In: TTXVN. https://bnews.vn/quy-dinh-sang-xe-doi-tai-xe-o-can-tho-dang-gay-kho-khan-cho-luu-thong-hang-hoa/209421.html. Accessed 23 Nov 2024
70. Thường trực Ủy ban Xã hội [Standing Committee for Social Affairs] (2021) Tác động của dịch COVID-19 đối với lĩnh vực xã hội và một số gợi ý về chính sách cho Chương trình phục hồi và phát triển kinh tế - xã hội [The impact of COVID-19 on the society and some policy implications for the Socio-Economic Recovery and Development P
71. Bấc G (2021) Ái nữ KTS Ngô Viết Thụ chết oan do cách ly, phong tỏa! [Architect Ngô Viết Thụ's beloved daughter tragically died under unfortunate circumstances related to quarantine and lockdown]. RFA Tiếng Việt
72. (2021) Which countries have rolled out COVID vaccine? In: Al Jazeera. https://www.aljazeera.com/news/2020/12/24/vaccine-rollout-which-countries-have-started. Accessed 27 Nov 2024
73. Feng E, Ruwitch J (2020) China is vaccinating thousands against COVID-19 with unapproved vaccines. Why? In: NPR. https://www.npr.org/sections/goatsandsoda/2020/11/12/933956247/china-is-inoculating-thousands-with-unapproved-covid-19-vaccines-why. Accessed 27 Nov 2024
74. Dũng V (2021) Thủ tướng yêu cầu Bộ Y tế báo cáo kế hoạch nhập khẩu và tiêm vaccine ngừa Covid-19 [Prime Minister requests Ministry of Health to report on plan to import and inject Covid-19 vaccine]. VOV
75. Ng E (2021) Other nations in South-east Asia ramping up Covid-19 vaccination drives. The Straits Times
76. Klobucista C (2021) What to know about the global COVID-19 vaccine rollout so far. In: Counc. Foreign Relations. https://www.cfr.org/in-brief/what-know-about-global-covid-19-vaccine-rollout-so-far. Accessed 27 Nov 2024
77. Low and Middle-Income Countries and Regions. In: ISI. https://isi-web.org/low-and-middle-income-countries-and-regions. Accessed 27 Nov 2024
78. Vietnam GDP - Gross Domestic Product 2020. In: Ctry. Econ. https://countryeconomy.com/gdp/vietnam?year=2020. Accessed 27 Nov 2024
79. Vietnam GDP - Gross Domestic Product 2021. In: Ctry. Econ. https://countryeconomy.com/gdp/vietnam?year=2021. Accessed 27 Nov 2024
80. (2021) Việt Nam là một trong 20 nước trên thế giới có số liều vaccine tiêm chủng nhiều nhất [Vietnam ranks among the world's top 20 countries for the number of COVID-19 vaccine doses delivered]. Cổng thông tin Bộ Y tế [Ministry Heal. Portal]
81. Lộc H, Anh L (2020) Văcxin COVID-19 của Việt Nam: chuẩn bị thử nghiệm lâm sàng trên người [Vietnam's COVID-19 vaccine: preparing for clinical trials on humans]. In: Tuổi Trẻ Online. https://tuoitre.vn/vacxin-covid-19-cua-viet-nam-chuan-bi-thu-nghiem-lam-sang-tren-nguoi-20201207082152099.htm. Accessed 27 Nov 2024

82. (2022) GDP năm 2021 của Việt Nam tăng 2,58% [Vietnam's GDP in 2021 increased by 2.58%]. In: Trang thông tin điện tử Bộ Tài chính. https://mof.gov.vn/webcenter/portal/vclvc-stc/pages_r/l/chi-tiet-tin?dDocName=MOFUCM222589. Accessed 13 May 2025
83. An N (2021) Chuyên gia đề nghị: Cần nhanh chóng tiêm vắc xin cho 70% dân số để đạt miễn dịch cộng đồng [Experts recommend: Need to quickly vaccinate 70% of the population to achieve community immunity]. Tuổi Trẻ Online
84. (2024) Top 10 tỉnh thành giàu nhất Việt Nam 2024: Tên gọi và xếp hạng [Top 10 Most Economically Prosperous Provinces and Provincial Cities in Vietnam, 2024: Names and Rankings]. In: Sputnik. https://kevesko.vn/20240608/top-10-tinh-thanh-giau-nhat-viet-nam-2024-ten-goi-va-xep-hang-29919109.html. Accessed 28 Nov 2024
85. (2021) Hà Nội: Hàng trăm người chen lấn để xét nghiệm COVID-19 [Hanoi: Hundreds of people jostle for COVID-19 testing]. Báo Tuổi Trẻ
86. (2021) Chống Covid-19 ở Việt Nam: Sai lầm, trách nhiệm và cách khắc phục [Vietnam's COVID-19 Fight: Mistakes Made, Who's Responsible, and What's Next]. BBC News Tiếng Việt
87. (2021) Thiếu trang thiết bị, vật phẩm chống dịch Covid-19: Do cơ chế? [Shortage of COVID-19 Equipment and Supplies: Fault of the System?]. Người Lao động [The Employee]
88. Liễu D (2022) "Đếm" ca mắc COVID-19 không còn nhiều ý nghĩa, Hà Nội cần tập trung vào ca bệnh nặng [As case counts lose relevance, Hanoi is advised to shift focus toward caring severe COVID-19 cases]. In: Tuổi Trẻ Online. https://tuoitre.vn/dem-ca-mac-covid-19-khong-con-nhieu-y-nghia-ha-noi-can-tap-trung-vao-ca-benh-nang-20220108084434245.htm. Accessed 29 Nov 2024
89. (2021) Một nữ Bí thư Quận ủy dũng cảm và quyết đoán [A brave and decisive female District Party Secretary]. Lao động và Công Đoàn
90. (2021) Bộ Y tế đề nghị ưu tiên đặc biệt việc vận chuyển oxy y tế cho điều trị COVID-19 [The Ministry of Health requests that the transportation of medical oxygen for COVID-19 treatment be given top priority]. In: Cổng thông tin Bộ Y tế. https://moh.gov.vn/thong-tin-chi-dao-dieu-hanh/-/asset_publisher/DOHhlnDN87WZ/content/bo-y-te-e-nghi-uu-tien-ac-biet-viec-van-chuyen-oxy-y-te-cho-ieu-tri-covid-19. Accessed 28 Nov 2024
91. Minh Đ (2021) Xe chở băng vệ sinh, tã bỉm bị chặn vì "không phải thiết yếu" [Truck carrying sanitary napkins and diapers was stopped because it was "not essential."]. VnExpress
92. Hai DT, Diep DTP, Duong NNH (2022) Can feminist legal theory transcend Vietnamese legal education and legal research towards the goal of addressing women's issues? Asian J Leg Educ:1–15. https://doi.org/10.1177/23220058221101984
93. (2021) Cả nước đã chi hơn 12.000 tỷ đồng nhập sản phẩm xét nghiệm COVID-19 [The country has spent more than 12,000 billion VND to import COVID-19 testing products]. Báo Phụ Nữ
94. Cao S (2020) COVID-19 vaccine prices revealed: Pfizer. Moderna and AstraZeneca, Observer
95. Hoàng T (2022) Bắt phó chủ tịch Hà Nội Chử Xuân Dũng liên quan vụ 'chuyến bay giải cứu' [Hanoi Vice Chairman Chử Xuân Dũng arrested for being involved in the "rescue flight" case]. In: Tuổi Trẻ Online. https://tuoitre.vn/bat-pho-chu-tich-ha-noi-chu-xuan-dung-lien-quan-vu-chuyen-bay-giai-cuu-20221222170547085.htm. Accessed 29 Nov 2024
96. Bổng Đ (2022) Loạt lãnh đạo, cán bộ bị khởi tố trong vụ án "chuyến bay giải cứu" [Series of leaders and officials prosecuted in "rescue flight" case]. In: Vietnamnet https://vietnamnet.vn/loat-lanh-dao-can-bo-bi-khoi-to-trong-vu-an-chuyen-bay-giai-cuu-2066710.html. Accessed 28 Nov 2022
97. (2021) Giá kit xét nghiệm Việt Á công bố 470.000 đồng, Bộ Y tế liên quan thế nào? [Việt Á test kits priced at 470,000 VND — What was the Ministry of Health's involvement?]. In: Lao Động. https://laodong.vn/y-te/gia-kit-xet-nghiem-viet-a-cong-bo-470000-dong-bo-y-te-lien-quan-the-nao-986962.ldo. Accessed 29 Nov 2024
98. Ủy ban Kiểm tra Trung Ương [Central Committee of Inspection] (2022) Ủy ban Kiểm tra Trung ương chỉ rõ nguyên nhân và bài học từ vụ Việt Á [The Central Inspection Commission pointed out the causes and lessons from the Việt Á incident]. VTV

99. Ủy ban Kiểm tra Trung Ương [Central Committee of Inspection] (2022) Nguyên nhân và bài học kinh nghiệm từ kiểm tra khi có dấu hiệu vi phạm đối với tổ chức đảng, đảng viên qua vụ kít xét nghiệm Covid-19 xảy ra tại Công ty Việt Á [Causes and lessons learned from testing when there are signs of violations against party organ. In: Trang Thông Tin Điện Tử Ủy Ban Kiểm Tra Trung Ương [Portal Cent. Comm. Insp. https://ubkttw.vn/danh-muc/nghien-cuu-trao-doi/nguyen-nhan-va-bai-hoc-kinh-nghiem-tu-kiem-tra-khi-co-dhvp-doi-voi-to-chuc-dang-dang-vien-qua-vu-kit-xet-nghiem-covid-19.html?fbclid=IwAR18jvvxqSUb49EuJbI-ggGWRSHlBxDN9pQnE0MwIl8ZSDoqKcKfvz-z6Ac. Accessed 3 Dec 2022
100. (2021) Chấm dứt ngay chuyến bay hồi hương "trục lợi", sớm nối thông đường bay thường lệ quốc tế [Immediately terminate the "profiteering" repatriation flight, soon connect the international regular route]. In: VnEconomy. https://vneconomy.vn/cham-dut-ngay-chuyen-bay-hoi-huong-truc-loi-som-noi-thong-duong-bay-thuong-le-quoc-te.htm. Accessed 28 Nov 2022
101. Childress JF, Faden RR, Gaare RD et al (2002) Public health ethics: mapping the terrain. J Law Med Ethics 30:170–178. https://doi.org/10.1111/j.1748-720X.2002.tb00384.x
102. Meiklejohn A (1948) Free speech and its relation to self-government. Harper & Brothers, New York
103. Mill JS (2012) On liberty. Cambridge University Press, New York, N.Y
104. European Commission (2000) Communication from the Commission on the precautionary principle
105. Doan H, Nie J, Fenton E (2023) The precautionary principle in public health emergency regime: ethical and legal examinations of Vietnamese and global response to COVID-19. Bioethics:11–23. https://doi.org/10.1111/bioe.13236
106. Sen A (2000) Development as freedom. Alfred A. Knopf, New York
107. Berlin I (2002) Two concepts of liberty. In: Hardy H (ed) Liberty. Oxford University Press, Oxford, pp 166–217
108. Keeney RL (1990) Mortality risks induced by economic expenditures. Risk Anal 10:147–159. https://doi.org/10.1111/j.1539-6924.1990.tb01029.x
109. Chapman KS, Hariharan G (1996) Do poor people have a stronger relationship between income and mortality than the rich? Implications of panel data for health-health analysis. J Risk Uncertain 12:51–63. https://doi.org/10.1007/BF00353330
110. Viscusi WK (2014) The value of individual and societal risks to life and health. Handb Econ Risk Uncertain 1:385–452. https://doi.org/10.1016/B978-0-444-53685-3.00007-6
111. Viscusi WK (1994) Risk-risk analysis. J Risk Uncertain 8:5–17
112. (2025) chauvinism. Britannica
113. Lêdunganh Cuộc Chiến Không Có Ai Bị Bỏ Lại! [A Battle Where No One Is Left Behind!]. In: Cờ Đỏ Tp. Hồ Chí Minh. https://www.facebook.com/codotphcm/posts/pfbid031k-V1aHpJE9pkfk3mKEhkcmDQtuYQoH5qGEV72LzyqG8eDnob1v8UYcP3KEbNB85Rl. Accessed 8 Nov 2024
114. Lêdunganh (2020) Cuộc Chiến Không Có Ai Bị Bỏ Lại! [A Battle Where No One Is Left Behind!]. In: Diễn Châu Style. https://www.facebook.com/photo.php?fbid=162524148500524&id=110573250362281&set=a.110759983676941&locale=vi_VN
115. (2020) Ngạo Nghễ Việt Nam? [The Majestic Spirit of Vietnam?]. BBC News Tiếng Việt
116. Anh M (2024) Thể hiện lòng yêu nước, tự hào dân tộc đúng đắn, phù hợp [Express patriotism and national pride properly and appropriately]. Tạp chí Tuyên giáo
117. Long T (2020) TP.HCM kỷ luật phó giám đốc sở ký văn bản gây hoang mang dư luận [HCM City disciplined deputy director of the department who signed the document causing public confusion]. In: Tuổi Trẻ Online. https://tuoitre.vn/tp-hcm-ky-luat-pho-giam-doc-so-ky-van-ban-gay-hoang-mang-du-luan-20200424184124308.htm. Accessed 13 Apr 2023
118. Rin T (2021) Xử lý nhóm "cò" hỏa táng hoạt động bát nháo ở Bình Hưng Hòa [Authorities take action against unregulated cremation middlemen at Bình Hưng Hòa]. Thanh Niên
119. Tiến T (2021) Bộ Tư lệnh TP Hồ Chí Minh thông tin vụ chở 46 thi thể mắc Covid-19 về Bến Tre hỏa táng [HCMC Command provides details on the transport of 46 bodies of COVID-19 patients to Bến Tre for cremation]. Kinh tế và Đô Thị

120. (2021) Dồn toàn lực, kiểm soát hoàn toàn dịch bệnh trước khi kết thúc Chỉ thị 16 ["All-out push to fully contain the epidemic before the end of Directive 16]. In: Cổng thông tin Bộ Y tế. https://moh.gov.vn/tin-noi-bat/-/asset_publisher/3Yst7YhbkA5j/content/don-toan-luc-kiem-soat-hoan-toan-dich-benh-truoc-khi-ket-thuc-chi-thi-16. Accessed 26 Nov 2024
121. David R, Brierley JEC (1978) Major legal systems in the world today: an introduction to the comparative study of law. Simon and Schuster
122. WHO, UN, UNICEF, et al (2020) Managing the COVID-19 infodemic: promoting healthy behaviours and mitigating the harm from misinformation and disinformation
123. Doan TH (2025) Social listening and its issues: what can the precautionary principle advice? Asian Bioeth Rev 17
124. Tse-tung M (1943) Some questions concerning methods of leadership. In: Sel. Work. Mao Tse-tung. https://www.marxists.org/reference/archive/mao/selected-works/volume-3/mswv3_13.htm. Accessed 3 Jun 2025
125. Hoa M (2017) Hơn 10.000 người trong "Lực lượng 47" đấu tranh trên mạng [More than 10,000 people in "Force 47" fight online]. Tuổi Trẻ Online
126. Hookway J (2017) Introducing force 47, Vietnam's new weapon against online dissent. Wall Str J
127. Huete-Alcocer N (2017) A literature review of word of mouth and electronic word of mouth: implications for consumer behavior. Front Physiol 8. https://doi.org/10.3389/fpsyg.2017.01256
128. Straubhaar J, LaRose R, Davenport L (2018) Media now: understanding media, culture, and technology, 10th edn. Gale Cengage Learning, Boston, MA
129. Littlejohn SW, Foss KA (2009) Two-step and multi-step flow. In: Encyclopedia of communication theory. SAGE Publications, Thousand Oaks, CA, pp 970–972
130. Hassan A, Barber SJ (2021) The effects of repetition frequency on the illusory truth effect. Cogn Res Princ Implic 6. https://doi.org/10.1186/s41235-021-00301-5
131. Mallinson DJ, Hatemi PK (2018) The effects of information and social conformity on opinion change. PLoS One 13:1–22. https://doi.org/10.1371/journal.pone.0196600
132. Solomon EA (1951) Effects of group pressure upon the modification and distortion of judgments. In: Groups, leadership and men; research in human relations. Carnegie Press, Oxford, UK
133. Asch SE (1955) Opinions and social pressure. Sci Am 193:31–35
134. Minh HC (1951) Báo cáo Chính trị tại Đại hội lần thứ II, tháng 2 năm 1951 của Đảng Lao động Việt Nam [Political Report at the Second Congress, February 1951 of the Labor Party of Vietnam]
135. Faden R, Bernstein J, Shebaya S (2025) Public health ethics. Stanford Encycl Philos
136. John-Stewart Gordon Bioethics. Internet Encycl. Philos
137. TTXVN (2020) Kiểm tra thân nhiệt tất cả hành khách nhập cảnh từ Vũ Hán [Temperature check of all passengers entering from Wuhan]. Báo Hòa Bình
138. Long X (2020) Vĩnh Phúc cách ly toàn xã Sơn Lôi, hỗ trợ mỗi người dân 40.000 đồng/ngày [Vĩnh Phúc places the entire Sơn Lôi commune under quarantine, providing each resident with 40,000 VND per day in support]. Tuổi Trẻ Online
139. Kiên M (2020) Huyện Mê Linh sẵn sàng thực hiện việc dừng cách ly y tế tại thôn Hạ Lôi [Mê Linh District set to lift medical quarantine in Hạ Lôi Commune]. In: Ban Chấp hành Đảng bộ TP Hà Nội. https://thanhuyhanoi.vn/tin-tuc/chung-tay-chong-dich/huyen-me-linh-san-sang-thuc-hien-viec-dung-cach-ly-y-te-tai-thon-ha-loi-10004819.html
140. Vỹ T (2020) Quảng Nam tiếp tục thực hiện cách ly xã hội tại 05 huyện, thị xã [Quảng Nam continues to implement social distancing in 05 districts and towns]. In: Diễn đàn Doanh nghiệp. https://diendandoanhnghiep.vn/quang-nam-tiep-tuc-thuc-hien-cach-ly-xa-hoi-tai-05-huyen-thi-xa-10056469.html. Accessed 19 Nov 2024
141. Thơ H (2020) Quảng Trị: TP.Đông Hà như thế nào sau chỉ thị giãn cách xã hội [Quảng Trị: What is the Life in Đông Hà City like after Social Distancing]. Lao Động [Labor Newspaper]
142. WHO Key planning recommendations for mass gatherings in the context of COVID-19 Interim guidance - First issued 14 February 2020, Third update 4 November 2021. https://iris.who.int/bitstream/handle/10665/348018/WHO-2019-nCoV-POE-mass-gathering-2021.1-eng.pdf?isAllowed=y&sequence=1&utm_source=chatgpt.com. Accessed 16 Jul 2025

143. Area, population and population density by locality. In: Gen. Stat. Off. Vietnam. https://www.gso.gov.vn/px-web-2/?pxid=V0201&theme=Dân số và lao động. Accessed 28 Nov 2024
144. Khá H (2016) 13 triệu người thường xuyên có mặt ở TP HCM [13 million people regularly present in Ho Chi Minh City]. In: Tuổi Trẻ Online. https://tuoitre.vn/13-trieu-nguoi-thuong-xuyen-co-mat-o-tp-hcm-1145929.htm. Accessed 28 Nov 2024
145. (2018) Những phận đời không giấy tờ [Lives Without Legal Documents]. In: VTV. https://vtv.vn/chuyen-dong-24h/nhung-phan-doi-khong-giay-to-20180911125319081.htm. Accessed 28 Nov 2024
146. (2023) 10 tỉnh, thành có dân số lớn nhất Việt Nam [10 provinces and provincial cities with the largest population in Vietnam. In: Cafef. https://cafef.vn/10-tinh-thanh-co-dan-so-lon-nhat-viet-nam-20230215133417338.chn. Accessed 28 Nov 2024
147. Chung T (2021) Sa thải giảng viên phát ngôn sai lệch về công tác phòng chống dịch COVID-19 [Lecturer Fired Over False Claims About COVID-19 Affairs]. Lao Động [Labor Newspaper]
148. Mudde C (2004) The populist zeitgeist. Gov Oppos 39:542–563. https://doi.org/10.1111/j.1477-7053.2004.00135.x
149. (2021) Giãn cách xã hội TPHCM; phong tỏa toàn bộ quận Gò Vấp; tạm dừng tuyển sinh lớp 10 [Social distancing in HCM City; blockade of the entire Go Vap District; suspending the grade-10th enrollment]. Báo Chính Phủ [Government's News]
150. (2021) New Zealand phong tỏa thành phố Auckland để kiểm soát Covid-19 [New Zealand locks down Auckland city to control Covid-19]. VOV
151. Trọng D (2021) Đăng tin "Hà Nội ngày đầu phong tỏa", 1 người bị phạt 12,5 triệu đồng [Posting "Hanoi on the first day of lockdown," one person fined 12.5 million VND]. Tuổi Trẻ Online
152. Đông S (2020) Chia sẻ thông tin "phong tỏa thành phố 14 ngày", một tài xế bị phạt 10 triệu đồng [Sharing information about "14-day city lockdown", a driver was fined 10 million VND]. Thanh Niên
153. Byă HH (2021) Công tác dân vận: "mưa dầm thấm lâu" [Mass mobilization: "a steady drizzle soaks deeply"]. Nhân Dân
154. Văn H (2020) Cẩn trọng với kiểu "mưa dầm thấm lâu" của các thủ đoạn xuyên tạc, bóp méo [Cautions against the "steady drizzle soaks deeply" approach of distortion and falsification]. VietNamNet

Case Law

Canada

Taylor v. Newfoundland and Labrador Taylor v. Newfoundland and Labrador, [2020] N.J. No. 191 2020 NLSC 125; [2020] N.J. No. 191

U.S.

Korematsu v United States, 323 US 214, 65 S Ct 193, 89 L Ed 194, 1944 US LEXIS 1341

Vietnam

High People's Court of Ho Chi Minh City, Judgment No.: 988/2023/HC-PT, Date: 20 Nov 2023

Part IV

Oceania

10 Holding the Frontline: The Success of the Cook Islands COVID-19 Response

Nuhisifa Seve, Bob Williams, Yin Yin May, Metua Bates, Garth Henderson, Carl Hunter, Teariki Faireka, Karen Ngamata, Ralema Makena Geno, and Parau Ngamata

Abstract

Border restrictions established on 25 March 2020 in the Cook Islands assisted to prolong the entry of the coronavirus (COVID-19 [COVID]) into the Cook Islands. On 16 April 2020, the Cook Islands was declared a COVID-free zone, signalling the shift from a public health emergency to a public health threat. The border closures were eased in 2021 when the New Zealand government made the COVID-19 vaccination available to Te Marae Ora Ministry of Health Cook Islands (Te Marae Ora [TMO]). The high rate of vaccinations of the Cook Islands population showed the trust they had in the mitigation approach to the pandemic that was implemented by the lead agency TMO who were supported by public

N. Seve (✉)
Auckland, New Zealand

B. Williams
Te Marae Ora Ministry of Health Cook Islands, Rarotonga, Cook Islands
e-mail: williams346@slingshot.co.nz

Y. Y. May
Hospital Health Services Te Marae Ora Ministry of Health Cook Islands, Rarotonga, Cook Islands

M. Bates · K. Ngamata · P. Ngamata
Public Health Te Marae Ora Ministry of Health Cook Islands, Rarotonga, Cook Islands

G. Henderson
Ministry of Finance and Economic Management, Rarotonga, Cook Islands

C. Hunter
The Office of the Public Service Commission, Rarotonga, Cook Islands

T. Faireka
Primary Healthcare Te Marae Ora Ministry of Health Cook Islands, Rarotonga, Cook Islands

R. M. Geno
ICT Te Marae Ora Ministry of Health Cook Islands, Rarotonga, Cook Islands

L. P. Marquez et al. (eds.), *COVID-19 in the Pacific Rim and Beyond*, Collaborative Bioethics 6, https://doi.org/10.1007/978-3-032-21218-4_10

and private sector agencies. The Cook Islands successfully remained COVID free until 13 February 2022, when the first positive case presented—2 weeks after the Cook Islands reopened its borders. In subsequent months, over 50% of the population (spread across six islands) would contract the virus. By the end of 2022, the numbers contracting the virus had dropped markedly, and there remain at the time of writing (2022) only two COVID-related fatalities. This chapter is a descriptive account of the successful Cook Islands response to the COVID-19 pandemic during the years 2020–2022. The narratives were written from the perspectives and experiences of the 10 named contributors who were in leadership roles in the public sector or as frontline staff or support personnel in the health sector.

Keywords

Public health · Success · Leadership · Frontliners · Ethics

Introduction

In the Pacific region, many Pacific nations remained relatively free from the direct impact of the COVID-19 virus due to border restrictions. Only Guam (north Pacific) and French Polynesia (south Pacific) reported large-scale community transmission in 2020, with the remaining Pacific nations reporting small numbers of localised community transmission or imported cases, or no cases at all [1]. At the end of 2021, the Pacific region was home to some of the last countries in the world to remain entirely COVID-free. The Cook Islands was one of those countries and successfully remained COVID free until 13 February 2022 when the first positive case presented 2 weeks after the Cook Islands reopened its borders.

Country Context: People and Place

The Cook Islands is in the middle of the South Pacific Ocean between Tonga to the west and the Society Islands to the east and comprises 15 islands (13 inhabited and two uninhabited islands) that are spread over 850,000 square miles (2.2 million square kilometres). The 15 islands are geographically divided into two groups, commonly referred to as the northern and southern group islands or the Pa Enua (outer islands). The northern group comprises Manihiki, Rakahanga, Pukapuka, Nassau, Penrhyn and Suwarrow, while the southern group comprises Aitutaki, Atiu, Mitiaro, Mauke, Mangaia, Palmerston, Takutea, Manuae and the main island of Rarotonga. Avarua is the nation's capital and is located on the main island of Rarotonga [2].

The total population of the Cook Islands, enumerated on 1 December 2021, was 15,040 people: 7392 men and 7648 women (this number included all persons present at census night in the Cook Islands). The 2021 Census count records a decrease of 2394 people since the 2016 Census (17,434). The decrease was predictable due to the effect of the pandemic. Many chose to seek employment overseas, and migrant workers who worked in the tourism industry lost their jobs and returned to their countries of origin.

Over 70% (10,898) of the total population live on Rarotonga, 20.2% (3040) in the southern group islands and 7.3% (1102) in the northern group islands. Rarotonga is the main island in the Cook Islands. The indigenous Cook Islands Māori population remains the largest ethnic group at 77% (n = 11,603), a slight decrease from 78% in 2016 and 81% 2011. The non-Cook Island Māori population accounts for 14% of the population [3].

The Cook Islands is a self-governing nation in free association with New Zealand since 1965. The government is an independent parliamentary democracy consisting of 25 elected members. A separate House of Ariki made up of 24 members acts in an advisory role to the government on various issues [4].

The Cook Islands is a small, open economy whose economic growth is heavily reliant on the export of services to several key partner countries—New Zealand, Australia and the United States of America in particular. The tourism industry dominates, with the tertiary or services sector accounting for three-quarters of the total economy in 2019–2020. The two largest tertiary industries, trade and accommodation services, accounted for nearly a quarter of total economic output. New Zealand remains the Cook Islands primary trading partner.

In January 2020, there were 8056 people in the Cook Islands labour force. About 69% are engaged in the private sector (employees and self-employed or sole proprietors), with 26% in the public sector [3].

Health Systems

Te Marae Ora is the main provider of healthcare in the Cook Islands and has a regulatory function through various legislation. Te Marae Ora provides health services to the population through its five directorates—Hospital Health Services (HHS), Public Health, Oral Health, Primary Healthcare and Funding and Planning. There are 300 plus staff that work at TMO at any given time [5].

Health services range from public health (inclusive of primary care) to secondary care. Overall, TMO is relatively well equipped to provide basic primary- and secondary-level care and able to deliver an adequate range of general clinical services in the core areas of surgery, medicine, anaesthesia, obstetrics, gynaecology, ophthalmology and paediatrics. These services are supplemented by visiting specialist teams from New Zealand, and access to tertiary services is through referral to New Zealand health providers. There are a small number of private health providers in Rarotonga.

The main health facilities are based in Rarotonga and include Rarotonga Hospital with 80 beds (provides health specialist visits and organises overseas referrals), Aitutaki Hospital with 26 beds (provides the basic primary health care services), Tupapa primary healthcare facility in Tupapa and one main oral health clinic in Tupapa. In each of the Pa Enua islands are small health clinics which are run by public health nurses on rotation from Rarotonga [5].

The Cook Islands COVID Response

Context

Prior to the pandemic announcement by World Health Organization (WHO), health systems preparedness had already commenced in the Cook Islands by TMO to respond to the virus should it present in the Cook Islands.

On 22 January 2020, TMO activated the health emergency response and incident management system following reports of rapidly increasing numbers of COVID cases and associated deaths in Wuhan, China. Two structures were established—the National Health Emergency Taskforce (NHET) and the Incident Management System (IMS). The NHET was the leadership and governance arm that was made up of senior staff from across public and private sector agencies. Their role was to direct the response activities through to the IMS and to keep the Minister of Health and the government of the Cook Islands informed. The IMS was the operational arm of NHET and was made up primarily of TMO staff and seconded staff from the public and private sector agencies. The NHET held its first meeting on 27 January 2020.

There were three phases to the Cook Islands COVID response over the three-year period (2020–2022), with each stage requiring modification of health systems strategies. The first phase was emergency responsiveness in 2020. This phase focused on preparing health systems and the country to mitigate against COVID should it present in the Cook Islands. The main mitigation strategy was to close the borders (air and sea). Although the border closure had an economic impact on a country reliant on tourism, it did allow TMO to prepare its health systems and communication framework. The communication framework was important to set up for the dissemination of public health information to the community. The information was disseminated through TMO's multimedia platforms. Public meetings were also held.

The second phase was Quarantine Free Travel (QFT) between the Cook Islands and New Zealand in 2021. This phase commenced on 17 May 2021 at the same time as the COVID vaccination roll-out for the Cook Islands population began. Travellers from New Zealand were permitted entry into the Cook Islands under restrictions (for example negative COVID test prior to entry and six-day testing after entry).

The third phase was the normalisation (living with COVID) of COVID in 2022. This phase involved the lifting of restrictions as there were high vaccination rates among Cook Islanders. However, quarantine facilities were set up during this period for positive COVID cases of travellers. Cook Islands residents who tested positive were quarantined in their homes.

Through the support of the Cook Islands government, a legislative framework to deliver a robust public health response was secured along with endorsement of the national emergency response plan (March 2020), border restrictions and budget. The border restrictions established on 25 March 2020 in the Cook Islands assisted to prolong the entry of the virus into the Cook Islands, and on 16 April 2020, the

Cook Islands was declared a COVID-free zone, signalling the shift of COVID from a public health emergency to a public health threat.

From the Frontline: The Lead Agency Te Marae Ora

As the primary healthcare provider in the Cook Islands, TMO was the lead agency responsible for preparing its health systems and administering the COVID vaccination. The high vaccination rates of the population ensured zero fatalities when COVID-19 initially presented in the Cook Islands on 13 February 2022. At the time of writing (2022), there have only been two COVID-related fatalities.

Informing the Operations

Legislation, plans, policies and guidelines are essential to pandemic response planning as the documents provide guidance to operational activities and clarify any issues that may arise that are critical to health and safety and legal and regulatory requirements of a country.

In 2020, it became apparent that the *Public Health Act* (PHA) *2004* [6] had major gaps. The Act did not provide legislative procedures for public health measures, such as supervised quarantine for inbound travellers, restrictions on gatherings and other physical distancing measures, and contact tracing, all of which were necessary to respond to the emerging threat. The Cook Islands government had to enact the *COVID-19 Act 2020* [7] on 25 March 2020 to provide a legal framework to temporarily overcome shortcomings in the PHA and implement necessary COVID measures. In early 2020, the reform of the *Public Health Act 2004* commenced, and the new *Public Health Act 2024* [8] was passed in 2024.

At the onset of the response to COVID, the TMO policy and planning team was integrated into the IMS Information and Planning team alongside staff from within TMO—Oral Health Directorate and from other public sector organisations (Office of the Public Service Commission, the Cook Islands Seabed Minerals Authority and the Financial Supervisory Commission). Although there was only one document that dealt with pandemics—*the Cook Islands Influenza plan* 2009 [9] at the time, over a period of 10 months in 2020 a substantial number of documents were written. The first *Cook Islands Emergency Response Plan to Coronavirus Disease 2019 (COVID-19)* [10] was also operationalised on 10 March 2020 to support the operational emergency framework that was established.

By the end of 2020, a total of 30 standard operating procedures (SOPs) and four policies were developed in consultation with respective TMO staff who would be operationalising them. Other plans like the *Containment and Mitigation Plan* [11], *Risk Communication and Community Engagement Plan* [12], *Easing Border Restrictions Plan* [13] and the *Public Health Protocol Coronavirus Disease 2019* [14] were also written. Te Marae Ora borrowed heavily from the New Zealand Ministry of Health documents and adapted many to suit their circumstances.

In October 2020, the IMS was stood down as there had been no COVID cases and the border remained closed except for the occasional repatriation flight of Cook Islands residents stranded in New Zealand.

The COVID-related documents were reviewed and updated to align with phase two and three by the TMO Policy and Planning Unit. During phase three, a new Cook Islands emergency plan was developed in collaboration with the Policy Directorate of the Office of the Prime Minister. Additionally, a further 20 SOPs were added by the TMO Policy and Planning Unit.

Information and Communication Technologies

Essential to health systems' response was the information and communication technologies (ICT) team, who were responsible for ensuring that ICT services were maintained and managed. Services like video conferencing with local and international health organisations like the WHO and the New Zealand Ministry of Health (MoH) were central in keeping TMO abreast with COVID response developments globally. Maintaining TMO's Wi-Fi connectivity throughout the period was critical for the dissemination of public health information to the Cook Islands community through various TMO multimedia platforms like Facebook and the TMO website. As ICT was an essential service during the response, the team would work in shifts in order to ensure 7-day and 24-h ICT coverage when required.

Reorientation of Health Systems

Primary Healthcare and Public Health

In 2020, the emergence of the pandemic globally impacted the Cook Islands, particularly in terms of its critical resources (human and monetary), pushing primary healthcare (PHC) to the forefront. Primary health care is defined as:

> ...the essential care based on practical, scientifically sound, and socially acceptable methods and technology, made universally accessible to individuals and families in the community through their full participation, and at a cost that the community and country can afford to maintain at every stage of their development in the spirit of self-reliance and self-determination [15].

The implementation of the PHC Directorate was a strategic response to health systems preparedness. Basing the PHC directorate in the Tupapa outpatient clinic was a planned move to minimise people from visiting Rarotonga Hospital for their healthcare needs. Additionally, ten puna (puna premises were formalised under the auspices of the Emergency Management Cook Islands structure in 2010 for disaster relief management purposes and are located around the island of Rarotonga) were seconded from the Emergency Management Cook Islands and turned into puna health clinics in order to provide non-emergency healthcare services closer to the community. The puna health clinics were staffed by public health nurses. While there were still some face-to-face consultations with patients by medical practitioners, phone consultations took precedence during this period.

In the Public Health Directorate during the QFT phase, the Health Intelligence Unit (HIU) was established alongside the QFT team made up of public health staff (public health specialist, QFT coordinator, technical adviser, health intelligence consultant/clinical coordinator, infection control coordinator and health intelligence manager). The QFT team was responsible for coordinating the two-way QFT arrangements between the Cook Islands and New Zealand, ensuring systems and protocols for testing, contact tracing and case management were in place, and staff were trained in delivering services safely.

As a newly created unit within the Public Health Directorate, the Health Intelligence Unit (HIU) was an integral part of the public health response to COVID. Although the unit faced several practical challenges in the beginning (for example, no designated space for operations), these were quickly set aside as there was a pressing need to compile the COVID statistics internationally and locally for government, TMO and the NHET daily reports. The HIU team also became responsible for contact tracing when COVID presented in the Cook Islands in February 2022. During the positive cases surge in the Cook Islands in 2022, HIU was responsible for positive cases reporting and contact tracing. The HIU worked with Red Cross to ensure seamless transition of information from HIU to the puna health clinics. Temporary staff were employed, and HIU expanded from 6 to 23 staff who were rostered on two shifts, 24 h a day, 7 days a week, for 4 months.

The Pa Enua

Travel from Rarotonga to the Pa Enua (outer islands) was restricted for 2 years. However, the restrictions were slowly eased for the island of Aitutaki (tourist destination) during phase three. In preparation for phase three in late 2021, two people were deployed from TMO in Rarotonga to work in Aitutaki to prepare and support Aitutaki's COVID response. Preparations included establishing an Emergency Operations Centre (EOC) in the main town area, stakeholder engagement and training with TMO staff and community volunteers (Au, Arenikau and Vaivaitau Taskforce teams) on public health measures, Rapid Antigen Testing (RAT) and community contract tracing. The 11 Healthcare Assistants (HCAs) that were employed for the brief period of the pandemic were later deployed to full-time roles in the health system and supported into further training or other career pathways. The COVID workforce and Taskforce teams also worked with the local Ministry of Internal Affairs (MIA) team to distribute welfare packs to affected households. The first COVID cases presented in the Pa Enua on 7 March 2022 were on the island of Aitutaki.

In the northern group of islands in Manihiki, the first positive cases of COVID occurred on 1 August 2022. On 12 August, a team of four public health nurses and a health protection officer was deployed from TMO Rarotonga to Manihiki.

Supported by the local TMO staff and the Manihiki Island Council, the team set up a quarantine facility for positive cases, started community screening for those who were symptomatic and issued quarantine orders. The team also visited households on the island to provide public health information on how they could contain the spread of the virus in their household.

Working alongside their local Manihiki counterparts, Police and staff from the MIA team also assisted in the distribution of welfare packs to households in need, refilled water bottles and fed pigs for those who were in quarantine. The coordinated efforts of TMO, Police, the Manihiki Island Council and the MIA assisted in containing the virus spread. As the community was highly vaccinated and followed the advice given by the TMO team, they were able to contain the spread of the virus over a four-week period. The total positive cases in Manihiki were 102 with no fatalities.

Hospital Health Services

Like the other two directorates (Primary Healthcare and Public Health), the HHS Directorate played a pivotal role in the Cook Islands COVID response. The Director of HHS provided leadership during the COVID vaccination information campaign and roll out. Alongside their TMO and public sector colleagues, HHS senior staff were involved in the governance and operational groups like the NHET and the IMS.

In the reorientation of the health services during the COVID period, the HHS implemented several preventative measures to protect high-risk inpatients in the Rarotonga Hospital in 2020. These measures included the establishment of the Cough and Cold Clinic and the relocation of Emergency services and Outpatients clinic from Rarotonga Hospital to Tupapa Primary Healthcare in March 2020. Rarotonga Hospital readiness included repurposing one of the hospital wards into the Te-Kou Ward for COVID positive patient needing ventilatory support. The ward had 32 beds including two negative pressure rooms that could accommodate ten patients.

The New Zealand MoH required several assurances that the Cook Islands health systems were prepared during the QFT period. Amongst the requirements was the establishment of a reverse transcription polymerase chain reaction (aka RT-PCR) laboratory, and electrical wiring and upgrading of the oxygen plant. The laboratory was essential as it meant that there was a fast turnaround for COVID results. The fast turnaround of results enabled better decision-making by both countries. The oxygen plant was to ensure TMO has adequate oxygen supply. The laboratory and oxygen plant were functional in May 2021.

Prior to the vaccination rollout that paralleled the QFT phase, several community meetings were held in Rarotonga and the outer islands to inform the community of the rollout and to answer any questions they had. The community meetings were chaired by the Secretary of Health supported by the Director HHS and Director PH. Staff from various directorates also attended to assist in the logistics of the public information meetings. At the meetings, pamphlets were given to those who required more information. In May 2021, the mass vaccination programme commenced in Rarotonga, followed by the Pa Enua. Leading the vaccination efforts was Director HHS with support from staff across all TMO directorates and staff from the public and private sector. The Pharmacy unit at HHS contributed significantly to the successful deployment of vaccine and the high immunisation rates in people on the 12 inhabited islands of the Cook Islands. The pharmacy team took on the responsibility of quality assurance for the nationwide deployment of the COVID-19 vaccine.

While there were several dissenting (anti-vaccination) voices at the public information meetings, the high vaccine coverage (eligible population 98% and eligible population 12+ years 94%) in a total of 128 working days had a positive impact on the community.

Multisectoral Support

The effects of border closures in New Zealand and Australia in March 2020 and the Cook Islands in April 2020 and the introduction of quarantine requirements in 2021 would have severe consequences on the Cook Islands economy reliant on the tourism industry.

Responsibility for managing government expenditure was the Ministry of Finance and Economic Management (MFEM). This ministry played a key role in the management of government expenditure during this period. The *Economic Response Plan* and the *Economic Recovery Roadmap* [2] directed the work of MFEM—supported payments to the private sector in form of wage support, grants and other strategies. The MFEM negotiated loans and grants to support the Cook Islands economy while also maintaining cash management. Customs border processes were developed and implemented to allow the safe continuance of trade and passenger routes. Working closely with TMO, the MFEM procured consumables like protective personal equipment (PPE) and equipment—oxygen plant. At the community level, the ministry funded welfare packs which were distributed by the MIA.

Although the Office of the Public Service Commissioner (OPSC) was initially uncertain of their role in the COVID response stage, they did host meetings for head of ministries to share information on local and international issues regarding COVID, and to clarify any work-related questions like leave entitlements and recruitment and secondment of staff. In 2022, the OPSC updated the *Cook Islands Government Leave Policy* [22] for public servants to provide pandemic leave.

Other public sector organisations like the Tourism Cook Islands Corporation and the MIA played an important role in communications and wellbeing of the community during this period. Private sector organisations like Red Cross provided practical support during the COVID cases surge and the vaccination period.

Conclusion: What We Learnt

The multisectoral (public and private sector) approach used to respond to COVID-19 minimised COVID-related fatalities when the virus presented in the Cook Islands on 13 February 2022. Firstly, a governance and leadership and operational framework was implemented in early 2020. The NHET was made up of senior staff from across public and private sector agencies who were able to make decisions at meetings and then convey the decisions and public health information to the staff of their respective organisations. The NHET was chaired by the Secretary of Health who was also responsible for updating the Minister of Health and the Cook Islands Cabinet. At the operational level, the IMS implemented operational decisions made at the NHET level. The IMS was primarily made up of TMO and seconded staff from private and public organisations. Collaboration with all sector agencies was an essential component evident in the nine contributor's narratives above. As one contributor stated: *Everyone needs to work together for the protection of our people.*

Secondly, essential to the response was funding. The response budget was administered by MFEM, and aid for the response primarily came from New Zealand. Consumables like PPE were donated from countries like China and Singapore. The funding received from New Zealand was used to purchase resources to set up the RT-PCR laboratory and refurbish the oxygen plant. Funding was also used to support amongst other things, overtime payments to frontline staff on call, welfare packs for the community and quarantine facilities for tourists who contracted COVID.

Thirdly, ensuring ICT infrastructure was maintained and operational throughout the response period was essential for the dissemination of information and telehealth consultations. Communications were key to the response, in particular to the Cook Islands communities, advising them of response measures. Working with the Tourism Cook Islands Corporation, the IMS Communications team used various multimedia platforms to disseminate key public health messages to the population. Social media (TMO Facebook page) was the main media platform evident in the rise of followers from 2000 in 2020 to 7000 in 2023.

Lastly, documents such as legislation, plans and policies are an essential part of the response because they directed the response work of TMO and other agencies within the parameters of legislative requirements.

This chapter has provided a descriptive account of the successful Cook Islands response to the COVID-19 pandemic during the years 2020–2022. The narratives were written from the perspectives and experiences of the nine contributors who were in leadership roles in the public sector or as frontline staff or support personnel in the health sector.

References

1. Bell L, van Gemert C, Merilles OE Jr et al (2022) The impact of COVID- 19 on public health systems in the Pacific Island countries and territories. Lancet Reg Health West Pac 25:100498
2. Government of the Cook Islands (2022) Te Kaveinga Tupu'anga iti Tangata: Cook Islands Population Policy 2022–2032. Retrieved from: https://www.pmoffice.gov.ck/wpcontent/uploads/2023/01/CI-Pop-Pol-highres- digital.pdf

3. Government of the Cook Islands (2022) The census of population and dwellings. Retrieved from: https://stats.gov.ck/2021-census-of-population-and-dwellings/Government of the Cook Islands https://parliament.gov.ck/about/
4. Te marae akarau vānanga o te kuki airani Parliament of the Cook Islands. Retrieved from: https://parliament.gov.ck/about/
5. Te Marae Ora Ministry of Health Cook Islands. Retrieved from https://www.health.gov.ck/wp-content/uploads/2024/01/CI_National_Health-Strategic_Plan_2023_2027.pdf
6. Te Marae Ora Ministry of Health Cook Islands. Retrieved from: https://www.health.gov.ck/public-health-act-2004/
7. Te Marae Ora Ministry of Health Cook Islands. Retrieved from: https://www.health.gov.ck/covid-19-act-2020/
8. Te Marae Ora Ministry of Health Cook Islands. Retrieved from: https://www.health.gov.ck/public-health-act-2024/
9. Te Marae Ora Ministry of Health (2009) Te Marae Ora Cook Islands Ministry of Health Influenza Pandemic Plan 2009. [Unpublished manuscript]
10. Government of the Cook Islands. Retrieved from https://samoa.un.org/en/43064-cook-islands-emergency-response-plan-coronavirus-disease-2019-covid-19. [Manuscript archived]
11. Te Marae Ora Ministry of Health (2020) Containment and mitigation plan; surveillance and testing plan [Manuscript archived]
12. Te Marae Ora Ministry of Health (2020) Risk communication and community engagement plan for coronavirus disease 2019 (COVID-19). Retrieved from: https://www.health.gov.ck/risk-communication-and-community-engagement-plan-for-coronavirus-disease-2019-covid-19/
13. Te Marae Ora Ministry of Health (2021). Easing border restrictions plan. [Manuscript archived]
14. Te Marae Ora Ministry of Health (2020) Public health protocol coronavirus disease 2019. [Unpublished manuscript]
15. World Health Organization (1978) The International Conference on Primary Health Care, AlmaAta, USSR, 6–12 September 1978. Retrieved from: https://cdn.who.int/media/docs/default-source/documents/almaata-declaration-en.pdf?sfvrsn=7b3c2167_2

Part V

North America

Ethical Considerations on the Response to the COVID-19 Pandemic in Costa Rica: Learnings and Challenges

11

Luis Fernando Solís-Calvo, Wendy López-Vargas, Ingrid Gómez-Duarte, Carlos F. Cáceres-Palacios, Jean-Paul Sarrazin, and Rocio Sáenz-Madrigal

Abstract

The COVID-19 pandemic, identified in late 2019 in Wuhan, China, spread rapidly and caused around 14.9 million deaths between January 2020 and December 2021. Deaths mainly affected middle-income countries and older people. The pandemic interrupted advances in life expectancy, affecting the Americas and Southeast Asia the most. It also exacerbated problems such as obesity and malnutrition, significantly impacting the health of the most vulnerable people. For the development of this chapter, the responses of the government, the private sector, and community organizations in Costa Rica were described. Then, an analysis was conducted from the perspective of public health ethics, keeping in mind the relationship between policies adopted in-country and guidelines disseminated internationally. This analysis aims to understand the implications of these responses and help formulate strategies for future health crises. Costa Rica's health crisis revealed structural weaknesses connected to pre-existing economic and social problems. The fiscal crisis intensified, and poverty and inequality increased, affecting the most vulnerable. However, the focus on fiscal stability weakened social investment and aggravated problems in education and public services. In any event, the response seemed to be closely aligned with uniform measures recommended by international agencies that did not address the local living conditions, population density, the status of health systems, the local

L. F. Solís-Calvo (✉) · W. López-Vargas · I. Gómez-Duarte · R. Sáenz-Madrigal
Center for Research in Nursing and Healthcare, University of Costa Rica,
San José, Costa Rica
e-mail: l.solis@raesequidadsalud.org

C. F. Cáceres-Palacios
Research Vice-Rectorate, Cayetano Heredia Peruvian University, Lima, Peru

J.-P. Sarrazin
Department of Sociology, University of Antioquia, Medellin, Antioquia, Colombia

L. P. Marquez et al. (eds.), *COVID-19 in the Pacific Rim and Beyond*,
Collaborative Bioethics 6, https://doi.org/10.1007/978-3-032-21218-4_11

economy, and people's livelihoods. There were limited attempts to consider local technical input alongside that advice. Moreover, public communication did not allow dissenting scientific views, treating them as misinformation. In the event of future pandemics, localized decision-making should be promoted, based on local data, and communication should reflect cultural competence. Appropriate economic support mechanisms for the most vulnerable should be put in place.

Keywords

COVID-19 · Ethics · Public health · Costa Rica · Equity

Introduction

Following the notification of a cluster of pneumonia cases in Wuhan, Hubei, China, on December 31, 2019, later identified as COVID-19, a new type of coronavirus, the situation spread worldwide [1].

The World Health Organization (WHO) reported that between January 1, 2020, and December 31, 2021, approximately 14.9 million people died from the COVID-19 pandemic, representing 9.5 million more than the 5.4 million deaths initially reported. Excess deaths mainly affected middle-income countries (81%), with lower percentages in high-income (15%) and low-income (4%) countries. Deaths were more common in older people and men [2]. The COVID-19 pandemic interrupted a decade of gains in life expectancy and healthy life expectancy (HLEs), reducing them by 1.8 and 1.5 years, respectively, between 2019 and 2021; this unequally affected different regions, with the Americas and Southeast Asia being the most impacted, while the Western Pacific suffered minimally [1]. WHO highlighted the urgency for countries to invest in resilient health systems capable of sustaining essential services during crises [2].

The pandemic also exacerbated problems such as obesity and malnutrition, affecting more than 1 billion people with obesity and 148 million children with stunting; people with disabilities, refugees, and migrants also face significant health inequalities [1]. In addition to these major setbacks experienced globally with the pandemic, there are challenges such as improving universal health coverage and pollution, which increase the challenges to accelerate progress and meet the health goals for 2030 [1].

In the case of Costa Rica, the COVID-19 pandemic in 2020 showed that the emergency preparations that had been planned were not sufficient to face a crisis of such magnitude due to its pre-existing structural fragility. According to data from the State of the Nation (2020) [3], before COVID-19, the country was already experiencing an economic slowdown, high levels of inequality, and high unemployment. On the other hand, the country's chronic fiscal crisis intensified due to reduced tax revenues and increased public spending to face the emergency. In social terms, poverty and inequality deepened rapidly, hitting the most vulnerable sectors harder.

Given this scenario, the government implemented mitigation measures to alleviate the impact on the most affected households [3].

In the economic sphere, there was a sharp drop in production and massive job losses, with a setback of almost five years in terms of productive capacity, highlighting the urgency of rethinking the country's development model. Tourism, trade, and transport sectors were the most affected, while others, such as digital services and real estate, showed greater resilience [3].

According to the 2024 State of the Nation Report, the country has seen a tendency to prioritize fiscal stability without a strategic vision over human development, which made it possible to achieve monetary and fiscal stability after the COVID-19 pandemic at the expense of the weakening of public social investment and capital spending on infrastructure, the deterioration of environmental management, the deepening of a severe educational crisis, and growing difficulties in assisting the most vulnerable populations [4].

This chapter was developed by reviewing the policies and practices of the government, private actors, and community organizations in response to the COVID-19 health crisis in Costa Rica from 2020 to 2025, focusing on ethical considerations aligned with principles such as obedience to moral obligations and laws, as well as the consequences of the measures adopted in light of information available about their impact.

Methodological Approach

This study was developed using a qualitative methodology based on data collection of various documentary sources available online and in open access. First, official policy documents, newsletters, interviews, press releases, web pages, news articles, and country reports were analyzed to understand the State's perspective during the response to the COVID-19 pandemic. Second, documents that reflected the reactions of the private sector and community organizations were examined.

In addition, available information on the impact of the measures implemented was incorporated, including post-crisis studies in Costa Rica (since 2020). This exploration covered key indicators in health, education, economy, and various social areas of an institutional and community nature. Finally, the practices adopted by the different actors and the most recent evidence of their effectiveness were contrasted, examining the current recognition of their impact on the country. A descriptive approach was considered from an ethical perspective, which would allow an understanding of the implications of these responses on social well-being and the formulation of future strategies for health crises at the national level.

The National Response to the Health Crisis

Policies in Dealing with the Pandemic and the Country's Normative-Institutional Perspective

In Costa Rica, according to the Ministry of Health (2020), the first suspected case of COVID-19 corresponded to a 52-year-old Costa Rican woman. A resident of Limón on the Caribbean coast of the country had returned on February 29, 2020, after her stay in Italy and Tunisia. Upon entry, she had been reported without symptoms that matched the case definition; however, she began to show symptoms in subsequent days [5]. Even before the first case, the country had an emergency plan that included the activation of an Emergency Operations Center (COE) that integrated different institutions to address the crisis from a technical, scientific, economic, and health perspective [6].

According to the United Nations Population Fund (UNFPA 2020, the first two cases were confirmed on March 6, and social distancing measures were adopted to keep the contagion at an average of 17 cases per day [7]. As of March 10, the Ministry of Health, the National Commission for Risk Prevention and Emergency Response (CNE), and the COE notified the first health measures, such as the suspension of mass events, although cinemas, theaters, bars, and restaurants remained enabled [7, 8]; they also published Presidential Directive No. 073-S-MTSS that instructed public institutions to adopt teleworking [8].

According to the State University at a Distance (UNED 2020) [8] of the country, the second phase of the yellow alert, including three new measures, was activated on March 12.

> (i) preventive closures of at least 14 days of educational centers at risk; (ii) reduction of capacity to 50% of public meeting places (with the interpersonal distancing of at least two meters); and (iii) cancellation of trips abroad by public servants. On 13 March (23 confirmed cases), telephone line 1322 was put into operation to receive remote medical advice. [8]

The United Nations Population Fund (2020) [7] noted that due to the failure of some establishments to comply with 50% capacity, the government decreed a total closure on March 15, responding with a strategy led by the Executive Branch. Under the direction of the Ministry of Health and the Ministry of the Presidency, measures were adopted to contain the virus, mitigate the economic effects, and support the most affected sectors, including implementing sectoral health protocols. Between March and December 2020, more than 170 specific protocols for different sectors were developed [9].

Within the scope of the decree, article 3 included all actions, works, and services necessary to safeguard the health and life of the inhabitants, to preserve public order and protect the environment, a plan prepared with the technical input of the Ministry of Health as the governing and key institution for addressing the declared emergency [7]. In addition, article 4 gave the Ministry of Health and the National Commission for Risk Prevention and Emergency Response responsibility for the planning, direction, control, and coordination of protection, rescue, care, and rehabilitation programs and activities in areas declared in a state of emergency and stated that.

> In conjunction with the National Commission for Risk Prevention and Emergency Response and the Ministry of Health, the Costa Rican Social Security Fund will establish the necessary contingency measures to keep health services operational throughout the country to ensure the preservation of the health and life of the population. (...) the Costa Rican Social Security Fund will act as the Executing Unit. The Ministry of Health and the National Vaccination and Epidemiology Commission will act as supervisory authorities. They will support the National Commission for Risk Prevention and Emergency Response in selecting the products, suppliers, and sanitary conditions necessary to protect people's health and lives. [9]

The contagion figures gradually increased until reaching 29 cases per day and 1744 cases in total by June 15, 2020. This motivated the Ministry of Health to reinforce measures such as hand washing, a correct cough and sneeze protocol, and self-responsibility to inform health authorities in case of presenting symptoms after a visit to countries of transmission in the last 14 days [9].

According to the State of the Nation Report (nd), the government declared the country in a state of national emergency in March 2020, measures that directly impacted social relations, daily activities, and the subsistence of people who were obliged to comply with the established norms, in addition to Costa Rican institutions [10]. The Office of the Executive Director of UNFPA instructed.

> Representatives of country offices should explore opportunities to reprogram regular resources to respond to requests from governments and other partners to support the emergency response, always within the framework of UNFPA's mandate. [10]

The Government of Costa Rica (2020) reported that on July 25, infections reached a peak, with 931 new cases on this day alone [11]. This month had the highest number of infections, hospitalizations, and deaths, with 80% of the total at the moment.

> Hospitalizations, which at the beginning of the month were only 50 patients, by July 31 had reached 270, that is, 5.4 times more than at the beginning of the month. The accelerated rate of growth of infections creates enormous pressure from the high number of patients in the limited Intensive Care Units (ICU). (...) The number of patients in the ICU increased from only six at the beginning of the month to 75 patients at the end of the month. This represents 12.5 more patients in ICU than at the beginning of the month. The increase in patients in ICU is greater than the proportion of the increase in infections and hospitalizations. [11]

Among the most important milestones of the government response is the creation of the Proteger Program by the Ministry of Labor and Social Security (MTSS) and the Ministry of Human Development and Social Inclusion (MDHIS), which provided economic support to 724.000 people who had a direct impact on reducing or suspending their working hours. 388,804 bonuses were given to women, 335,484 to men, and 42 to intersex people.

> Other positive impacts were achieved, such as containing poverty by 3.4 percentage points in the face of the shock caused by the pandemic and allowing more than 200,000 people to access banking services. [6]

On Wednesday, April 7, 2020, 43,200 doses of COVID-19 vaccines were received through the COVAX Mechanism, which was a global solidarity effort between the Coalition for the Promotion of Innovations for Epidemic Preparedness (CEPI), the Global Alliance for Vaccines and Immunization (Gavi), the United Nations Children's Fund (UNICEF), the Pan American Health Organization (PAHO), and the WHO (Pan American Health Organization, 2020), and on May 6, 2020, the Ministry of Health identified epidemiological links by conducting tests at sentinel sites, as well as on repatriated people, potential outbreaks, and transporters entering Costa Rica, allowing for timely identification and breaking transmission chains, especially in risk areas. The name "sentinel sites" comes from the number of population or border components, for a total of 33 zones with 30 tests per week in each one [12].

A year and a half after the confirmation of the first case and death from COVID-19 in the country, there were 530,113 infections, 6349 deaths, and a fatality rate per hundred positive cases of 1.19, the lowest in Central America at that time. In a comparative analysis of the Central American subregion, Panama, Costa Rica, and Belize had the highest rates of infections per 100 thousand inhabitants (10,764, 10,023 and 4754, respectively), and also the highest death rates per million inhabitants, with Panama in first place (1663), followed by Costa Rica (1189) and Belize (993) [13].

From a state perspective, Costa Rica's response to COVID-19 was highly effective, mainly due to its robust and consolidated health system. The Comprehensive Basic Health Care Teams (EBAIS), which include more than 1000 facilities across the country, were a key element in the early detection and monitoring of cases, enabling timely identification and control of virus transmission. The low fatality rate (six deaths in two months) and the low number of hospitalizations (16 patients) in the first months of the pandemic demonstrate the effectiveness of management in this health crisis [14]. In August 2024, Costa Rica had a total of 1884.63 deaths per million, far surpassed by countries such as Peru with 6601.52, the United States with 3510.88, and the United Kingdom with 3404.43, respectively [15].

A multidisciplinary research group from the University of Costa Rica (UCR) and the Inciensa Foundation (2023) conducted a study to analyze the impact of COVID-19 on mortality in Costa Rica between 2020 and 2021. It revealed that in the first stage of the pandemic, deaths decreased by 9% due to strict health measures, such as social distancing and the use of masks. However, between July 2020 and March 2021, mortality increased by 15% due to the spread of the first variant of the virus and the lack of vaccines [16]. Priority vaccination for older people was essential to reduce deaths, especially among this high-risk group. It was also shown that deaths were not only caused directly by COVID-19 but also by chronic diseases exacerbated by the pandemic. Some causes of death, such as respiratory diseases unrelated to SARS-CoV-2, decreased due to hygiene measures [16].

The preventive approach adopted by the authorities in Costa Rica, which included closing borders, suspending mass activities, and promoting social distancing, access to timely information, and handwashing campaigns, had a high level of compliance by the population [14]. Experts pointed out that investment in health, which represents more than 6% of the Gross Domestic Product, and in health infrastructure, which includes an adequate number of hospital beds, respirators, and the Sentinel surveillance model, also facilitated the early detection of cases and rapid response [14].

Practices and Responses of Other Actors to Measures Against the Pandemic

As mentioned, Costa Rica joined the multilateral mechanism COVAX-Facility to secure vaccines against COVID-19. This program, coordinated by the WHO and supported by organizations such as GAVI-Alliance, UNICEF, the Gates Foundation, and the World Bank, sought to accelerate global development and equitable vaccine access through funds and negotiations with manufacturers. With Costa Rica's accession, the country had access to vaccines as they became available, and the agreement guaranteed 20% coverage of the population. The State made a first advance payment of USD 6,316,560 during the first month to have access through COVAX to the vaccines developed against COVID-19 [17].

According to the Ministry of Health (2020), this payment corresponded to 30% of the "weighted average" price per dose set by COVAX at USD 10.55. This price included factory costs, the access/speed premium, and financing/risk mitigation and operating costs [17]. With a population projection in Costa Rica of 5,163,021 inhabitants in 2021, the Costa Rican Social Security Fund (CCSS) reported achieving coverage of 90.1% of the population vaccinated with the first dose, 84.6% with the second dose, 57.8% with the third dose, 21.3% with the fourth dose, and 0.5% with the fifth dose [18]. In May 2022, 4,042,081 people had two doses, 78.3% of the total population. Meanwhile, 2,018,556 people had the third booster shot, 39.1% of the country's total population [19].

The effectiveness and safety of COVID-19 vaccines, particularly those using mRNA or recombinant DNA, remains disputed; official discourse has always maintained that the vaccine is "safe and effective" [20]; likewise, some critics argue that vaccination would have had a positive and negligible association with fatality rates and a moderate association with deaths/million inhabitants in America so that vaccination advances would not have been effective in reducing COVID-19 deaths [20]. The majority of the population would remain susceptible to SARS-CoV-2 infection. However, vaccines would provide limited protection against infection, mitigating future outbreaks, hospitalizations, ICU admissions, disease burden, and the economic impact of COVID-19. However, others consider that the consequences of vaccines outweigh their eventual benefits [21].

In the case of health services at the primary care level of the CCSS, they experienced interruptions during the first eight months of the COVID-19 pandemic, although they were not severe according to WHO parameters; the most affected production indicators included cervical cancer detection, anemia screening in children under two years of age, and high blood pressure control [22]. There was a significant decrease in the number of medical consultations, including those for prenatal care, postpartum care, and care for chronic diseases such as diabetes and hypertension. However, no negative change was observed in the early recruitment of children under one year of age or in prenatal and postpartum check-ups [22].

With the government's declaration of a state of emergency, the health system adapted, highlighting the reconversion of the National Rehabilitation Center (CENARE) into the new anti-COVID-19 hospital named the Specialized Center for the Care of Patients with COVID-19 (CEACO). This center, managed by the CCSS, was equipped with advanced technology and could simultaneously care for 88 patients with mild or moderate complications, with more than 1350 workers operating in 24-hour shifts [19].

The journey of COVID-19 patients in Costa Rica began with the attending physician contacting the patient management center, who, together with the COVID-19 management team, decided whether the patient should be transferred to CEACO or another center, depending on the severity of the case. If the transfer was necessary, the modality was coordinated, and the medical and nursing team was prepared to receive the patient, assigning a bed according to the severity [23]. The hospital was divided into wards with different levels of complexity. In addition, important data such as the patient's arrival time, name, and ID had to be collected, and laboratory tests and X-rays had to be performed before admission. The cubicles had to be ready to receive the patient, and poor prognostic factors were assessed to adjust the treatment if necessary. The protocol was adjusted according to the clinical evolution and updated evidence on COVID-19 [23].

Government Health Policies and the Unequal Impact on the Private Sector

Government actions focused on reducing the effects of the pandemic, prioritizing public health and the collective interest over the individual. In the health sector, Costa Rican regulations, specifically the General Health Law through its article 386, allowed the Executive Branch to issue a decree that authorized the use of intensive care beds in private centers for the admission of non-COVID-19 patients, allowing the CCSS to free up more than 400 beds by transferring patients with low complexity to shelters, clinics, health areas, and public and private hospitals, as well as by keeping all victims of traffic accidents in the Trauma Hospital of the National Insurance Institute (INS) [24].

On the other hand, the Costa Rican Chamber of Health diagnosed the private sector in the country and pointed out that the role of private laboratories during the COVID-19 pandemic was key, as these centers carried out 2 of the three million total tests carried out in the country [25].

Regarding fiscal and labor regulations, the government allowed the application of tax moratoriums on the payment of value-added tax (VAT) and income tax. The reduction of working hours and the temporary suspension of employment contracts were also permitted, which offered companies legal tools to adjust to the reduction in economic activity [26]. Likewise, support was provided to strategic sectors; for example, resources were allocated to the cultural, agricultural, and manufacturing sectors. The *Alivio Program* invested 5.6 million dollars in 200 small- and medium-sized enterprises (SMEs), ensuring operational continuity amid the crisis. In addition, mechanisms were established to guarantee essential services, such as drinking water and electricity access, especially in vulnerable communities [27].

Additionally, the Legislative Assembly approved international loans and redirected institutional surpluses to guarantee the government's liquidity in managing the crisis. These actions supported social programs such as pensions for impoverished older adults and subsidies for vendors affected by the pandemic. At the same time, public universities contributed with technological innovations, such as producing respirators and masks, demonstrating the value of an inter-institutional response in times of crisis [27].

As the COVID-19 pandemic progressed, challenges arose in various areas and sectors in Costa Rica, which sometimes generated tensions between the private sector and the government [28]. In the economic and social spheres, the pandemic highlighted inequalities exacerbated by the implementation of measures taken by the government to mitigate the spread of COVID-19 infections [27]. Rural communities and peripheral regions faced a disproportionate impact during the pandemic; precarious conditions on agricultural plantations, characterized by overcrowding and lack of access to essential services, facilitated the spread of the virus; the limited institutional capacity to monitor compliance with labor and health regulations exacerbated these inequalities [28].

On the other hand, the application of the protocols generated by the government caused the temporary closure of companies that did not comply with the measures, as occurred in pineapple and banana packing plants and other plantations in the northern part of the country, where significant outbreaks were detected; one of the main problems arose from the pressures exerted by business chambers, private associations, and small- and medium-sized entrepreneurs who demanded the economic reopening [28]. This resulted in a conflict between collective well-being and economic priorities; the pressures exerted by these sectors increased the risks associated with COVID-19 as the state strategy for managing the health crisis was made more flexible in favor of economic opening [28].

These measures also significantly impacted the decline in employment, which was unevenly distributed across economic sectors. At the same time, the education sector was better able to adapt to teleworking and activities such as accommodation and food, which faced drastically reduced demand. This situation reflected the Costa Rican labor market's structural difficulties in adapting quickly to economic shocks [29]. Pineapple monoculture in rural areas, for example, reflected the disproportionate impact of the crisis on vulnerable sectors, exacerbating problems such as overcrowding, poverty, and lack of access to essential services [28].

On the other hand, the regions furthest from the country's center lacked access to technological infrastructure and educational services, which affected education and adaptation to teleworking. According to data from the State of the Nation 2020 [3], while 67% of students in the Greater Metropolitan Area (GMA) had access to the internet at home, only 40% of students in regions such as Brunca had this possibility. This made implementing remote educational and work strategies complex during the pandemic, deepening existing gaps [28].

Digitalization and teleworking were promoted during the pandemic, especially in the public sector, where it reached 51.19% of employees in 2020. However, its adoption in the private sector was limited (10%), reflecting inequalities in access to technology and digital skills. This adjustment made combining mobility restrictions with job continuity possible, although unevenly between sectors and regions. Regions with less access to technological infrastructure lagged in adapting to new labor and educational dynamics, putting students and workers in these areas at a disadvantage [29].

Although the government demonstrated the ability to mobilize resources and coordinate a comprehensive response, the pandemic highlighted the country's structural limitations. The need to strengthen mechanisms for dialogue and cooperation between sectors is evident to ensure a more effective response in future crises. In addition, the pandemic underlined the importance of reducing territorial and digital inequalities. Investing in technological infrastructure and strengthening institutional capacity to monitor compliance with labor rights and access to health are key priorities to promote an inclusive recovery [29].

Protests, Discontent, and Other Community Responses to the Pandemic

During the first year of the pandemic, the project "Community participation for people-centered health in the context of the COVID-19 pandemic" was carried out, proposed by the PAHO Office in Costa Rica jointly with the Mixed Institute of Social Assistance (IMAS), and promoted by the Social Protection Roundtable, the Ministry of Health, the CCSS, and the CNE [30].

According to the University of Costa Rica (2020), this initiative aimed to promote community networks, both formal and informal, based on shared responsibility for controlling the pandemic, develop empowerment processes from local governments and Municipal Emergency Committees, and support community surveillance and risk communication capacities [31].

However, despite the restrictions of social distancing and confinement, there was the presence of different social movements around three themes related to (1) health conditions, (2) working conditions, and (3) the economic situation in response to the different measures (health, social, economic) taken by the government and the consequences on the functioning of the national economy [32].

Some of the protests were by community sectors and groups of students, mothers, and fathers, whose claims focused on the lack of water in marginalized urban communities. Other groups, together with unions, requested the temporary closure of institutions as a measure to contain the spread of the disease.

> In contrast, in rural areas of the country, specifically in tourist attraction areas, groups of neighbors carried out direct actions (road blockades) to stop the arrival of tourists, mainly from the Greater Metropolitan Area, and thus prevent the spread of COVID-19 in their communities. [32]

Public sector workers also questioned health measures such as restrictions on movement, given the effects on mainly informal work associated with the immediate service economy. Unemployed people recorded actions demanding government attention to the issue. In addition, requests were made to the government to facilitate payroll cuts and define conditions for non-payment or deferral of taxes during the pandemic [32].

According to the report "A Year of Protests and COVID-19 in Costa Rica" by the Institute of Social Research (IIS) of the UCR, it reported global data from March 2020 to May 2021 showing that 72.2% of the 1542 protest actions recorded were related to the social situation caused by the pandemic [33].

Following the decisions taken by the government, various sectors of Costa Rican society reacted critically to the measures implemented during the pandemic, as evidenced by the demonstration on July 27, 2020, in San José, led by the National Association of Public Employees (ANEP). This protest reflected the discontent of trade unionists, farmers, and workers, who demanded an emergency fund to support approximately one million people economically affected by the crisis, such as the unemployed, workers with reduced hours, or those with suspended contracts [34].

Among the main demands was the creation of an emergency fund by imposing taxes on large fortunes and incomes from high-profit sectors. In addition, the protesters requested the elimination of the value-added tax (VAT) on essential basket products, the strengthening of the CCSS with the payment of outstanding debts by the government, and the approval of laws to combat tax fraud. The movement also rejected government proposals such as reducing working hours in the public sector. It demanded specific measures to support farmers, small- and medium-sized businesses, and sectors severely affected by the restrictions, such as bars, restaurants, and shops [34].

The number of protests in Costa Rica increased significantly during 2020, from 29 in January to 88 in July of the same year. Between March and July, 74% of the demonstrations, 227 in total, were linked, directly or indirectly, to the effects of the pandemic. The main factors behind these protests included the economic crisis, social inequality, and a growing distrust in the legitimacy of government policies. This was reflected in a drastic drop in the president's approval rating, which reached just 15% in November 2020 [35].

Another particular element in the protests during the pandemic was some people's skepticism about the pandemic's existence. However, not all protests represent an explicit rejection of the existence of the pandemic; many of them were associated with the discontent of the sectors most economically affected by the health restrictions imposed by government institutions. This is also combined with the influence of pre-existing beliefs in certain groups, considered in some analyses as subscription to conspiracy theories, distrust in science, or adherence to pseudoscience ideas [35, 36].

At a demonstration held on July 13, 2020, in front of the Presidential House, a group of around 30 people expressed skepticism about the pandemic, stating that the virus was not as severe as claimed and that the measures imposed were part of an alleged "setup" intended to control the population. This event gave rise to various debates on social media, polarizing opinions; on the one hand, those who supported the health measures appealed to collective solidarity and accused skeptics of putting public health at risk, and on the other, those who rejected them and argued that they were excessive, damaged the economy, and represented an abuse of the government. This exchange of opinions highlighted the underlying social tensions and the diversity of perceptions regarding the health crisis, showing how the pandemic not only had an impact on health and the economy, but also on the social cohesion of the country [35].

However, in November 2020, the Center for Political Research and Studies (CIEP) of the UCR conducted its periodic sociopolitical study. The results showed that, despite the multiple difficulties generated by the pandemic, the protests and social tensions originated from the crisis. The actions to contain the virus were perceived as necessary and effective, so the general support of the Costa Rican population toward the measures implemented by the government reflected a high degree of confidence in public health institutions. According to the data collected, more than 70% of the people interviewed expressed their support for government decisions to confront COVID-19, reaching approximately 94% approval, with the Ministry of Health and the CCSS standing out, especially as the best-evaluated entities, with 8.99 and 8.54, respectively [37].

The positive assessment of the government can also be explained by the institutional response to the health emergency; the coordination between institutions such as the Ministry of Health, the CCSS, the Ministry of Security, and other key entities allowed confidence to be generated in the management of the crisis, which contrasts with the protest movements and demands of specific affected sectors and suggests that the perception of the measures varied significantly between different groups in society [37].

Consequences of the Response to COVID-19 and Widening Social and Economic Inequalities in Costa Rica

The national impact of the crisis unleashed by the pandemic was not minor. Despite the various containment measures implemented by the government the intersectoral coordination for a joint approach, there were direct impacts on people, their families, and communities at the economic, labor and occupational, social, mobility, and protection of people in the most vulnerable conditions levels, among others, which added to the discontent of certain groups of the population.

> Several significant events have had a powerful impact on the country's course and have created muscular social and political tensions. It is worth highlighting the outbreak of a severe crisis in public finances, the health, production, social, and environmental shock of the COVID-19 pandemic, and a profound crisis of political representation of citizens. [4]

The UCR CIEP was surveyed in November 2020, and 48% of the people surveyed reported having been affected by job loss. This segregated data show that 19% of those who participated in the survey suffered job loss directly, while 29% attributed the impact to a member of their household; by sex, men between 35 and 54 years old and women over 55 years old experienced the greatest direct impact on their families [38].

In an analysis of the impact by province, job losses were highest in Guanacaste, Puntarenas, and Limón. Although 69% of citizens still supported the health measures, and there was an increase in the assessment of economic actions (43%), the administration of then-President Carlos Alvarado received only 15% positive opinions, placing him as the leader with the least support since 2018, while negative public opinion regarding the country's economic situation remained latent, with 87% of people considering the national outlook as bad [38].

Such discontent was linked to unemployment and the government's handling of social tensions. However, social protection based on principles of freedom, peace, and protection of biodiversity, as well as a widespread rejection of anti-democracy, allowed citizens to continue to lean toward dialogue and negotiation to face the challenges and demonstrations of October 2020, which allowed Costa Rican democracy to remain strong [38].

According to García-Marín and García Fernández (2022) [39], from a social perspective, a link has been assessed between the characteristics of the distribution of contagion and death from COVID-19 with a possible inequitable social structure and persistent economic disparities, as well as other gender, territorial, age, health, labor and socio-occupational disparities, beyond the role of health risk factors that was part of the predominant discourse. The authors, citing the Economic Commission for Latin America and the Caribbean (ECLAC), report that mobility restrictions exacerbated the economic crisis, unemployment, and poverty, especially in countries such as Latin America, whose economy depends mainly on tourism. Unemployment in Costa Rica rose from 12% to more than 24% [40].

At the national level, the government implemented a local program for the distribution of food and supplies and another aimed at job reintegration; however, there was also an orientation toward business stabilization, with the State providing training to unemployed workers so that they could be inserted into the companies that demanded it [41].

At the labor level, some government measures affected the working class's historical achievements, such as the suspension of wage increases and the flexibility of payments of employers' debts to the Costa Rican Social Security Fund [41]. Salary reductions of up to 50% were also allowed for workers in companies that showed losses. These actions reflect an approach that, although aimed at mitigating the crisis, favored business sectors to the detriment of labor rights [41].

The health measures in response to the COVID-19 pandemic caused a 5% reduction in economic output, equivalent to a loss of USD 3 billion; women and low-wage workers were the most affected. Poverty in the country also increased considerably, with extreme poverty rising from 5.7% to 7% in 2020. Before the pandemic, Costa Rica had made progress in the fight against climate change with policies such as the National Adaptation Policy and the National Decarbonization Plan, which aim to achieve carbon neutrality by 2050. However, it is now necessary to explore possible recovery trajectories and the effects of an investment in decarbonization on employment and economic activity to overcome the impacts of the pandemic [40].

The annual survey *Actualidades 2021* by the UCR showed that the population valued citizen security with a score of 6.2, while job creation, the fight against poverty, corruption, and the economic crisis were rated with lower grades, between 4.1 and 4.7; the general average rating on the handling of the country's problems was 5.2, which reflects a negative perception of government management (Universidad de Costa Rica, 2021). When segmenting the results, it was found that people over 50 and foreigners most valued the health measures, while women aged 30 to 49 and Costa Ricans gave the worst ratings. The proportion between those who consider poor management and those who support the government's actions was almost equal, with 23.5% negative and 29.2% positive opinions [42].

A study on mental health in times of COVID-19 in Costa Rica conducted by the Universidad Estatal a Distancia (UNED) and the Universidad Nacional (UNA) in 2020, with support from the Ministry of Health and the CCSS, revealed a significant increase in mental health disorders in the Costa Rican population due to the pandemic [43]. Sixty-one percent of participants showed symptoms of depression, an increase of 50% compared to March 2020; 43.7% presented severe symptoms of anxiety, and more than half of the population reported low resilience, indicating that some coping strategies were not effective. Emotional reactions such as fear of contracting COVID-19 increased from 53.8% in March to 88.3% in October. The mental impact was more significant in women, especially between 36 and 40 years old, and in people with a lower educational level, single, divorced, and unemployed people. People with greater job stability reported less affected mental health [43].

Regarding physical impact, 74.6% of the population reported feeling more tired, 67.8% experienced muscle pain, and 49.4% suffered from headaches. In addition, 45% reported symptoms such as a racing heart and shortness of breath. In cognitive terms, most participants had trouble sleeping (91.3%) and difficulty relaxing or switching off (90.7%); they also experienced concentration problems (77%) and frequent forgetfulness (71.4%). Forty-two percent increased their consumption of substances such as caffeine, nicotine, and alcohol, and 25% engaged in dangerous actions such as reckless driving. Emotionally, 85.3% of those affected experienced mood swings and high irritability (69.6%), while 70.4% felt they could not handle the situations they faced. It was found that generalized anxiety affected 43.7% of the population, and the highest prevalence was found among private sector workers (53.8%) and informal workers (78.5%), respectively [43].

The COVID-19 pandemic also brought uncertainty and multiple changes in daily life, limiting social and leisure activities. As the pandemic progressed, measures became more specific and adapted to the most affected areas, with more substantial restrictions in cantons that had a high impact [44]. The most significant closures occurred during Holy Week, Father's Day, and July when an "epidemiological fence" was implemented in the GMA. Although there were moments of severe restrictions, these were limited, representing only 17% of the days, and for the most part, the measures were less strict after those initial moments [44].

Ethical Considerations on the Response to the COVID-19 Pandemic in Costa Rica: Lessons and Challenges

Ethical analysis of the response to COVID-19 in any population involves estimating the magnitude of the risks to which some of these measures could have exposed those who undertook them. The growing recognition that the SARS-CoV-2 virus could be a virus modified in the laboratory using "gain of function" procedures [45, 46] raises questions about its purpose and the circumstances of its release.

There is now growing doubt as to whether lockdowns were effective in delaying the spread of the virus, and there is also growing fear that they may have caused a range of deleterious health consequences for those affected, which could even have resulted in their death [47, 48]. The radical lockdown adopted in many countries isolated those who needed help, did not allow treatment to be followed, increased contagion in overcrowded homes, left informal workers without income, and caused damage to the mental health of many [49]. It is also now clear that there was no evidence of the usefulness of masks in preventing the user's acquisition of infection; they served to prevent transmission of the user's infection to others, and their use decreased the concentration of oxygen that the person breathed; they are only considered helpful in hospital settings [50, 51].

Due to the decision to apply mRNA (Pfizer, Moderna) or recombinant DNA (Astra Zeneca, Janssen) vaccines to the entire population, the possibility of comparing the effectiveness and safety of the vaccine between those who received it and those who received placebo were lost; and in the last three years, a growing number of reports have appeared recognizing the side effects of these products, in the face of effectiveness that was also declining [52, 53]. Regarding global vaccination against COVID-19, it has been argued that the use of non-sterilizing vaccines (i.e., incapable of eliminating the virus) such as these during the development of an epidemic eventually favors the natural selection of versions of the virus resistant to the immune response generated by said vaccine [54].

The ethical analysis of what happened in any country in response to the COVID-19 crisis is highly complex for several reasons: First, because of the significant role that global actors would have had in promoting the measures adopted in such a homogeneous way at the country level throughout the world, which would imply a re-revaluation of the responsibility of each national government. The response to COVID-19 in Costa Rica was similar to that in other countries [55], simply changing actors, population sizes, and geography. This finding turns out to be an important discovery since it turns the discussion on the ethics of the response in Costa Rica into a particular version of the discussion on the ethics of the global response, differentiating between the responsibility of the supranational actors (who would have imposed a response model) and that of the countries—which to a greater or lesser extent would have acted with submission, without a critical attitude, out of fear and thinking about access to vaccines, supposedly "doing what science indicated"; although here it should also be highlighted how the dynamics of hegemonic scientific production centered on a few countries contributes to a kind of scientific colonialism. Thus, the conduct of supranational actors requires a specific ethical analysis.

Second, since the emergence of COVID-19, the nature of SARS-CoV-2 was controversial, as many disputed its natural origin, posing instead that it was the result of genetic engineering ("gain of function") [45, 46]. Furthermore, it has transpired that the publication in which an elite of scientists attacked the hypothesis of a non-natural origin of the virus was published at the behest of the Director of the NIAID and that those scientists had originally stated that said hypothesis was probably true [56]. Responding to a natural pandemic resulting from a zoonosis would not be the same as responding to one produced by an altered virus, as the latter scenario opened the possibility of an epidemic resulting from the purposeful release of a bioweapon—rather than from the fortuitous escape of a virus engineered with benevolent intentions. If that possibility existed, global authorities were ethically required to conduct a serious investigation, and both global and local authorities should have taken that hypothesis into account when implementing control measures. However, they not only failed to do so, but the hypothesis of a non-natural origin was ridiculed.

Third, the fact that panic prevailed [57] created a state of emergency, with many rules suspended and many deaths occurring. While assessing the response may lead to identifying errors that could eventually be considered the cause of a more significant number of deaths, it will be prudent to remember that the attribution of responsibility is problematic and to concentrate such inquiries on probable cases of intent.

Fourthly, the ethical analysis of the response to COVID-19—which would generally correspond to a response within the framework of public health—implies an ethical analysis of the operating bases of public health. Here is a complex ethical dilemma: As Bayer and Fairchild (2004) [58] said, public health differs significantly from clinical medicine due to its strongly paternalistic features, affecting people's privacy and freedom.

Public health proposes measures that in a given situation would be necessary to achieve the collective good, and that would have to be adopted by everyone, which is why they are presented as necessary measures and, usually, as mandatory measures; the authority that public health claims for this are assumed to be derived from the best scientific evidence available. However, science does not produce absolute or definitive truths; all scientific truth is unstable and temporary, which is why public health imposes the assumption of a collective risk—the risk that the promoted measure, based on what is at that time the "scientific truth," will not be effective, when new evidence is discovered. In such a situation, such a measure would be harmless in the best case, and in the worst case, it would be harmful. In the case of the response to COVID-19, there are some aggravating factors since the measures promoted were never considered the "state of the art" from a scientific point of view:

- Measures such as the lockdown and the use of masks, whose effectiveness was not proven, were imposed as "necessary" measures and were presented simply as state of the art [49]. However, when the measures were terminated, it was known that problems had occurred, and the balance of costs and benefits was uncertain.
- Active efforts were reportedly made to discredit the use of well-known drugs, for which there was growing evidence of effectiveness; this is now explained by the fact that, for the emergency authorization of new vaccines, there should not be any drug that could be considered effective [59].
- The formula that vaccines were safe and effective was accepted without question, despite early reports of ineffectiveness and adverse effects; and it was also known that the studies had been stopped early, without being able to collect information on longer-term effectiveness; anyone who established a different perspective was censored, labeled as "anti-vaccine," and its use was established as almost obligatory, even though in some places users were made to sign a "voluntary informed consent" [60].

The Costa Rica case follows a pattern familiar to most Latin American countries. In mid-March 2020, the declaration of a pandemic by the World Health Organization was received and accepted, immediately entering a moment of significant influence by the latter and its regional office, the Pan American Health Organization, for the reception of recommendations and the entry into mechanisms for access to vaccines. The response of all social sectors (government, private sector, community) was marked by fear, constantly reinforced by the media; local media would echo international media and social networks to consolidate a single discourse [37].

Implementing a major public health intervention—as occurred in Costa Rica—is an important decision that should be made after careful and rigorous analysis, with the most extensive possible basis in evidence suggesting that such an intervention will be effective and safe in the various sectors of the population. However, when the media and social networks generate panic in public opinion, demands for immediate action can prevent such analysis. Governments are compelled to "do something" as soon as possible without waiting for the necessary studies and debates to be carried out. Although the media and governments presented health policies as the result of "Science" [61], in reality, in many cases, not all the protocols of the scientific method were respected.

For implementing a health intervention to be considered appropriate, its positive effects must outweigh its adverse effects. For example, the decision to confine the Costa Rican population should have been evaluated according to studies or reviews of evidence on its possible adverse effects on society. However, it was adopted simply because it was a WHO recommendation, supposedly based on evidence and maintained for long periods. In contrast, the frightened population seemed to accept the confinements as a common-sense measure to prevent the transmission of the virus. Little was known in the first half of 2020 about the multiple negative consequences of the measure.

When the media justified the lockdown by saying that it was a matter of choosing between life and the economy ("lives *vs.* the economy"), implying that many lives would inevitably be lost if the measure was not implemented—something that was never confirmed [62]—it was assumed that "the economy" was simply a matter of financial figures that only matter to some politicians and managers of large companies, and by separating "life" from "the economy" it was made to forget that the economy is part of life, that it is necessary to live; sadly it was forgotten that impoverishment could also cause deterioration of health and death. Finally, by presenting the economy as the only thing that could seriously question the lockdown, the multiple social, cultural, emotional, and, in short, human aspects that would be negatively affected were excluded from the public debate, thus going against general well-being and not only against economic stability.

Not only were critical voices that attempted to present their analyses from the health sciences ignored or excluded [61], but studies or analyses from the human sciences that could have shed some light on the effects of health policies on specific and local population groups were practically absent. On the contrary, the "government of microbiology" prevailed [63], and the social sciences came to occupy a very marginal (if not wholly non-existent) place in advising governments on decision making. This marginalization was particularly regrettable in low- and middle-income countries such as Costa Rica, since measures defined in other countries with social, political, economic, and cultural situations very different from local ones were imposed there. Thus, it could have been predicted that European countries with a better standard of living and greater possibilities for teleworking, for example, were better adapted and prepared for mobility restrictions than countries with a greater number of households characterized by overcrowding and poverty.

In any case, the lockdown was not implemented based on evidence of its usefulness in Costa Rica, and there was no evidence of its usefulness elsewhere either. In the end, its implementation in Costa Rica does not seem to have stopped the spread of the epidemic. Therefore, it would be concluded that said implementation was not based on technical criteria and did not show any usefulness, which should be considered for future pandemics.

However, some people claimed that such considerations were impossible due to the coronavirus's novel nature and that the situation's urgency did not allow for the usual development of studies. Such claims are hazardous and must be discussed now.

The sociopolitical importance of the state of exception has been studied notably by Agamben (2004) [64]. The state of exception (a general concept that in countries such as Costa Rica was called a "state of emergency") is justified by stating that there is an exceptional, unique situation that prevents us from using the models we have previously constructed to understand the world and respond to it. When a government declares a state of exception (as occurred in 2020), it acquires unrestricted powers and faculties; it suspends constitutional rights and citizen oversight mechanisms created to prevent extreme drift and abuses by rulers [64].

Consequently, the declaration of a state of exception is a powerful and dangerous weapon of governmentality since those who, at a given moment, decide that we are facing an exceptional event and manage to convince the population of this have absolute power beyond the limits of law and ethics. Therefore, declaring that we are facing an exceptional situation is a political act that should always be subject to broad debate. Was Sars-Cov-2 such a lethal virus that it merited the restriction of freedoms and the impossibility of disagreeing [61]? Likewise, it is worth remembering that the situations or events humans must respond to are always different from what happened previously ("No one bathes twice in the same river," said Heraclitus). From the law (which is based on a thousand-year-old tradition of ethical reflections), it has been established that the decisions made by judges in previous cases must be used to evaluate new cases. Of course, no case is identical to another.

In the case of this pandemic, governments such as Costa Rica claimed to rely on scientific studies to affirm that we were facing an exceptional situation that warranted declaring a national emergency and, therefore, taking exceptional measures. The State legitimized its actions by presenting them as a "scientific approach to the emergency" [9]. Therefore, we must remain vigilant regarding the political uses of scientism since, as occurred with this pandemic, uncertainty is inherent to science, and naturally, controversies abound within the scientific community. Therefore, part of the lessons of this crisis is that there must be greater transparency regarding the link that exists between science and power so that citizens can judge whether a situation merits the declaration of a state of exception and whether we accept its consequences even if it goes against our ethical principles.

Conclusion

An ethical analysis of the response to the COVID-19 pandemic in Costa Rica cannot be separated from ethical considerations about the technical advice that came from international agencies, which had a notable influence worldwide. It also affects the problems generated by the centralization of scientific production in some countries and the tendency to assume supranational recommendations without a minimum of technical considerations of their own, which was accentuated by the prevailing fear. Finally, the COVID-19 epidemic shows the temporary and unstable nature of scientific truth, which forces us to normalize the possibility of disagreement in the context of science and also to remember that health interventions must take into account with sufficient care the possibility of evidence emerging contrary to that which guides specific measures, always alerting the public about such uncertainty, and always respecting the right of people to dissent in an informed manner from specific interventions, based on their judgment about what may or may not be helpful to them.

References

1. World Health Organization (2024) COVID-19 has wiped out a decade of progress in life expectancy worldwide. Available on: https://www.who.int/en/news/item/24-05-2024-covid-19-eliminated-a-decade-of-progress-in-global-level-of-life-expectancy. Accessed 3 Feb 2025
2. United Nations (n.d.) COVID-19 deaths are expected to reach 15 million between 2020 and 2021. Available on: https://www.un.org/es/desa/las-muertes-por-covid-19-sumar%C3%ADan-15-millones-entre-2020-y-2021. Accessed 3 Feb 2025
3. State of the Nation Program (2020) State of the Nation Report 2020. Available on: https://estadonacion.or.cr/wp-content/uploads/2020/11/RESUMEN-EN-26-2020-WEB.pdf. Accessed 3 Feb 2025
4. National Council of Rectors (2024) State of the Nation Report 2024. Available on: https://estadonacion.or.cr/?informes=informe-estado-de-la-nacion-2024. Accessed 23 Jan 2025
5. Ministry of Health of Costa Rica (2020a) First suspected case of COVID-19 in Costa Rica. Ministry of Public Health of Costa Rica. Available on: https://www.ministeriodesalud.go.cr/index.php/prensa/42-noticias-2020/397-primer-caso-sospechoso-por-covid-19-en-costa-rica. Accessed 23 Jan 2025
6. Judicial Branch of Costa Rica (2020) Laws and Regulations: No. 90737. Judicial Branch of Costa Rica. Available on: http://www.pgrweb.go.cr/scij/Busqueda/Normativa/Normas/nrm_texto_completo.aspx?nValor1=1&nValor2=90737. Accessed 23 Jan 2025
7. United Nations Population Fund (UNFPA) (2020) COVID-19 in Costa Rica. United Nations Population Fund. Available on: https://costarica.unfpa.org/es/covid-19. Accessed 23 Jan 2025
8. State University at a Distance (2020) Draft Costa Rica: 5-month COVID-19 situation. *State University at a Distance (UNED)*. Available on: https://www.uned.ac.cr/ocex/index.php/capsulas/592-borrador-costa-rica-estado-de-situacion-de-5-meses-de-covid-19. Accessed 23 Jan 2025
9. Ministry of Health of Costa Rica (2020) Two months since the first case of COVID-19 in Costa Rica: 60 days of a scientific approach to the emergency. Ministry of Public Health of Costa Rica. Available on: https://www.ministeriodesalud.go.cr/index.php/prensa/42-noticias-2020/494-dos-meses-desde-el-primer-caso-de-covid-19-en-costa-rica-60-dias-de-un-abordaje-cientifico-de-la-emergencia. Accessed 23 Jan 2025
10. National Council of Rectors (n.d.) Health restrictions. State of the Nation Report Costa Rica. Available on: https://estadonacion.or.cr/restricciones-sanitarias/. Accessed 23 Jan 2025
11. Government of Costa Rica (2020) COVID-19 Pandemic. Government of Costa Rica. Available on: https://gobiernocarlosalvarado.cr/pandemia-covid-19/. Accessed 23 Jan 2025
12. Delfino.cr (2020) Costa Rica confirms its first imported case of COVID-19. Delfino.cr. Available on: https://delfino.cr/2020/03/costa-rica-confirma-su-primer-caso-importado-de-covid-19. Accessed 23 Jan 2025
13. Pérez-Laurrabaquio O, Rizo-Amezquita JN (2022) The response of American social security systems to the COVID-19 pandemic, assessment more than a year after the emergency: the case of Costa Rica. Inter-American Conference on Social Security. Available on: https://ciss-bienestar.org/wp-content/uploads/2022/02/cuadernoCovid2CostaRica.pdf. Accessed 22 Jan 2025
14. Brooks D (2020) Coronavirus in Costa Rica: What is the effective formula in the Latin American country where fewer COVID-19 patients die? BBC Mundo. Available on: https://www.bbc.com/mundo/noticias-america-latina-52480615. Accessed 22 Jan 2025
15. Our World in Data (2024) Total deaths and confirmed cases of COVID-19 per million people. Available on: https://ourworldindata.org/grapher/total-covid-cases-deaths-per-million?tab=table&time=2020-01-04..2024-08-31. Accessed 22 Jan 2025
16. University of Costa Rica (2023) The pandemic altered the number and causes of deaths in Costa Rica. Available on: https://www.ucr.ac.cr/noticias/2023/11/27/la-pandemia-altero-el-numero-y-las-causas-de-muertes-en-costa-rica.html. Accessed 22 Jan 2025

17. Ministry of Health of Costa Rica (2020d) Costa Rica joins a mechanism to obtain the COVID-19 vaccine for more than one million inhabitants. Available on: https://www.ministeriodesalud.go.cr/index.php/prensa/42-noticias-2020/727-costa-rica-se-adhiere-a-mecanismo-para-obtener-vacuna-contra-covid-19-para-mas-de-un-millon-de-habitantes. Accessed 22 Jan 2025
18. Costa Rican Social Security Fund (2023) Campaign against COVID-19. Available on: https://www.s.sa.cr/web/coronavirus/vacunacion. Accessed 22 Jan 2025
19. Alvarado Quesada Administration (2020) COVID-19 Pandemic. Available on: https://gobiernocarlosalvarado.cr/pandemia-covid-19/. Accessed 22 Jan 2025
20. Alvarado DM (2023) Impact of complete vaccination against COVID-19 on fatality rates and deaths per million inhabitants in America as of 06/05/2022. Tecnología en Marcha 36(5):96–106. Available on: https://revistas.tec.ac.cr/index.php/tec_marcha/article/view/6498/6498. Accessed 22 Jan 2025
21. Singh P et al (2024) Impact of COVID-19 vaccination: a global perspective. Front Public Health 11:1272961. Available on: https://www.frontiersin.org/journals/public-health/articles/10.3389/fpubh.2023.1272961/full. Accessed 22 Jan 2025
22. Barrantes-Arroyo A, Cortés-Ruiz A, Molina-Granados JM et al (2022) Impact on primary health care services by the COVID-19 pandemic in Costa Rica: a count of the interruptions. Health Soc Secur Manage 1(1):e185. Available on: https://doi.org/10.62999/gestion.v1i1.185. Accessed 22 Jan 2025
23. Costa Rican Social Security Fund (2021) COVID-19 management protocol for the CEACO internal medicine service. Available on: https://www.cendeisss.sa.cr/wp/wp-content/uploads/2021/09/Actualizacio_n-de-nuevo-protocolo-CEACO-VERSION-4.pdf. Accessed 22 Jan 2025
24. Ministry of Health of Costa Rica (2020e) Government promotes 5 areas of action to strengthen pandemic care. Available on: https://www.ministeriodesalud.go.cr/index.php/prensa/43-noticias-2021/1031-gobierno-impulsa-5-areas-de-acciones-para-fortalecer-la-atencion-de-la-pandemia
25. El Mundo CR (2024) Diagnosis shows the positive impact of the private sector on the national health system. Available on https://elmundo.cr/costa-rica/diagnostico-muestra-el-impacto-positive-del-sector-privado-en-el-sistema-nacional-de-salud/#google_vignette
26. Martínez Franzoni J, Sánchez-Ancochea D (2022) COVID-19 response policies in Costa Rica. Available on https://repositorio.cepal.org/server/api/core/bitstreams/a78d487a-b9c7-4f50-9e33-39a324441784/content
27. Chavarría Chinchilla SD (2021) Government management during the COVID-19 pandemic in Costa Rica. Available on https://repositorio.cepal.org/server/api/core/bitstreams/a78d487a-b9c7-4f50-9e33-39a324441784/content
28. Vargas Solís LP (2022) Working conditions and occupational health in agricultural plantations. Available on: https://cicde.uned.ac.cr/images/cicde/recursos/Crisis_de_la_democracia_en_los_tiempos_del_COVID.pdf. Accessed 22 Jan 2025
29. Ulate A, Hall B, Segura B, Pastrana G, Araya A (2020) COVID-19 Pandemic: a push for teleworking in Costa Rica. Available on: https://ecodatos.fce.ucr.ac.cr/s/xzsHpa6gn3m2aMn. Accessed 22 Jan 2025
30. Pan American Health Organization (2020) PAHO and IMAS promote a pilot plan for community participation in the face of the pandemic. Pan American Health Organization. Available on: https://www.paho.org/en/news/10-19-2020-paho-and-imas-promote-pilot-plan-community-participation-in-the-face-of-the-pandemic. Accessed 23 Jan 2025
31. University of Costa Rica (2020a) Expert voice: social protest in Costa Rica in times of COVID-19. University of Costa Rica. Available on: https://www.ucr.ac.cr/noticias/2020/6/01/voz-experta-la-protesta-social-en-costa-rica-en-tiempos-del-covid-19.html. Accessed 23 Jan 2025
32. University Seminar (2020) Social protest moves to the rhythm of pandemic restrictions. University Seminar. Available on: https://semanariouniversidad.com/pais/protesta-social-se-mueve-al-ritmo-de-las-restricciones-de-la-pandemia/. Accessed 22 Jan 2025

33. Institute for Social Research (2021) A year of protests and COVID-19 in Costa Rica: partial results report (January–May 2021). Available on: https://www.kerwa.ucr.ac.cr/items/2830b13c-1104-4679-8ed2-f6ec5d2f3c73/full. Accessed 22 Jan 2025
34. Deutsche Welle (DW) (2020). Costa Rica: government measures protested due to the pandemic. Available on: https://www.dw.com/en/unions-protesting-in-costa-rica-against-government-economic-measures-during-pandemic/a-54341864. Accessed 22 Jan 2025
35. Cambronero Quesada M (2020) «Which pandemic?»: experiences of people skeptical of COVID-19 in Alajuela, Costa Rica. Reflex Mag 101(1):1–21. https://doi.org/10.15517/rr.v101i1.46531. Accessed 22 Jan 2025
36. Sarrazin JP, Cáceres CF (2025) Disputes over the figures of the COVID-19 pandemic: epistemic diversity, dissemination of science, and political opposition. Sociol Health Illn 47(1):e13833. https://doi.org/10.1111/1467-9566.13833
37. Center for Political Research and Studies (CIEP) (2020) Report on the results of the sociopolitical opinion study. November 2020. University of Costa Rica. Available on: https://ciep.ucr.ac.cr/estudios-deopinion/. Accessed 22 Jan 2025
38. University of Costa Rica (2020b). Costa Ricans prioritize everyday problems in the face of COVID-19. Available on: https://www.ucr.ac.cr/noticias/2020/11/11/costarricenses-priorizan-los-problemas-cotidianos-frente-al-covid-19/imprimir.html?empotrar=true&ruta=https%3A%2F%2Fwww.ucr.ac.cr%2Fnoticias%2F2020%2F11%2F11%2Fcostarricenses-priorizan-los-problemas-cotidianos-fr. Accessed 22 Jan 2025
39. García-Marín A, García Fernández R (2022) COVID-19 in Costa Rica: a longitudinal analysis of infection and death rates from a social inequality perspective. Popul Health Mesoam 20(1):244–280. Available on: https://www.scielo.sa.cr/scielo.php?script=sci_arttext&pid=S1659-02012022000200244. Accessed 22 Jan 2025
40. Groves DG, Molina-Perez E, Syme J, Alvarado G, Denegri FDL, Román JDA, Rojas AJ (2022) A green recovery from COVID-19 in Costa Rica. Available on: https://cambioclimatico.minae.go.cr/wp-content/uploads/2022/05/PNUD-Una_recuperacion_verde_del_COVID-19_en_Costa-Rica-Espanol.pdf. Accessed 22 Jan 2025
41. Fallas-Jiménez YI (2023) State actions to confront COVID-19 in Costa Rica. Reflexiones Magazine, 103. Available on: https://doi.org/10.15517/rr.v102i00.54350. Accessed 22 Jan 2025
42. University of Costa Rica (2021) Costa Ricans are happy with handling the pandemic and dissatisfied with employment, corruption, and the economy. Available on: https://www.ucr.ac.cr/noticias/2021/12/22/costarricenses-a-gusto-con-el-manejo-de-la-pandemia-e-insatisfechos-con-empleo-corrupcion-y-economia.html. Accessed 22 Jan 2025
43. National University of Costa Rica (2020) Depression and anxiety increased due to the pandemic. Available on: https://www.unacomunica.una.ac.cr/index.php/dicembre-2020/3297-aumentan-depresion-y-ansiedad-ante-la-pandemia. Accessed 22 Jan 2025
44. Gómez-Campos S, Aguilar-Gómez C (2023) Were the health measures very restrictive during the pandemic? State of the Nation. Available on: https://estadonacion.or.cr/restricciones-sanitarias/. Accessed 22 Jan 2025
45. Looi M (2024) Will we ever know where covid-19 came from? BMJ 386:q1578. https://doi.org/10.1136/bmj.q1578
46. Honderich H (2025) CIA says lab leak most likely source of Covid outbreak. BBC, 25 January. Available on: https://www.bbc.com/news/articles/cd9qjjj4zy5o. Accessed 7 Feb 2025
47. Joffe AR (2021) COVID-19: rethinking the lockdown groupthink. Front Public Health 9. Available on: https://www.frontiersin.org/articles/10.3389/fpubh.2021.625778/full. Accessed 7 Feb 2025
48. John S (2020) The ethics of lockdown: communication, consequences, and the separateness of persons. Kennedy Inst Ethics J 30(3):265–289. https://doi.org/10.1353/ken.2020.0015
49. Bardosh K (2023) How did the COVID-19 pandemic response harm society? A global evaluation and state of knowledge review (2020-21). Available on: https://ssrn.com/abstract=4447806 or https://doi.org/10.2139/ssrn.4447806. Accessed 7 Feb 2025
50. Elisheva R (2020) Adverse effects of prolonged mask use among healthcare professionals during COVID-19 [Details missing—check source for journal/book information

51. Intemann K, de Melo-Martín I (2023) On masks and masking: epistemic harms and science communication. Synthese 202:93. https://doi.org/10.1007/s11229-023-04322-z
52. Rhodes P, Parry P (2024) Pharmaceutical product recall and educated hesitancy towards new drugs and novel vaccines. Int J Risk Safe Med 35(4):317–333. https://doi.org/10.1177/09246479241292008
53. Faksova K et al (2024) COVID-19 vaccines and adverse events of special interest: a multinational Global Vaccine Data Network (GVDN) cohort study of 99 million vaccinated individuals. Vaccine 42(9):2200–2211. https://doi.org/10.1016/j.vaccine.2024.01.100
54. Raszek M, Cowley D, Rubio-Casillas A (2024) Considerations for improving future pandemic responses. J Vacc Immunol 10(1):001–005. https://doi.org/10.17352/jvi.000059
55. Tabari P, Amini M, Moghadami M, Moosavi M (2020) International public health responses to COVID-19 outbreak: a rapid review. Iran J Med Sci 45(3):157–169. https://doi.org/10.30476/ijms.2020.85810.1537
56. Moffit R (2023, updated 2024) How Fauci and NIH Leaders Worked to Discredit COVID-19 Lab Leak Theory. The Heritage Foundation, 18 July. Available at: https://www.heritage.org/public-health/commentary/how-fauci-and-nih-leaders-worked-discredit-COVID-19-lab-leak-theory. Accessed 7 Feb 2025
57. Luqmani YA, El Hashim A (2022) The COVID-19 pandemic: a health crisis managed or a panic response with disastrous future consequences? Med Princ Pract 31(1):1–10. https://doi.org/10.1159/000520258
58. Bayer R, Fairchild A (2004) The genesis of public health ethics. Bioethics 18(6):473–492. https://doi.org/10.1111/j.1467-8519.2004.00412.x
59. Cáceres CF (2022) Unresolved COVID-19 controversies: "Normal science" and potential non-scientific influences. Glob Public Health 17(4):622–640. https://doi.org/10.1080/17441692.2022.2036219
60. Blaylock R (2022) COVID-19 Update: What is the truth? Surg Neurol Int 13:167. Available on: https://surgicalneurologyint.com/surgicalint-articles/COVID-19-19-update-what-is-the-truth/. Accessed 7 Feb 2025
61. Sarrazin JP, Aguirre F (2023) Governmentality, science and the media: examining the "Pandemic reality" with Foucault, Lyotard Baudrillard. Foucault Stud 35:21–45. https://doi.org/10.22439/fs.i35.7073
62. Bhattacharya J, Packalen M (2020) Lives vs. lives—the global cost of lockdown. The Spectator, 16 May. Available on: https://www.spectator.co.uk/article/lives-vs-lives-the-global-cost-of-lockdown. Accessed 25 Apr 2021
63. Basile G (2020) The government of microbiology in the response to Sars-Cov-2. Salud Problema 14(27):14–35
64. Agamben G (2004) State of exception. Homo Sacer II, 1. Pre-textos, Valencia

Mexico's COVID-19 Response: An Analysis of Policies, Practices, and Ethical Considerations

12

Mauricio Torres-Martínez, Diego Sepúlveda-González, Carlos A. Gonzalez-Assad, Miguel A. Walwyn-Hernández, Francisco E. Herrera-Carrillo, Ricardo J. Estrada-Mendizabal, Constantine S. Psimopoulos, and Diego Ramonfaur

Abstract

This chapter explores the ethical dimensions of Mexico's COVID-19 response from November 2019 to December 2024. Using Beauchamp and Childress's principles of biomedical ethics as a guide, we analyze the interplay between public health policies, practices, and societal impacts. Key interventions, including vaccination strategies, travel restrictions, mask mandates, hospital management, and public communication campaigns are evaluated not only for their effectiveness but also for their ethical implications concerning respect for

M. Torres-Martínez
Department of International Health, Johns Hopkins Bloomberg School of Public Health, Baltimore, MD, USA

D. Sepúlveda-González · C. A. Gonzalez-Assad · M. A. Walwyn-Hernández
Tecnológico de Monterrey, Escuela de Medicina y Ciencias de la Salud, Monterrey, Nuevo León, Mexico

F. E. Herrera-Carrillo
Department of Pediatrics, Boston Children's Hospital, Harvard Medical School, Boston, MA, USA

R. J. Estrada-Mendizabal
Division of Allergy, Asthma, and Clinical Immunology, Mayo Clinic, Scottsdale, AZ, USA

C. S. Psimopoulos
Department of Epidemiology, Harvard T.H. Chan School of Public Health, Boston, MA, USA

Department of Global Health and Social Medicine, Harvard Medical School, Boston, MA, USA

Human Flourishing Program, Harvard University, Cambridge, MA, USA

D. Ramonfaur (✉)
Department of Internal Medicine, Cleveland Clinic, Cleveland, OH, USA

L. P. Marquez et al. (eds.), *COVID-19 in the Pacific Rim and Beyond*, Collaborative Bioethics 6, https://doi.org/10.1007/978-3-032-21218-4_12

autonomy of the patient as a person, beneficence, nonmaleficence and distributive justice. We explore how government actions, private-sector contributions, and community engagement collectively shaped the pandemic trajectory, while highlighting the influence of cultural and socioeconomic factors. By reflecting on both successes and shortcomings, the chapter provides normative insights and adaptive recommendations for creating equitable, ethically sound and informed frameworks that enhance preparedness for future public health emergencies.

Keywords

Mexico · COVID-19 · Emergency response and preparedness · Biomedical ethics · Beauchamp and Childress

Introduction

The coronavirus disease 2019 (COVID-19) pandemic was a defining moment for public health, challenging the resilience of healthcare systems worldwide. As the severe acute respiratory syndrome coronavirus 2 (SARS-COV-2) spread rapidly across borders, governments were confronted with the daunting dual responsibility of containing its transmission while addressing the immediate and long-term health needs of their populations [1]. These challenges were particularly acute in developing countries, where the rapid surge in cases unmasked the longstanding structural weaknesses in health infrastructure. In this context, the inadequacies of public health infrastructure were exacerbated, leading to devastating consequences [2]. This was particularly evident in Latin America where countries were profoundly affected by COVID-19, bearing a disproportionate share of its human and societal costs.

Mexico's COVID-19 experience illustrates the devastating impact of the pandemic on nations with systemic vulnerabilities. Although representing only 1.6% of the global population, Mexico accounted for 7% of global COVID-19 deaths and had the second highest case-fatality rate at 4.5% [3]. Additionally, the country ranked ninth globally in COVID-19 mortality, recording 260 deaths per 100,000 population [3]. Beyond estimates, these figures tell the story of a healthcare system pushed to its limits, compounded by a heavy burden of chronic diseases such as diabetes, obesity, and hypertension [4]. Furthermore, they underscore the influence of socioeconomic disparities that deepened the crisis for underserved populations, where access to adequate care and resources was often insufficient. Against this challenging backdrop, Mexico's public health response was characterized by both notable efforts and significant policy missteps that diverged from international best practices and often contradicted domestic guidelines [5, 6].

For reference, this chapter includes comparative figures that pinpoint Mexico alongside a carefully selected group of countries and benchmarks: the Organization for Economic Cooperation and Development (OECD) average, the United States, Croatia, Chile, and Malaysia. The latter three countries were chosen because they share a comparable Gross Domestic Product (GDP) per capita ($15,564; $14,496, and $10,920 USD, respectively) with Mexico ($10,370 USD), creating a more equitable basis for evaluating demographic baselines, healthcare spending, and subsequent outcomes [7]. These nations provide insights into how countries with similar economic resources addressed the shared challenges of COVID-19.

The OECD's average serves as a valuable benchmark, offering a lens through which to evaluate Mexico's pandemic outcomes against broader international standards of healthcare performance and socioeconomic context. Similarly, while the United States does not share a similar GDP per capita with Mexico, its inclusion remains critical. As Mexico's largest trading partner with deeply intertwined labor markets, significant cross-border mobility, and extensive bilateral cooperation in health and security, the United States represents a relevant comparator. Moreover, its proximity and influence often shape public perception and policy discourse within its southern neighboring country, making it an essential frame of reference for understanding Mexico's relative outcomes and challenges. Together, these comparisons offer a comprehensive and multidimensional perspective, balancing the global, regional, and economic contexts that shaped Mexico's pandemic experience.

This chapter examines the complexities of Mexico's response to the COVID-19 pandemic, unraveling the policies, practices, and societal implications that shaped the country's public health trajectory. By adopting the framework of Beauchamp and Childress [8], which employs the bioethical principles of respect for autonomy of the patient as person, beneficence, nonmaleficence, and distributive justice, the analysis delves into the ethical dimensions of key public health interventions. These include the implementation of quarantines, the enforcement and eventual lifting of mask mandates, and the development and rollout of vaccination strategies. Each of these measures carried significant ethical and societal ramifications, influencing not only public health outcomes but also the trust and subsequent behavior of Mexico's population during the pandemic.

Background

Mexico, the 15th largest country by land area and the 10th most populous nation globally, is home to over 127 million people [9]. This vast and diverse population contends with deep-rooted socioeconomic disparities that heavily influence access to and the quality of healthcare. In 2020, approximately 56 million people, 44% of the country's population, lived in poverty, while 11 million, 8.5% of the population, were classified as living in extreme poverty [10]. These inequities, rooted in long-standing structural imbalances, result in stark differences in healthcare access, with wealthier citizens enjoying far greater opportunities for quality care compared to their impoverished counterparts. Figure 12.1 shows the distribution of healthcare affiliation across Mexico in 2020, illustrating that 27% of the population lacked any healthcare coverage at the start of the pandemic. (Note: The figure's percentages exceed 100% because many individuals in Mexico are affiliated with multiple healthcare systems.)

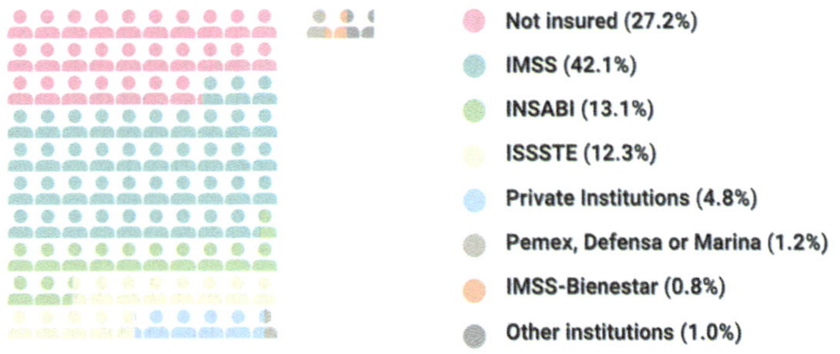

Fig. 12.1 Different healthcare affiliations in Mexico, 2020

Healthcare in Mexico is delivered through a fragmented system that offers varying levels of access and quality depending on employment status and income [11, 12]. Coverage options include private insurance, employer-based social security programs, and publicly funded alternatives for those outside the formal labor market. For over two decades, the Seguro Popular program served as a crucial safety net for uninsured citizens. However, in January 2020, Seguro Popular was replaced by the National Institute of Health for Well-Being (INSABI), an initiative introduced by President Andrés Manuel López Obrador to centralize healthcare under federal control [5]. Although this reform was intended to address longstanding inequalities, its implementation faced significant challenges, ultimately leaving millions without coverage. Following the transition, enrollment in INSABI dropped by 16.8% compared to the levels achieved under Seguro Popular [13], and by 2020, only 13.1% of the population was enrolled in INSABI [10], while 27.2% remained uninsured (see Fig. 12.1) [11].

The fragmentation of Mexico's healthcare system was further complicated by chronic underfunding. In 2019, the country allocated just 2.8% of its GDP to healthcare, the lowest proportion among OECD countries [5]. Figure 12.2 compares health expenditure per capita and as a percentage of GDP between Mexico and selected countries, illustrating the limited financial resources dedicated to healthcare in Mexico relative to its peers [14]. This underinvestment left significant gaps in medical infrastructure and human resources [15]. As shown in Fig. 12.3, Mexico lagged far behind the OECD average, with only 1 hospital bed, 2.5 doctors, and 2.9 nurses per 1000 population, compared to 4.4 beds, 3.6 doctors, and 8.8 nurses per 1000 population in OECD countries [14, 16, 17].

These challenges were compounded by austerity measures implemented in the years leading up to the pandemic. In 2019, Mexico's Ministry of Health reduced healthcare staffing by 30% as part of broader cost-cutting policies [5]. By 2020, per capita health spending decreased by 26.4% compared to 2015 levels [18]. This erosion of healthcare capacity left the system ill-equipped to handle the demands of the COVID-19 pandemic.

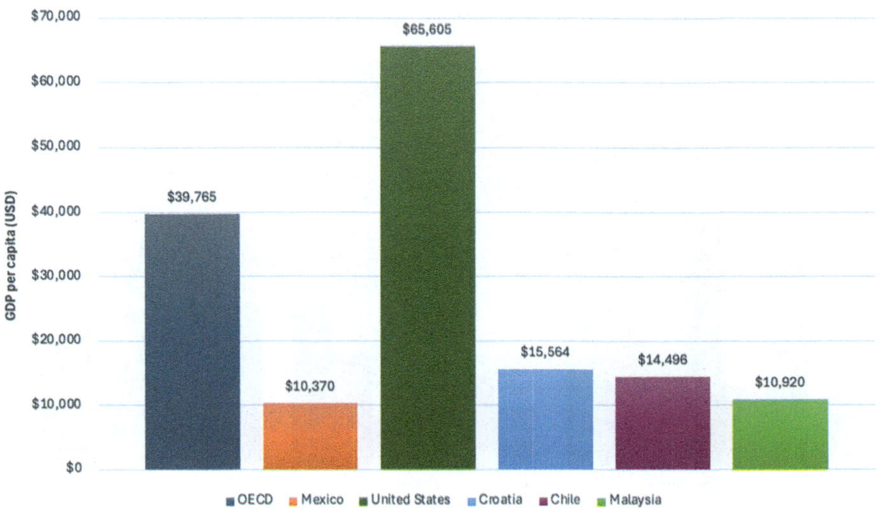

Fig. 12.2 Countries' GDP per capita

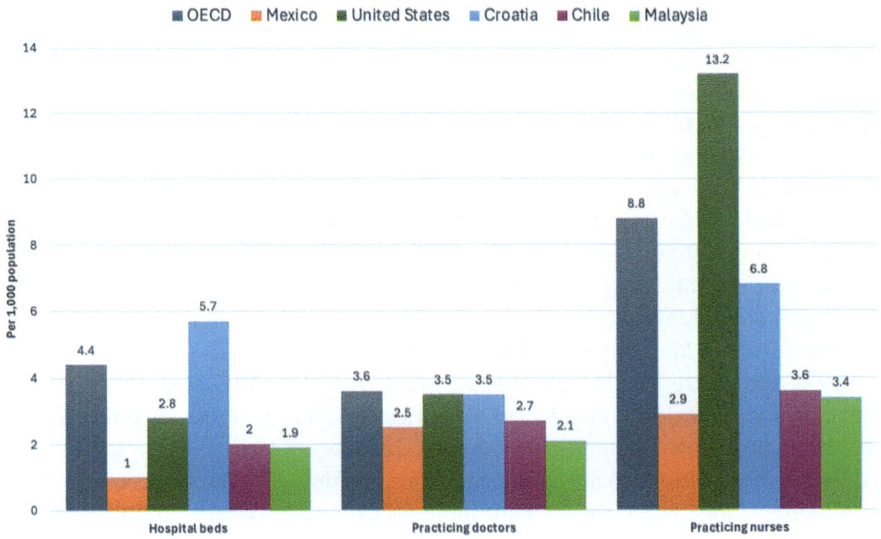

Fig. 12.3 Comparative health system performance across selected countries

Resource distribution in Mexico reflects broader urban-rural disparities, further limiting access for vulnerable populations. Urban centers like Mexico City and Nuevo León benefit from disproportionately higher concentrations of medical personnel, while rural states remain critically underserved. For example, in 2020, half of the country's 700 pulmonologists were concentrated in Mexico City, leaving 24 of Mexico's 32 states with fewer than 10 pulmonologists [5]. Similarly, 56% of

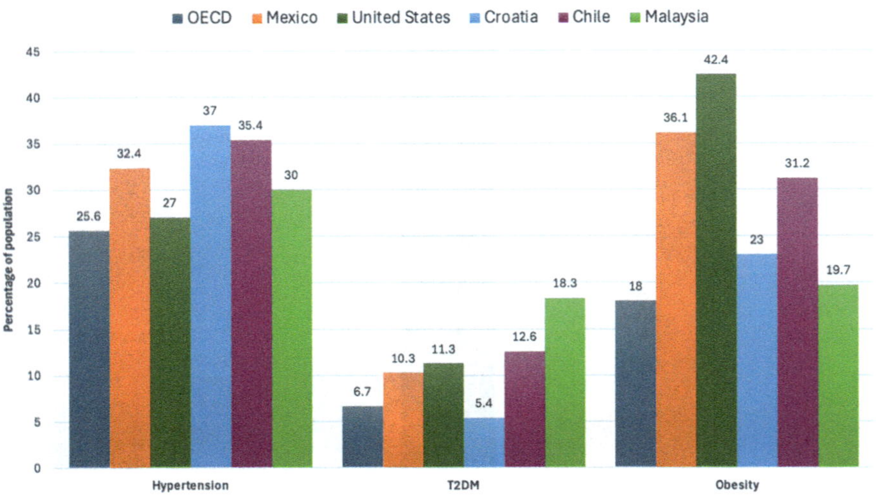

Fig. 12.4 Baseline demographic and comorbidity profiles across selected countries

critical care specialists were based in Mexico City, with 15 states having fewer than 10 specialists each [5].

Mexico's health system also faced the additional burden of addressing high rates of chronic diseases. As of 2020, 24.9% of men and 26.1% of women had hypertension [19], while 15.7% of the population had diabetes mellitus [20]. Figure 12.4 compares baseline demographic and comorbidity profiles across Mexico and selected countries, emphasizing the prevalence of hypertension [21–25], type 2 diabetes mellitus [26–31], and obesity [32–36], which significantly contributed to severe COVID-19 outcomes [16, 17]. Overweight and obesity rates were particularly concerning, affecting 31.5% of men and 40.2% of women [37]. These underlying health conditions compounded the strain on an already fragile system, leading to Mexico's high mortality rates during the pandemic.

To contextualize Mexico's healthcare challenges, Fig. 12.5 compares GDP per capita among Mexico and selected countries, offering a frame of reference for understanding how economic constraints influenced healthcare systems [38]. Although these nations share similar economic profiles with Mexico, countries such as Croatia, Chile, and Malaysia built more robust health systems, highlighting the importance of strategic investments in healthcare infrastructure and resource allocation.

Timeline of Mexico's COVID-19 Response

The first reports of SARS-CoV-2 emerged in December 2019 from Wuhan, China, marking the beginning of the global health crisis [39]. Mexico confirmed its first case on February 27, 2020, a citizen returning from Italy with mild symptoms [40].

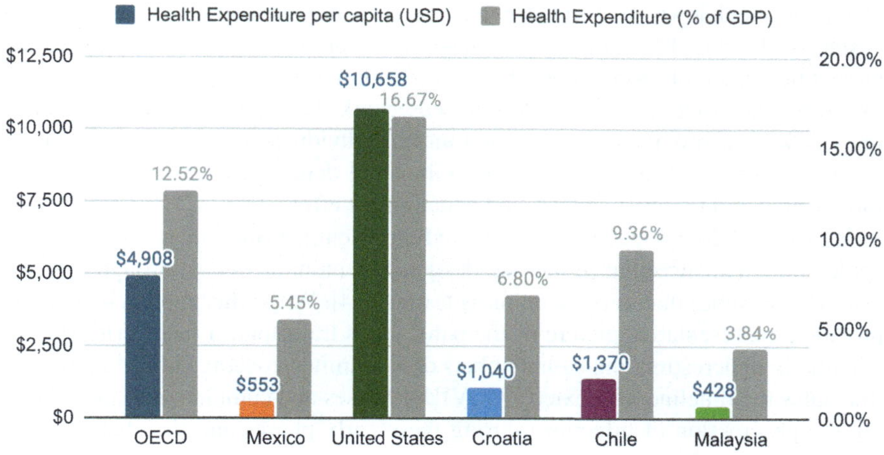

Fig. 12.5 Health expenditure per capita and as a percentage of GDP across selected countries

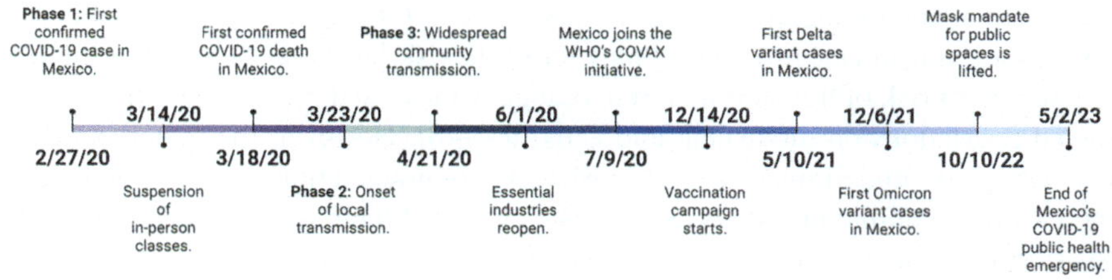

Fig. 12.6 Key milestones in Mexico's COVID-19 public health emergency response

By the following day, two additional cases linked to international travel were reported [5, 41]. Despite observing the devastating impact of the virus in countries like China and Italy, Mexican authorities delayed implementing significant preventive measures [42]. Early guidance from public health officials remained advisory, lacking the enforcement necessary to curb the virus's spread.

As cases began to rise, the private sector took a proactive role in mitigating the pandemic's impact. On March 14, 2020, Mexico's largest private university, Tecnológico de Monterrey, suspended in-person classes nationwide, setting a precedent for educational institutions across the country [43]. Just four days later, Mexico reported its first COVID-19-related death, with confirmed cases rising to 118 [41]. The federal government introduced a three-phase framework to manage the pandemic: the importation phase, the person-to-person transmission phase, and the community transmission phase [44]. Figure 12.6 provides an overview of the public health emergency's key moments, including the declaration of these phases and major policy shifts, illustrating how the country's response evolved over time.

The onset of local transmission was marked by Phase 2, which was declared on March 23, 2020 [43]. The government launched the "National Healthy Distancing Protocol" (Jornadas Nacionales de Sana Distancia), which implemented a nationwide lockdown, suspended non-essential activities, and closed schools and businesses [5, 45]. The campaign emphasized social distancing, hand hygiene, and protecting vulnerable populations. However, compliance varied, as many citizens in lower-income groups could not afford to cease work. Testing capacity also remained limited, with efforts focusing on symptomatic individuals and their close contacts [5]. Challenges such as logistical constraints and delays in obtaining testing kits further hindered timely diagnosis and containment efforts [5].

By April 2020, the government adopted the "Sentinel Surveillance System," an epidemiological surveillance method designed to monitor trends through selected health units rather than relying on mass testing [44]. While this approach aimed to provide a representative picture of the pandemic's trajectory, it faced criticism for potentially underestimating case numbers due to limited coverage [46]. Figure 12.7 illustrates the timeline of Mexico's COVID-19 cases and cumulative deaths, showing the progression of infections during these early phases and the challenges of suppressing the virus's spread [47].

The widespread community transmission was acknowledged by Phase 3, which was declared on April 21, 2020 [44]. To manage this phase, the government introduced the Epidemiological Spotlight System (*Semáforo Epidemiológico*), a color-coded mechanism categorizing regions into four risk levels, in order from highest risk to lowest risk of transmission: red, orange, yellow, and green. These tiers determined restrictions on movement and activities [48]. However, the system's effectiveness was undermined by inconsistent communication, enforcement and increasing economic pressures. Some states lowered their alert levels prematurely to prioritize economic recovery, leading to further transmission [44]. Political alignment with the federal government also influenced enforcement, eroding public trust in the system's objectivity [49].

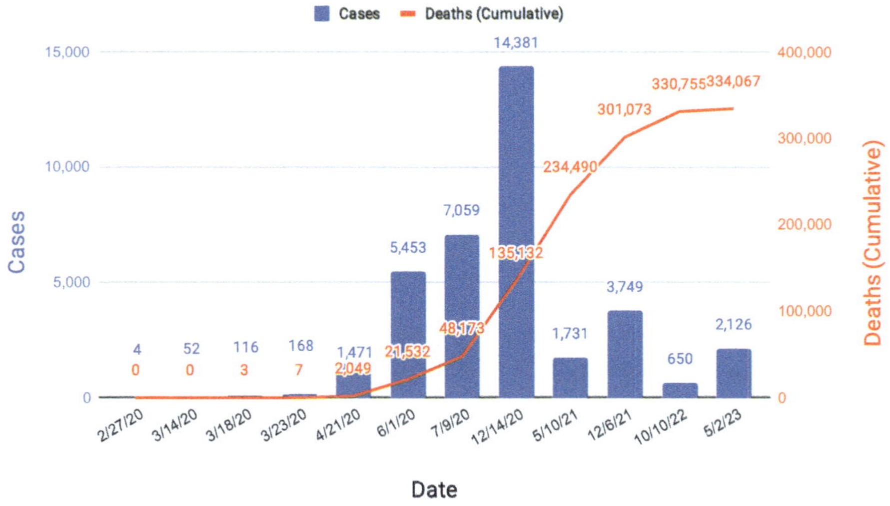

Fig. 12.7 Timeline of COVID-19 cases and deaths in Mexico

In June 2020, the federal government began easing restrictions to address mounting economic concerns despite escalating cases and limited healthcare resources [44]. Essential industries such as mining, construction, and manufacturing were prioritized for reopening, sparking debates over the balance between economic recovery and public health [50]. During this period, hospitals underwent significant reconversion to accommodate the surge in COVID-19 patients, expanding intensive care units and repurposing facilities [51]. Yet, these efforts were hampered by shortages of medical personnel and equipment, including oxygen tanks and other critical supplies [52].

Mexico's engagement with international organizations marked a turning point in its pandemic response. On September 7, 2020, the country joined the World Health Organization's COVAX initiative to secure equitable vaccine access [53]. By December 14, 2020, the vaccination campaign commenced, prioritizing healthcare workers and high-risk populations [44]. The government procured vaccines from multiple sources, including Pfizer-BioNTech, AstraZeneca, Sinovac, and Sputnik V, to diversify supply and mitigate shortages [54]. However, logistical challenges and vaccine hesitancy slowed initial progress, leaving Mexico behind countries like Chile in vaccination rates [5]. Figure 12.8 illustrates the progression of Mexico's vaccination coverage throughout the pandemic [55].

In early 2021 new challenges emerged with the spread of the Delta variant, which strained the Mexican healthcare system further [56]. Vaccination efforts intensified during this period, accompanied by public campaigns to combat misinformation and vaccine hesitancy. Collaborations with community leaders and public service announcements aimed to rebuild trust in the vaccination program. By August 2021 schools began reopening under hybrid models, combining in-person and remote learning with sanitation protocols in place to minimize risks [57].

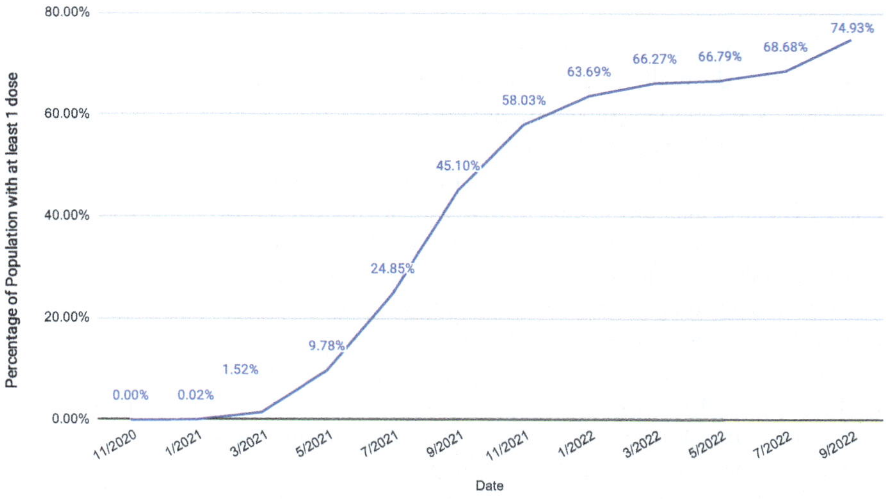

Fig. 12.8 Vaccination coverage in Mexico between 2020 and 2024

The Omicron variant brought additional challenges in late 2021, prompting the government to launch booster vaccination campaigns [58]. These efforts targeted healthcare workers, the elderly, and immunocompromised individuals to bolster immunity against the highly transmissible strain. Gradually, vaccination rates improved, and by March 2022, all states reached the green tier in the epidemiological spotlight system [59]. This milestone reduced transmission risks and the easing of restrictions nationwide.

On October 10, 2022, mask mandates for public spaces were lifted as transmission risks declined further [60]. Finally, on May 4, 2023, President Andrés Manuel López Obrador officially declared the end of Mexico's COVID-19 public health emergency [5, 41].

Bioethical Analysis

The COVID-19 pandemic introduced complex ethical challenges in public health decision-making. Beauchamp and Childress' 8th edition of the seminal work *Principles of Biomedical Ethics*, originally developed in 1979 to address clinical ethical dilemmas, offers a robust framework for navigating these issues [8]. Drawing from common morality and the study of moral philosophy, and centered on the four core principles of autonomy (respecting individual decision-making), beneficence (promoting well-being), nonmaleficence (avoiding harm), and justice (ensuring fairness), this framework has since evolved to guide responses to large-scale public health crises. Doing so by balancing individual rights, a person's own autonomous choices pertaining to their health, along with collective duties and responsibilities. It was notably applied during the HIV/AIDS epidemic, when difficult decisions regarding confidentiality, mandatory testing, and equitable access to treatment highlighted the need to balance individual rights with collective health responsibilities [61].

In examining Mexico's COVID-19 response, this framework provides a valuable lens for assessing public health strategies such as quarantines and lockdowns, mask mandates, vaccination campaigns, and hospital reconversion efforts. By emphasizing the balance between individual rights and collective welfare, it helps evaluate whether these interventions were ethically justified, proportionate, and equitably implemented. It also draws attention to critical challenges, including vaccine hesitancy, inconsistent communication, and the implementation of policies like the epidemiological spotlight system and travel restrictions. Through this analysis, the inherent ethical compromises of these measures are revealed, particularly regarding their effects on public trust, equity, and community well-being. It has been well documented in the literature by leading bioethicists that the lack of trust originated in the lack of trustworthiness within certain non-white or western communities, pertaining to the clinical trials process of the vaccines themselves [62].

By systematically evaluating how these principles balanced, were upheld, or compromised throughout México's response, this chapter highlights the ethical nuances and complexities embedded in public health interventions. In doing so, it contributes to a deeper understanding of the moral dimensions that must inform public health decision-making, particularly in times of crises, when the tension between protecting public welfare and respecting individual freedoms becomes most pressing.

Quarantines and Lockdowns

During the COVID-19 pandemic Mexico implemented quarantines and lockdowns as key components of its public health strategy to mitigate viral spread. These measures included mandatory stay-at-home orders, restrictions on non-essential activities, and the closure of schools and businesses. The rationale behind these policies stemmed from the highly contagious nature of SARS-CoV-2, the uncertainty of its incubation period (5–12 days) [63], and the potential for asymptomatic carriers to unknowingly spread the virus [64]. These factors rendered quarantines for symptomatic individuals insufficient, necessitating broader measures aligned with international practices.

While these interventions were essential in reducing transmission and alleviating pressure on the healthcare system they also raised significant ethical concerns. Quarantines, by their very nature, restrict individuals' physical liberty and often create a conflict between public health objectives and personal autonomy. In Mexico, these measures were enforced through mandatory orders that were largely effective and supplemented by efforts to encourage voluntary compliance. Daily press conferences known as "*las mañaneras*" were used to communicate detailed risk levels and healthcare capacity data, thereby empowering citizens to make informed decisions and preserving a measure of individual autonomy even amid restrictions [65].

From the perspective of beneficence, quarantines aimed to reduce viral spread, protect vulnerable populations, and provide critical time to bolster healthcare resources. These measures, however, had to be weighed against the inadvertent harms incurred: severe economic repercussions, deteriorating mental health due to prolonged isolation [66], and disrupted educational opportunities [67]. Mitigating these harms through robust support systems would have strengthened adherence to the principle of nonmaleficence by ensuring that measures designed to save lives did not inadvertently exacerbate other vulnerabilities.

Finally, the principle of justice is central to evaluating these policies, as the burdens of lockdowns were not evenly distributed. Economically disadvantaged populations, often employed in precarious jobs and residing in crowded conditions, experienced disproportionate hardship [68]. This inequity underscores the need for targeted interventions, such as direct financial assistance and community-specific support measures, to distribute the burdens and benefits of public health strategies more equitably.

Mask Compliance

Mask compliance during the COVID-19 pandemic in Mexico was marked by significant inconsistencies shaped by political dynamics, logistical challenges, and public perception. Between March and June 2020, national health authorities refrained from recommending face masks as a key preventive measure, a decision that likely delayed their widespread adoption [5]. Concurrently, states governed by the president's party often implemented more lenient restrictions, projecting an impression of control that downplayed the pandemic's severity [18, 49]. This divergence in policy and messaging contributed to uneven mask usage across the country, with compliance varying markedly across regions [49].

As the pandemic progressed, public mask usage gradually increased, rising from 45% in August 2020 to 66% by February 2021 [5]. However, during the critical early months, shortages of personal protective equipment (PPE) hindered adoption, particularly in lower-income areas [5]. Although masks are widely recognized as an effective tool for reducing viral transmission [69], compliance was influenced by a mix of political, social, and logistical factors, underscoring the complexities inherent in achieving uniform public health measures [5].

Ethically, mask compliance raises important questions related to beneficence, autonomy, and justice. In high-risk indoor settings, mask-wearing is often regarded as an ethical obligation under the principle of beneficence, as it protects vulnerable populations, such as the elderly and immunocompromised. Conversely, in lower-risk outdoor settings, except during large gatherings, the ethical imperative is less clear, suggesting that public health recommendations should be tailored to specific contexts to maximize effectiveness while respecting individual freedoms.

The principle of autonomy, which upholds individuals' rights to make personal health decisions, was invoked by some citizens in Mexico to justify non-compliance with mask mandates, framing such policies as infringements on personal liberty [70]. However, in public health emergencies, autonomy may be justifiably curtailed to protect the wider community [71]. Ensuring that citizens understand the collective benefits of mask-wearing through clear, consistent, and science-based communication is essential to foster voluntary compliance while respecting individual rights.

Finally, the principle of justice demands that all citizens be protected equitably, yet the unequal distribution of masks and inconsistent enforcement of mandates in Mexico exposed systemic inequities [70]. Vulnerable populations, including those in crowded living conditions or precarious employment, faced increased risks due to their limited access to protective resources. Addressing these disparities requires targeted interventions, such as subsidizing masks for low-income communities and ensuring uniform enforcement of public health guidelines.

Moreover, misinformation and conflicting government messaging compounded these ethical challenges. Contradictory statements from officials, some endorsing mask use, others questioning its efficacy, undermined public trust and eroded confidence in the overall public health strategy [65]. Effective mask policies, therefore, depend not only on ensuring access to PPE but also on delivering transparent, consistent messaging that aligns with scientific evidence.

Vaccination Strategies

Mexico's COVID-19 vaccination strategy was pivotal in combating the pandemic, yet its significant deviation from published guidelines raised serious ethical concerns. Initially, the guidelines prioritized vaccinating 70% of the population and ensuring 100% coverage for medical personnel, a strategy consistent with utilitarian principles that aim to maximize benefits by protecting frontline workers and those at highest risk. However, the rollout soon shifted focus toward rural areas with low transmission rates. This reallocation conflicts with the utilitarian objective of directing vaccines to areas where they could most effectively reduce severe outcomes and curb transmission.

This shift, while controversial, appeared to reflect a commitment to equity by addressing longstanding healthcare disparities in underserved communities. Rural areas in Mexico have historically faced limited access to medical resources, and prioritizing these regions may be seen as an effort to correct entrenched systemic inequalities. However, the absence of transparent communication regarding the rationale behind this shift undermined public trust. Critics argued that the reallocation of vaccines to rural areas was politically motivated, potentially aimed at garnering support for the president's party in these regions [72]. This perception of favoritism fueled confusion and skepticism regarding the fairness of the vaccination campaign, underscoring the vital role of clear and consistent communication.

Despite these challenges, Mexico implemented several measures to incentivize vaccination. Access to certain public spaces and services was contingent on proof of full vaccination, and international travel requirements further motivated individuals to complete their vaccination requirements [44]. These targeted incentives effectively boosted vaccine uptake. Nevertheless, for such measures to be truly effective, they must be paired with transparent, evidence-based communication to ensure they are perceived as fair and legitimate.

Ultimately, balancing equity and utilitarianism remains a central ethical challenge in public health. While addressing disparities in rural areas is commendable, diverging from established guidelines without clear public justification can erode trust in the vaccination campaign's integrity. Effective public health strategies require not only equitable distribution but also transparent reasoning that helps the public understand the ethical trade-offs involved. Building trust through honest communication is as essential as the policies themselves for the success of such initiatives.

Hospital Reconversion

As COVID-19 cases surged, Mexico's health authorities introduced a "hospital reconversion" strategy to mitigate systemic deficiencies in healthcare capacity [51]. This approach involved designating entire hospitals exclusively for the treatment of COVID-19 patients, while non-COVID patients were redirected to a limited number of facilities [51]. The primary goal was to increase the availability of critical resources, such as ICU beds and ventilators, for patients with severe COVID-19 symptoms [18]. Despite these efforts, the strain on the healthcare system remained immense. Between February and June 2020, 45% of hospitalized COVID-19 patients who died did not receive full support due to a lack of ICU bed availability, underscoring the challenge of meeting overwhelming demand [18].

While the hospital reconversion strategy sought to prioritize COVID-19 patients, it led to unintended consequences that exposed underlying systemic inequities. Healthcare utilization decreased significantly as many patients deferred or avoided care out of fear of contracting COVID-19 in medical facilities [6]. This trend disproportionately affected vulnerable populations with chronic conditions, contributing to a lower in-hospital mortality rate from non-communicable diseases during the pandemic [6]. However, the broader implications were stark: the disruption of routine care highlighted gaps in Mexico's health coverage, a factor that contributed to the country's disproportionately high COVID-19 mortality rate [73].

Further illustrating the pandemic's wider health effects, a study by Palacios-Mejía compared cause-specific mortality during 2020–2021 with that from 2015–2019. COVID-19 emerged as the leading cause of death in Mexico, surpassing ischemic heart disease, while mortality from other conditions, such as diabetes mellitus (up 36.8%), respiratory infections (up 33.3%), and hypertensive diseases (up 25%), also increased significantly [6]. In contrast, some causes of death experienced notable declines, including infectious diseases, skin diseases, non-traffic-related accidents, and malignant neoplasms [6]. These trends reveal how reallocating resources under the reconversion strategy benefited COVID-19 patients but inadvertently exacerbated other health crises.

Viewed through the lens of the bioethical principle of justice, the hospital reconversion strategy represented an effort to direct crucial resources where they were needed most during an unprecedented crisis. By dedicating entire hospitals to COVID-19 care, the government sought to address the needs of the majority. However, achieving justice for all patients proved challenging. The redirection of non-COVID patients to a limited number of facilities created significant barriers to care for those with chronic or acute conditions unrelated to the virus, highlighting the inherent difficulties of resource allocation in a system with limited capacity.

Vaccine Hesitancy

Over the past few decades, vaccine hesitancy has emerged as a focal point for both the healthcare community and the public, a trend that was further intensified during the COVID-19 pandemic. As public discourse surrounding vaccination grew, so did the hesitancy among the concerned population. According to the Strategic Advisory Group of Experts on Immunization (SAGE) Working Group on Vaccine Hesitancy, hesitancy is defined as "a delay in acceptance or refusal of vaccination despite the availability of vaccination services" [74]. This definition highlights the multifaceted and context-specific factors, from individual beliefs to broader societal dynamics, that shape vaccine attitudes.

In Mexico, vaccine hesitancy posed a significant challenge during the pandemic. Data from the National Health and Nutrition Survey (ENSANUT) revealed that 28.2% of Mexicans refused vaccination, while an additional 9.5% expressed hesitancy [75]. Combined with other literature, these figures suggest that nearly one-third of the Mexican population was either unwilling or uncertain about receiving the vaccine [76]. Such widespread hesitancy impeded efforts to achieve herd immunity and highlighted the need for targeted public health interventions.

The principle of autonomy asserts that individuals have the right to self-determination, provided they have access to sufficient information to make informed decisions. However, during the pandemic, this autonomy was heavily influenced by factors such as media narratives, historical and familial influences, perceptions of the healthcare system, geographic disparities, and cultural beliefs. In Mexico, conflicting messages from government officials further complicated public health efforts. Although national health sub-secretary Dr. Hugo López-Gatell provided daily updates intended to inform the public, these briefings became increasingly politicized [65]. President López Obrador's tendency to downplay the severity of the virus and question the importance of vaccination further fueled mistrust, undermining the government's credibility and its capacity to manage the crisis effectively [65].

The interplay between autonomy and vaccine hesitancy presents a profound ethical dilemma. While it is crucial to respect an individual's rights to make personal health decisions, this autonomy must be balanced against the collective need to safeguard public health. In Mexico, misinformation, distrust, and fear often compromised informed decision-making, creating barriers to vaccination uptake. Overcoming these challenges required transparent and consistent communication strategies that prioritized clarity and accuracy. Public health campaigns needed to foster trust by engaging with communities directly, addressing their concerns, and tailoring messages to their specific cultural and social contexts.

In addition to autonomy, the principles of beneficence and justice are essential in addressing vaccine hesitancy. Beneficence demands that public health efforts maximize the benefits of vaccination while minimizing harm, which includes ensuring equitable access to vaccines and actively combating misinformation. Meanwhile, the principle of justice highlights that the burdens of vaccine hesitancy often fall disproportionately on marginalized communities with limited access to reliable information and healthcare resources. Together, these ethical principles underscore the importance of designing public health strategies that not only encourage vaccine uptake but also foster trust and fairness across all segments of the population.

Epidemiological Spotlight System

The Epidemiological Spotlight System (*Semáforo Epidemiológico*) was introduced in Mexico in June 2020 as a strategy to guide the country's public health response and manage restrictions during the COVID-19 pandemic. This four-color coded system communicated the severity of the pandemic in each state and outlined the measures required to mitigate contagion [77]. The colors ranged from:

1. *Green:* Indicating the lowest risk, allowing for minimal restrictions.
2. *Yellow:* Advising precautionary hygiene measures when going out.
3. *Orange:* Suggesting limited movement and the avoidance of non-essential outings.
4. *Red:* Signaling the highest risk and the implementation of strict lockdown measures.

The system relied on a series of key indicators to determine each state's risk level. These included the effective reproduction number, the incidence rate of active cases, mortality rates, hospitalization rates, hospital occupancy (including ventilator-equipped beds), positivity rates, and trends in COVID-19 syndromes and mortality [48]. By incorporating these metrics, the system aimed to provide an objective and data-driven assessment of the pandemic's progression across regions. In a country as geographically and demographically diverse as Mexico, this localized decision-making framework was critical in addressing regional differences in population density, healthcare capacity, and socioeconomic vulnerability.

While the Spotlight System provided structure and clarity to Mexico's pandemic response, it also faced several challenges. One significant issue was the reliance on the reported data from states, which sometimes resulted in inconsistencies or delays in accurately reflecting local conditions [3, 78]. Additionally, the system's dependence on hospital capacity as a key metric highlighted inequities in Mexico's healthcare infrastructure. States with fewer healthcare resources were often flagged as high-risk, even when their case numbers were relatively low. This dynamic perpetuated restrictions in resource-limited regions, exacerbating existing socioeconomic disparities and placing additional burdens on already vulnerable populations [79].

By tailoring restrictions and allocating healthcare resources based on local epidemiological conditions, the system aimed to address the varying needs of different populations. For example, regions experiencing higher infection rates received stricter restrictions and increased access to critical resources, reflecting an attempt to prioritize those at greater risk. However, the challenges of accurately assessing risk and the uneven healthcare infrastructure across states revealed systemic inequities that complicated the system's ability to achieve true bioethical justice.

This inequity underscores the ethical responsibility of public health systems to not only address immediate risks but also to work toward reducing structural disparities that undermine equitable healthcare delivery. While the epidemiological spotlight system represented an innovative and adaptive strategy, its reliance on imperfect data and pre-existing inequalities highlighted the broader need for systemic reforms.

Travel Restrictions

The COVID-19 pandemic prompted countries worldwide to implement various travel restrictions based on their local epidemiological situation. In Mexico, the government advised its population to avoid nonessential international travel, with a primary focus on assisting Mexicans abroad who wished to return while limiting non-essential departures whenever possible [80]. However, these measures were rapidly de-escalated despite rising cases and deaths caused by SARS-CoV-2, which placed Mexico among the top 10 countries with the highest number of cases in 2020 and the top 5 in COVID-related deaths in 2020 and 2021 [81].

Mexico adopted a relatively lenient approach to international travel by keeping its borders open throughout most of the pandemic and refraining from imposing strict quarantine or testing requirements for incoming travelers [81]. This strategy likely reflected the country's economic reliance on its tourism sector, which was particularly vulnerable during the global crisis [82]. While this decision aimed to mitigate the pandemic's economic impact, it raised concerns about its potential contribution to the virus's rapid spread.

From an ethical perspective, Mexico's approach emphasized the principle of autonomy, respecting an individual's right to make personal decisions about travel. However, the relatively lax travel policies posed significant challenges to balancing this autonomy with the principles of beneficence and nonmaleficence, as unrestricted travel increased the risk of viral transmission, endangering public health and potentially exacerbating harm to vulnerable populations.

While the federal government avoided closing international borders, localized travel restrictions were implemented at state and municipal levels to reduce risks and uphold the principles of beneficence and nonmaleficence. Measures included limitations on nonessential travel within certain regions, reduced hotel occupancy, restrictions on guest numbers, and the requirement for travelers to demonstrate engagement in essential activities. These policies were guided by the Epidemiological Spotlight System, which categorized states by their COVID-19 risk level [83].

Airports also played a critical role in monitoring travelers for respiratory symptoms, such as fever, cough, and difficulty breathing. Suspected cases were reported to international health teams under the guidance of Mexico's Ministry of Health, ensuring continuous surveillance at points of entry [80].

Analyzing Mexico's decision-making on travel restrictions reveals a tension between prioritizing individual freedoms and safeguarding public health. While the principle of autonomy was respected, the lack of stringent federal measures may have compromised beneficence and nonmaleficence by allowing increased mobility, which ultimately contributed to the virus's spread. In contrast, local governments attempted to strike a balance by implementing targeted restrictions to protect some of their vulnerable populations.

Communication with the Public

Effective communication emerged as one of the most significant challenges during the COVID-19 pandemic, and in many cases, it was mishandled. In Mexico, Subsecretary of Health Hugo López-Gatell became a central figure in the country's pandemic response, delivering daily press conferences alongside President López Obrador [65]. These briefings aimed to provide guidance and reassure the public, but they often presented contradictory messages that generated confusion. López-Gatell's updates were rooted in data and science, yet they frequently clashed with the president's more relaxed approach to health policies, such as downplaying the severity of the virus or dismissing the use of masks [65]. This disconnection between scientific messaging and political rhetoric raised ethical concerns related to transparency, accountability, and public trust.

The role of social media further compounded these issues. Platforms like Facebook, WhatsApp, and YouTube served both as channels for disseminating critical public health information and as conduits for spreading misinformation [84, 85]. In Mexico, false narratives regarding the virus's origins, vaccine safety, and supposed miracle cures gained traction, even influencing some healthcare professionals [86]. These misleading narratives further eroded public trust, complicating efforts to achieve widespread compliance with public health measures.

In the absence of consistent federal messaging, local news outlets and independent media filled the gaps by reporting the realities unfolding on the ground, such as overwhelmed hospitals and medical supply shortages [87]. However, the lack of centralized, transparent data and a defensive governmental response to criticism made accountability difficult [87]. This patchwork of information intensified public confusion and highlighted the urgent need for clear, consistent, and trustworthy communication during crises.

From an ethical standpoint, clear communication is essential for upholding the bioethical principles of justice, autonomy, beneficence, and nonmaleficence. Justice demands that all citizens, particularly marginalized populations with limited resources, have equitable access to reliable and culturally relevant information. Autonomy relies on individuals receiving accurate, transparent, and timely data to make informed decisions about their health. However, political interference and widespread misinformation compromised these ideals in Mexico. Beneficence requires that the government act in the best interest of public health by communicating effectively, yet inconsistent messaging hindered adherence to measures like mask mandates and vaccination campaigns. Finally, the principle of nonmaleficence compels leaders to avoid actions that foster confusion or mistrust, yet defensiveness and opacity in reporting pandemic data only deepened public uncertainty.

Moving forward, Mexico's experience highlights the urgent need to combat misinformation, rebuild public trust in health institutions, and strike a better balance between government messaging, media reporting, and independent voices. In crises, transparent, consistent, and community-engaged communication is as critical as any policy measure in safeguarding public health.

Healthcare Workforce

The healthcare workers in Mexico were the backbone of the nation's COVID-19 response, managing surging patient volumes amid systemic deficiencies and often at great personal and professional cost. A significant number were employed under temporary or subcontracted arrangements, receiving lower pay and fewer benefits compared to their permanently employed counterparts [88]. Despite facing elevated risks, many did not receive hazard pay or other financial incentives, which exacerbated feelings of frustration and burnout. Furthermore, the government's decision to hire Cuban doctors instead of adequately addressing the needs of its own healthcare workers intensified perceptions of neglect and further eroded morale [89]. Such systemic injustices not only compromised the well-being of these essential workers but also revealed a failure to uphold the ethical principle of justice, which demands fair and equitable treatment for those risking their lives for the public good.

Healthcare workers in Mexico faced disproportionately high risks of COVID-19 infection and death [90]. By the end of 2021, Mexico had one of the highest healthcare worker death rates globally, with over 4000 fatalities and a mortality rate of 10.5 deaths per 1000,000 population (see Fig. 12.9) [91]. Contributing factors included overcrowded facilities, limited infection control measures, and inadequate access to PPE, conditions that violated the ethical principle of nonmaleficence by exposing frontline workers to preventable harm.

The widespread PPE shortage, particularly during the early pandemic, further exacerbated these risks. Numerous reports detailed instances of doctors and nurses reusing masks, improvising protective gear, or even purchasing PPE out-of-pocket due to hospitals' inability to provide adequate supplies [92]. These failures not only undermined the principles of beneficence and justice but also left healthcare

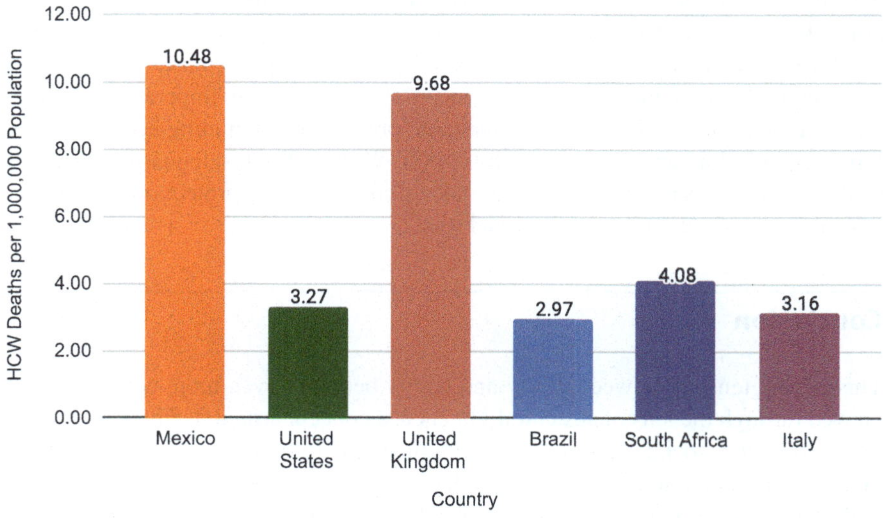

Fig. 12.9 Healthcare worker deaths

workers without the essential tools needed to safeguard their own health and that of their patients.

The psychological toll on healthcare workers during the pandemic was immense. Possible repercussions of pandemic care on healthcare providers include depression, sleep disruption, anxiety, and post-traumatic stress disorder [93]. The systemic lack of mental health support compounded these challenges, leaving many to cope in isolation. In this context, the ethical principle of beneficence was neglected, as the mental health needs of healthcare workers went largely unaddressed.

Medical students and *pasantes del servicio social* (medical interns completing their mandatory community service) also played a significant role in Mexico's pandemic response, particularly in rural and underserved areas. These individuals often worked with limited supervision and inadequate resources, facing significant risks without fair compensation, proper training, or sufficient PPE [94]. While their contributions were invaluable, the principle of justice raises serious concerns about the fairness of placing such a heavy burden on inexperienced individuals, and nonmaleficence underscores the ethical responsibility to protect them from unnecessary harm.

Additionally, workplace violence against healthcare workers, already a persistent issue in Mexico, worsened during the pandemic [95]. Reports of verbal abuse, physical assaults, and threats from frustrated patients and their families became increasingly common, often fueled by misinformation circulated on social media. For example, false narratives alleging that doctors were "killing patients to harvest synovial fluid" further undermined trust and heightened the risk of both psychological and physical harm [96]. This hostile environment compromised the worker's ability to perform their duties safely, undermining the principles of autonomy and nonmaleficence [95].

The pandemic also raised an important ethical question: do healthcare workers have the right to refuse work in unsafe conditions? The principle of autonomy supports their right to prioritize personal safety, while the principle of beneficence emphasizes their moral obligation to care for patients, particularly during crises when their skills are most needed. Balancing these competing imperatives requires that healthcare institutions provide adequate protections, including access to PPE, fair compensation, and mental health support. Without these safeguards, expecting healthcare workers to put their lives at risk without sufficient protection undermines ethical fairness and erodes institutional trust.

Conclusion

This ethical tension between utility and equity becomes even more complex when viewed through the lens of structural violence, a concept articulated by Paul Farmer to describe how social, economic, and political structures systematically disadvantage certain populations. Mexico's pandemic response exposes the need to reckon with the reality of structural violence. Health inequities, whether regarding access to vaccines, hospital beds, or protective equipment are not random or incidental, but foreseeable consequences of longstanding social, economic, and political exclusion. Farmers' call for structural justice and the protection of vulnerable groups provides an essential perspective for understanding why certain communities remain disproportionately burdened by health crises. Recognizing these patterns invites not only critical reflection but also concrete action to address the roots of health inequities, rather than treating their consequences alone.

The ethical dimensions analyzed in this chapter invite what some have called a moral awakening in healthcare: a call for deeper reflection on the values that guide public health action. Inspired by the ethics of sympraxis, which emphasizes ethical action through partnership and shared practice, such an approach challenges systems to move beyond transactional interactions toward genuine relational engagement. Sympraxis asks not simply how policies can be implemented, but how they can be co-created with the people most affected [97]. In this sense, accountability and empathy are not ancillary features of public health, but they are its ethical foundation.

At the same time, the pandemic has reaffirmed the deeply interconnected nature of global health. The most pressing global health challenges of our time, pandemics, climate change, and health inequities, know no borders and demand coordinated and unified responses. While domestic capacity and self-reliance remain a vital component for timely and effective responses, they must be balanced with a commitment to global solidarity. Collaborative frameworks, shared resources, and unified strategies are not just aspirational ideals but essential tools for addressing the complexities of our interdependent world.

The legacy of the COVID-19 pandemic in Mexico and around the world must be one of profound transformation. The collective pain and loss endured by millions should not simply be remembered; they must become a catalyst for bold, inclusive, and ethical public health reforms. This is an opportunity to reimagine systems that not only respond to crises but also actively work to dismantle structural inequities and foster authentic international cooperation. The next global health crisis is not a question of if, but when. The decisions we make now will shape the effectiveness, equity, and resilience of our future responses. Confronting these challenges with unwavering compassion, scientific integrity, and an unyielding commitment to the dignity and well-being of every individual in our communities is not merely a policy choice, but one of the most urgent and shared ethical obligations of all who work in public health.

References

1. Shvetsova O (ed) (2023) Government responses to the COVID-19 pandemic: between a rock and a hard place. Springer International Publishing, Cham
2. Garcia PJ, Alarcón A, Bayer A, Buss P, Guerra G, Ribeiro H, Rojas K, Saenz R, Salgado De Snyder N, Solimano G, Torres R, Tobar S, Tuesca R, Vargas G, Atun R (2020) COVID-19 response in Latin America. Am J Trop Med Hyg 103(5):1765–1772. https://doi.org/10.4269/ajtmh.20-0765

3. Johns Hopkins University Coronavirus Resource Center (2023) Mortality analyses. In: Coronavirus Resour. Cent. https://coronavirus.jhu.edu/data/mortality
4. Secretaría de Salud (2022a) Panorama epidemiológico de las enfermedades no transmisibles en México, CIERRE 2021. México
5. Sánchez-Talanquer M, González-Pier E, Sepúlveda J, Abascal-Miguel L, Fieldhouse J, del Río C, Gallalee S (2021) Mexico's response to Covid-19: a case study | Global Health Sciences. In: UCSF Inst. Glob. Health Sci. https://globalhealthsciences.ucsf.edu/sites/globalhealthsciences.ucsf.edu/files/mexico-covid-19-case-study-english.pdf. Accessed 24 Oct 2023
6. Palacio-Mejía LS, Hernández-Ávila JE, Hernández-Ávila M, Dyer-Leal D, Barranco A, Quezada-Sánchez AD, Alvarez-Aceves M, Cortés-Alcalá R, Fernández- Wheatley JL, Ordoñez-Hernández I, Vielma-Orozco E, Muradás-Troitiño MDLC, Muro-Orozco O, Navarro-Luévano E, Rodriguez-González K, Gabastou JM, López-Ridaura R, López-Gatell H (2022) Leading causes of excess mortality in Mexico during the COVID-19 pandemic 2020–2021: a death certificates study in a middle-income country. Lancet Reg Health Am 13:100303. https://doi.org/10.1016/j.lana.2022.100303
7. World Bank Group (2019) GDP per capita (current US$). In: data.worldbank.org. https://data.worldbank.org/indicator/NY.GDP.PCAP.CD?end=2023&most_recent_year_desc=true&start=2023&utm_source=chatgpt.com&view=map&year=2019
8. Holm S (2002) Principles of biomedical ethics, 5th edn.: Beauchamp T L, Childress J F. Oxford University Press, 2001, £19.95, pp 454. ISBN 0-19-514332-9. J Med Ethics 28(5):332.2–332. https://doi.org/10.1136/jme.28.5.332-a
9. WorldData (2024) The 30 largest and most populous countries. In: Worlddata.info. https://www.worlddata.info/the-largest-countries.php. Accessed 24 Oct 2023
10. CONEVAL (2021) Nota técnica sobre la carencia por acceso a los servicios de salud, 2018–2020
11. Gómez-Dantés O, Sesma S, Becerril VM, Knaul FM, Arreola H, Frenk J (2011) The health system of. Salud Publica Mex 53(Suppl 2):s220–s232
12. González-Block MA, Reyes-Morales H, Cahuana-Hurtado L, Balandrán A, Méndez-Hernández E, Allin S (2021) Health Systems in Transition: Mexico. University of Toronto Press
13. Instituto Nacional de Estadística y Geografía [INEGI] (2021) Encuesta Nacional de Ingresos y Gastos de los Hogares 2020. In: inegi.org.mx. https://www.inegi.org.mx/programas/enigh/nc/2020/. Accessed 31 Jan 2025
14. World Bank Group (2021) Nurses and midwives (per 1,000 people). In: World Bank Open Data. https://data.worldbank.org. Accessed 31 Jan 2025
15. Zamora-Ramos E, Alcántara-Rico D, Cabrera-Sánchez EG, Moreno-Marín TC (2016) Informe sobre la Salud de los Mexicanos 2016, 1st edn. Secretaría de Salud, México
16. OECD, European Observatory on Health Systems and Policies (ed) (2021) Croatia: country health profile 2021. OECD Publishing, Paris
17. Organisation for Economic Co-operation and Development [OECD] (2021) Health at a glance 2021: OECD indicators. OECD
18. Olivas-Martínez A, Cárdenas-Fragoso JL, Jiménez JV, Lozano-Cruz OA, Ortiz-Brizuela E, Tovar-Méndez VH, Medrano-Borromeo C, Martínez-Valenzuela A, Román-Montes CM, Martínez-Guerra B, González-Lara MF, Hernandez-Gilsoul T, Herrero AG, Tamez-Flores KM, Ochoa-Hein E, Ponce-de-León A, Galindo-Fraga A, Kershenobich-Stalnikowitz D, Sifuentes-Osornio J (2021) In-hospital mortality from severe COVID-19 in a tertiary care center in Mexico City; causes of death, risk factors and the impact of hospital saturation. PLoS One 16(2):e0245772. https://doi.org/10.1371/journal.pone.0245772
19. Secretaría de Salud (2023) En México, más de 30 millones de personas padecen hipertensión arterial: Secretaría de Salud. In: gobmx http://www.gob.mx/salud/articulos/en-mexico-mas-de-30-millones-de-personas-padecen-hipertension-arterial-secretaria-de-salud. Accessed 25 Oct 2023
20. Secretaría de Salud (2021a) Sistema de vigilancia epidemiológica hospitalaria de diabetes mellitus TIPO 2 | Boletín de Cierre Anual 2020. Dirección General de Epidemiología, México
21. Eurostat (2021) 22% of people in the EU have high blood pressure

22. Lanas F, Garcia MS, Paccot M, Beaney T, Xia X, Poulter NR (2021) May measurement month 2019: blood pressure screening results from Chile. Eur Heart J Suppl 23(Suppl_B):B40–B42. https://doi.org/10.1093/eurheartj/suab020
23. Palomo-Piñón S, Antonio-Villa NE, García-Cortés LR, Álvarez-Aguilar C, González-Palomo E, Bertadillo-Mendoza OM, Figueroa-Suárez ME, Vargas-Hernández F, Herrera-Olvera IG, Cruz-Toledo JE, Cruz-Arce MA, Serafín-Méndez B, Muñoz-Cortés G, Morfin-Macias CJ (2022) Prevalence and characterization of undiagnosed arterial hypertension in the eastern zone of Mexico. J Clin Hypertens 24(2):131–139. https://doi.org/10.1111/jch.14414
24. Ismail R, Ismail NH, Md-Isa Z, Mohd-Tamil A, Ja'afar MH, Mat Nasir N, Abdul-Razak S, Zainol Abidin N, Ab Razak NH, Joseph P, Yusof KH (2023) Prevalence and factors associated with prehypertension and hypertension among adults: baseline findings of PURE Malaysia cohort study. Am J Med Open 10:100049. https://doi.org/10.1016/j.ajmo.2023.100049
25. Khalid N, Haider S, Hasnat M, Abdullah M, Asghar S, Shamoon Y, Elkattawy S, Vasudev R, Fayez SE (2024) Trends and disparities in prevalence of diagnosed hypertension among U.S. Adults from 2019 to 2022. Curr Probl Cardiol 49(10):102750. https://doi.org/10.1016/j.cpcardiol.2024.102750
26. OECD (2015) Cardiovascular disease and diabetes: policies for better health and quality of care
27. L S-L, A C-C, Ag A, Mn C, Ma E, Lg G-O, M G, A P-S, S San M, A C, Jj G, Iberobdia PG (2019) Diabetes in Latin America: prevalence, complications, and socio-economic impact. Int J Diabetes Clin Res 6(3). https://doi.org/10.23937/2377-3634/1410112
28. Akhtar S, Nasir JA, Ali A, Asghar M, Majeed R, Sarwar A (2022) Prevalence of type-2 diabetes and prediabetes in Malaysia: a systematic review and meta-analysis. PLoS One 17(1):e0263139. https://doi.org/10.1371/journal.pone.0263139
29. Gallardo-Rincón H, Ríos-Blancas MJ, Montoya A, Saucedo-Martínez R, Morales-Juárez L, Mujica R, Cantoral A, Suarez-Idueta L, Lozano R, Tapia-Conyer R (2023) Evaluation of effective coverage for type 2 diabetes in Mexican primary care health information systems: a retrospective registry analysis. Int J Equity Health 22(61). https://doi.org/10.1186/s12939-023-01878-7
30. Centers for Disease Control and Prevention [CDC] (2024) National Diabetes Statistics Report. In: cdc.gov. https://www.cdc.gov/diabetes/php/data-research/index.html#:~:text=During%202001%E2%80%932020%2C%20the%20age,1;%20Appendix%20Table%202
31. Trading Economics (2025) Croatia—diabetes prevalence (% of population ages 20–79). In: tradingeconomics.com. https://tradingeconomics.com/croatia/diabetes-prevalence-percent-of-population-ages-20-to-79-wb-data.html
32. Rodríguez Osiac L, Cofré C, Pizarro T, Mansilla C, Herrera CA, Burrows J, Castillo C (2017) Using evidence-informed policies to tackle overweight and obesity in Chile. Rev Panam Salud Publica:1–5. https://doi.org/10.26633/RPSP.2017.156
33. Barquera S, Rivera JA (2020) Obesity in Mexico: rapid epidemiological transition and food industry interference in health policies. Lancet Diabetes Endocrinol 8(9):746–747. https://doi.org/10.1016/S2213-8587(20)30269-2
34. Mohd-Sidik S, Lekhraj R, Foo CN (2021) Prevalence, associated factors and psychological determinants of obesity among adults in Selangor, Malaysia. Int J Environ Res Public Health 18(3):868. https://doi.org/10.3390/ijerph18030868
35. Restrepo BJ (2022) Obesity prevalence among U.S. adults during the COVID-19 pandemic. Am J Prev Med 63(1):102–106. https://doi.org/10.1016/j.amepre.2022.01.012
36. OECD (2023) Health at a glance 2023: OECD indicators. OECD
37. Shamah-Levy TT, Romero-Martínez M, Barrientos-Gutiérrez T, Cuevas-Nasu L, Bautista-Arredondo S, Colchero MA, Gaona-Pineda EB, Lazcano-Ponce E, Martínez-Barnetche J, Alpuche-Arana C, Rivera-Dommarco J (2021) Encuesta Nacional de Salud y Nutrición 2020 sobre Covid-19. Resultados nacionales. Instituto Nacional de Salud Pública, Cuernavaca, México
38. World Bank Group (2024) Current health expenditure per capita (current US$). In: data.worldbank.org. https://data.worldbank.org/indicator/SH.XPD.CHEX.PC.CD

39. Malik YS, Sircar S, Bhat S, R VO, Tiwari R, Sah R, Rabaan AA, Rodriguez-Morales AJ, Dhama K (2020) Emerging coronavirus disease (COVID-19), a pandemic public health emergency with animal linkages: current status update. Indian J Anim Sci 90(3):303–317. https://doi.org/10.56093/ijans.v90i3.102316
40. Hamui-Sutton L, Arciniega-Martínez SM, Carmona-Zamudio I, Vázquez-de-Anda GF, Rodríguez-Hernández MF (2024) In retrospect: intensivists and COVID-19 pandemic on day zero. An analysis of communication exchanges on WhatsApp. Horiz Med Lima 24(1). https://doi.org/10.24265/horizmed.2024.v24n1.03
41. Suárez V, Suarez Quezada M, Oros-Ruiz S, Ronquillo-de-Jesús E (2020) Epidemiology of COVID-19 in Mexico: from the 27th of February to the 30th of April 2020. Rev Clínica Esp 220(8):463–471. https://doi.org/10.1016/j.rce.2020.05.007
42. Knaul FM, Touchton MM, Arreola-Ornelas H, Calderon-Anyosa R, Otero-Bahamón S, Hummel C, Pérez-Cruz P, Porteny T, Patino F, Atun R, Garcia PJ, Insua J, Mendez O, Undurraga E, Boulding C, Nelson-Nuñez J, Velasco Guachalla VX, Sanchez-Talanquer M (2022) Strengthening health systems to face pandemics: subnational policy responses to COVID-19 in Latin America: study examines policy responses to COVID-19 in Latin America. Health Aff (Millwood) 41(3):454–462. https://doi.org/10.1377/hlthaff.2021.00981
43. Aristegui Noticias (2020) Tec de Monterrey suspende clases por Covid-19
44. Bautista-Reyes D, Werner-Sunderland J, Aragón-Gama AC, Duran JRC, Medina KDC, Urbina-Fuentes M, Bautista-González E (2023) Health-care policies during the COVID-19 pandemic in Mexico: a continuous case of heterogeneous, reactive, and unequal response. Health Policy OPEN 5:100100. https://doi.org/10.1016/j.hpopen.2023.100100
45. Bautista-González E, Werner-Sunderland J, Pérez-Duarte Mendiola P, Esquinca-Enríquez-de-la-Fuente CJ, Bautista-Reyes D, Maciel-Gutiérrez MF, Murguía-Arechiga I, Vindrola-Padros C, Urbina-Fuentes M (2021) Health-care guidelines and policies during the COVID-19 pandemic in Mexico: a case of health-inequalities. Health Policy OPEN 2:100025. https://doi.org/10.1016/j.hpopen.2020.100025
46. Azanza-Ricardo CL, Hernandez-Vargas EA (2020) Epidemiological characteristics of COVID-19 in Mexico and the potential impact of lifting confinement across regions. Front Phys 8:573322. https://doi.org/10.3389/fphy.2020.573322
47. Consejo Nacional de Humanidades, Ciencias y Tecnologías [CONAHCYT] (2023) COVID-19 Tablero México. In: COVID—19 Tablero México. https://datos.covid-19.conacyt.mx/index.php. Accessed 31 Jan 2025
48. Secretaría de Salud (2020) 220. Semáforo de Riesgo Epidémico por COVID-19 se define con 10 indicadores. In: gob.mx. http://www.gob.mx/salud/prensa/nuevo-coronavirus-en-el-mundo-covid-19-comunicado-tecnico-diario-257735?idiom=es. Accessed 31 Jan 2025
49. Knaul F, Arreola-Ornelas H, Porteny T, Touchton M, Sánchez-Talanquer M, Méndez Ó, Chertorivski S, Ortega S, Chudnovsky M, Kuri P, the group from the Observatory for the Containment of COVID-19 in the Americas (2021) Not far enough: public health policies to combat COVID-19 in Mexico's states. PLoS One 16(6):e0251722. https://doi.org/10.1371/journal.pone.0251722
50. Pagés C, Aclan C, Alfonso M, Arroio R, Irigoyen JL, Mejía I, Mendieta C, Moreno S, Muente A, Peñaherrera S, Pombo C, Regalía F, Savedoff B, Stein E, Tejerina L (2020) Del confinamiento a la reapertura: Consideraciones estratégicas para el reinicio de las actividades en América Latina y el Caribe en el marco de la COVID-19. Inter-American Development Bank
51. Perez-Tezoco JY, Aguilar-Lasserre AA, Moras-Sánchez CG, Vázquez-Rodríguez CF, Azzaro-Pantel C (2023) Hospital reconversion in response to the COVID-19 pandemic using simulation and multi-objective genetic algorithms. Computers Ind Eng 182:109408. https://doi.org/10.1016/j.cie.2023.109408
52. Kitroeff N, Lopez O (2021) The death market': oxygen shortage leaves Mexicans to die at home. N. Y. Times
53. Coalition for Epidemic Preparedness Innovations [CEPI] (2020) More than 150 countries engaged in COVID-19 Vaccine Global Access Facility | CEPI. https://cepi.net//more-150-countries-engaged-covid-19-vaccine-global-access-facility. Accessed 31 Jan 2025

54. Secretaría de Relaciones Exteriores (2021) Mexico has one of the broadest vaccine portfolios in the world. In: gob.mx. http://www.gob.mx/sre/prensa/mexico-has-one-of-the-broadest-vaccine-portfolios-in-the-world. Accessed 31 Jan 2025
55. Mathieu E, Ritchie H, Rodés-Guirao L, Appel C, Gavrilov D, Giattino C, Hasell J, Macdonald B, Dattani S, Beltekian D, Ortiz-Ospina E, Roser M (2024) Coronavirus (COVID-19) Vaccinations. Our World Data
56. De La Cruz-Hernández SI, Barrera-Badillo G (2022) The Delta variant triggers the third wave of COVID-19 in Mexico. Disaster Med Public Health Prep 16(6):2225–2227. https://doi.org/10.1017/dmp.2022.49
57. Zapata-Garibay R, González-Fagoaga JE, Meza-Rodríguez EB, Salazar-Ramírez E, Plascencia-López I, González-Fagoaga CJ (2021) Mexico's higher education students' experience during the lockdown due to the COVID-19 pandemic. Front Educ 6:683222. https://doi.org/10.3389/feduc.2021.683222
58. De La Cruz-Hernández SI, Álvarez-Contreras AK (2022a) Omicron variant in Mexico: the fourth COVID-19 wave. Disaster Med Public Health Prep 16(6):2260–2261. https://doi.org/10.1017/dmp.2022.160
59. Bello-Chavolla OY, Antonio-Villa NE, Valdés-Ferrer SI, Fermín-Martínez CA, Fernández-Chirino L, Vargas-Vázquez A, Ramírez-García D, Mancilla-Galindo J, Kammar-García A, Ávila-Funes JA, Zúñiga-Gil CH, García-Grimshaw M, Ceballos-Liceaga SE, Carbajal-Sandoval G, Montes-González JA, Zaragoza-Jiménez CA, García-Rodríguez G, Cortés-Alcalá R, Reyes-Terán G, López-Gatell H, Gutiérrez-Robledo LM (2023) Effectiveness of a nationwide COVID-19 vaccination program in Mexico against symptomatic COVID-19, hospitalizations, and death: a retrospective analysis of national surveillance data. Int J Infect Dis 129:188–196. https://doi.org/10.1016/j.ijid.2023.01.040
60. Reuters (2022) Mexican state brings back mask mandate as COVID numbers rise | Reuters. In: reuters.com. https://www.reuters.com/world/americas/mexican-state-brings-back-mask-mandate-covid-numbers-rise-2022-12-12/. Accessed 31 Jan 2025
61. Varkey B (2021) Principles of clinical ethics and their application to practice. Med Princ Pract 30(1):17–28. https://doi.org/10.1159/000509119
62. Warren RC, Forrow L, Hodge DA, Truog RD (2020) Trustworthiness before trust—Covid-19 vaccine trials and the black community. N Engl J Med 383(22). https://doi.org/10.1056/NEJMp2030033
63. McAloon C, Collins Á, Hunt K, Barber A, Byrne AW, Butler F, Casey M, Griffin J, Lane E, McEvoy D, Wall P, Green M, O'Grady L, More SJ (2020) Incubation period of COVID-19: a rapid systematic review and meta-analysis of observational research. BMJ Open 10(8):e039652. https://doi.org/10.1136/bmjopen-2020-039652
64. Zhao H, Lu X, Deng Y, Tang Y, Lu J (2020) COVID-19: asymptomatic carrier transmission is an underestimated problem. Epidemiol Infect 148:e116. https://doi.org/10.1017/S0950268820001235
65. Agren D (2020) Mexican President López Obrador draws doctors' ire. Lancet 395(10237):1601. https://doi.org/10.1016/S0140-6736(20)31198-3
66. Hernández-Díaz Y, Genis-Mendoza AD, Ramos-Méndez MÁ, Juárez-Rojop IE, Tovilla-Zárate CA, González-Castro TB, López-Narváez ML, Nicolini H (2022) Mental health impact of the COVID-19 pandemic on Mexican population: a systematic review. Int J Environ Res Public Health 19(11):6953. https://doi.org/10.3390/ijerph19116953
67. Monroy-Gómez-Franco L, Vélez-Grajales R, López-Calva LF (2022) The potential effects of the COVID-19 pandemic on learnings. Int J Educ Dev 91:102581. https://doi.org/10.1016/j.ijedudev.2022.102581
68. Parolin Z, Lee EK (2022) The role of poverty and racial discrimination in exacerbating the health consequences of COVID-19. Lancet Reg Health Am 7:100178. https://doi.org/10.1016/j.lana.2021.100178
69. Hemmer CJ, Hufert F, Siewert S, Reisinger E (2021) Protection from COVID-19: the efficacy of face masks. Dtsch Arztebl Int. https://doi.org/10.3238/arztebl.m2021.0119

70. Ibarrola-Peña JC, Barbosa-Camacho FJ, Almanza-Mena YL, Chejfec-Ciociano JM, Reyes-Elizalde EA, Romero-Limón OM, Zaragoza-Organista R, Cervantes-Pérez E, Sapién-Fernández JH, Guzmán-Barba JA, Flores-Becerril P, Ochoa-Rodríguez I, Nájar-Hinojosa R, Cueto-Valadez AE, Cueto-Valadez TA, López-Zendejas M, Fuentes-Orozco C, Cervantes-Guevara G, Miranda-Ackerman RC, González-Ojeda A (2022) Preventive measures against the COVID-19 pandemic in Mexico: a cross-sectional study. Front Public Health 10:932010. https://doi.org/10.3389/fpubh.2022.932010
71. Biggar N (2020) Are there absolute rights? In: What's wrong with rights? 1st edn. Oxford University Press, Oxford, pp 167–189
72. Legler TF (2025) The politics of COVID-19 in Mexico: governance meets government, 1st. Routledge, London
73. Johns Hopkins Coronavirus Resource Center (2023) COVID-19 dashboard by the Center for Systems Science and Engineering (CSSE). In: Johns Hopkins Coronavirus Resour. Cent. https://coronavirus.jhu.edu/. Accessed 31 Jan 2025
74. MacDonald NE (2015) Vaccine hesitancy: definition, scope and determinants. Vaccine 33(34):4161–4164. https://doi.org/10.1016/j.vaccine.2015.04.036
75. Carnalla M, Basto-Abreu A, Stern D, Bautista-Arredondo S, Shamah-Levy T, Alpuche-Aranda CM, Rivera-Dommarco J, Barrientos-Gutiérrez T (2021) Acceptance, refusal and hesitancy of Covid-19 vaccination in Mexico: Ensanut 2020 Covid-19. Salud Pública México 63(5):598–606. https://doi.org/10.21149/12696
76. Ramonfaur D, Limaye RJ, Hinojosa-González DE, Barrera FJ, Rodríguez-Gómez GP, Castillo-Salgado C (2024) COVID-19 vaccine hesitancy prevalence in Mexico: a systematic review and metanalysis. Vaccine X 18:100488. https://doi.org/10.1016/j.jvacx.2024.100488
77. Secretaría de Salud (2022b) Semáforo COVID-19. In: coronavirus.gob.mx. https://coronavirus.gob.mx/semaforo/. Accessed 31 Jan 2025
78. Gutierrez E, Rubli A, Tavares T (2022) Information and behavioral responses during a pandemic: evidence from delays in Covid-19 death reports. J Dev Econ 154:102774. https://doi.org/10.1016/j.jdeveco.2021.102774
79. Cuesta-Aguirre JP, Ahmed-Hannan S (2020) A Mexican state-level perspective on Covid-19 and its economic fallout. International Monetary Fund, Washington, DC
80. Secretaría de Relaciones Exteriores (2020) Mexicans Advised to Avoid Nonessential International Travel. In: gob.mx. http://www.gob.mx/sre/en/articulos/mexicans-advised-to-avoid-nonessential-international-travel-237981. Accessed 31 Jan 2025
81. De La Cruz-Hernández SI, Álvarez-Contreras AK (2022b) COVID-19 pandemic in Mexico: the response and reopening. Disaster Med Public Health Prep 16(6):2264–2266. https://doi.org/10.1017/dmp.2022.177
82. Chapa-Cantú JC, Ayala-Gaytán EA, Vazquez-Acevedo A (2024) The impact of COVID-19 on international tourism in Mexico. J Tour Dev. Páginas:243–255. https://doi.org/10.34624/RTD.V45I0.32084
83. TROCHTAT (2020) Health alert—Mexico COVID-19 update. In: US Embassy Consulates Mex. https://mx.usembassy.gov/health-alert-mexico-covid-19-update-15/. Accessed 31 Jan 2025
84. Gabarron E, Oyeyemi SO, Wynn R (2021) COVID-19-related misinformation on social media: a systematic review. Bull World Health Organ 99(6):455–463A. https://doi.org/10.2471/BLT.20.276782
85. Chejfec-Ciociano JM, Martínez-Herrera JP, Parra-Guerra AD, Chejfec R, Barbosa-Camacho FJ, Ibarrola-Peña JC, Cervantes-Guevara G, Cervantes-Cardona GA, Fuentes-Orozco C, Cervantes-Pérez E, García-Reyna B, González-Ojeda A (2022) Misinformation about and interest in chlorine dioxide during the COVID-19 pandemic in Mexico identified using Google trends data: Infodemiology study. JMIR Infodemiology 2(1):e29894. https://doi.org/10.2196/29894
86. Wirtz VJ, Millán-Garduño G, Hegewisch-Taylor J, Dreser A, Anaya-Sanchez A, González-Vázquez TT, Escalera R, Torres-Pereda P (2023) Misinformation messages shared via

WhatsApp in Mexico during the COVID-19 pandemic: an exploratory study. Health Promot Int 38(3):daad041. https://doi.org/10.1093/heapro/daad041
87. Garza-Ramos J (2020) Pandemic toll: challenges for journalists in Mexico and beyond | ReVista. Rev Harv Rev Lat Am
88. Esquivel-Chirino C, Valero-Princet Y, Gaitán-Cepeda LA, Hernández-Hernández C, Hernández AM, Laparra-Escareño H, Ventura-Gallegos JL, Montes-Sánchez D, Lopéz-Macay A, Hernández-Sánchez F, Oliveira WAD, Morales-González JA, Carmona-Ruiz D, Rosen-Esquivel K, Zentella-Dehesa A (2021) The effects of COVID-19 on healthcare workers and non-healthcare workers in Mexico: 14 months into the pandemic. Medicina (Mex) 57(12):1353. https://doi.org/10.3390/medicina57121353
89. Martínez-Chacón M (2021) Médicos cubanos enviados a México sí estaban acreditados. AP News
90. Secretaría de Salud (2021b) Informes sobre el personal de Salud COVID19 en México. In: gob.mx. http://www.gob.mx/salud/documentos/informes-sobre-el-personal-de-salud-covid-19-en-mexico. Accessed 31 Jan 2025
91. McCarthy N (2020) Infographic: where most health workers have died from Covid-19. In: Stat. Dly. Data. https://www.statista.com/chart/22795/highest-number-of-health-worker-deaths-from-covid. Accessed 31 Jan 2025
92. Infobae (2020) Las indignantes fotos de la improvisación contra el coronavirus: equipan a doctores con guantes de limpieza y caretas para albañil. In: infobae. https://www.infobae.com/america/mexico/2020/03/02/las-indignantes-fotos-de-la-improvisacion-contra-el-coronavirus-equipan-a-doctores-con-guantes-de-limpieza-y-caretas-para-albanil/. Accessed 31 Jan 2025
93. Lucas-Hernández A, González-Rodríguez VDR, López-Flores A, Kammar-García A, Mancilla-Galindo J, Vera-Lastra O, Jiménez-López JL, Peralta Amaro AL (2022) Stress, anxiety, and depression in health workers during the COVID-19 pandemic. Rev Medica Inst Mex Seguro Soc 60(5):556–562
94. López MJ (2020) Pasantes de medicina carecen de insumos para atender Covid-19—El Sol de Hermosillo | Noticias Locales, Policiacas, sobre México. Sonora y el Mundo, El Sol Hermosillo
95. Mento C, Pandolfo G, Spatari G, Silvestri MC, Formica I, Bruno A, Falgares G, Muscatello MRA, La Barbiera C, De Carlo A (2025) The impact of workplace violence on healthcare workers during and after the COVID-19 outbreak. Public Health Pract 9:100572. https://doi.org/10.1016/j.puhip.2024.100572
96. Indigo Staff (2020) Robo de líquido de rodilla a muertos por Covid-19 es fake news; AMLO se pronuncia. https://www.reporteindigo.com/reporte/robo-de-liquido-de-rodilla-a-muertos-por-covid-19-es-fake-news-amlo-se-pronuncia/. Accessed 31 Jan 2025
97. Singh G, Owens J, Cribb A (2017) What are the professional, political, and ethical challenges of co-creating health care systems? AMA J Ethics 19(11):1132–1138. https://doi.org/10.1001/journalofethics.2017.19.11.msoc1-1711

COVID-19 in the United States: A Relational Justice Approach to Public Health Ethics

13

Marielle Antoinette H. Zosa and Mellissa Withers

Abstract

The COVID-19 pandemic exposed and intensified systemic issues in the United States, including racial health disparities, leadership failures, and an ingrained individualistic culture. In this chapter, we examine the reasons behind the United States' failed pandemic response through a relational justice perspective. Relational justice requires social justice to address structural racism, relational solidarity to reinforce effective crisis leadership, and relational autonomy to challenge rugged individualism. To counter systemic racism, we advocate a shift from individual biological explanations to upstream social determinants of health, alongside improved data collection on marginalized communities. Immediate policy reforms should ensure fair compensation for essential workers, expanded healthcare access, culturally competent care, and language-inclusive health communication. Long-term social justice efforts must address structural inequities in education, employment, housing, and community engagement. Second, effective public health leadership must emphasize trustworthy communication, strategic partnerships, healthcare and information systems thinking, equity-focused training, and emotional resilience to instill social solidarity. Lastly, we highlight the need to challenge culturally dominant hyper-individualism by promoting relational autonomy, which recognizes human interdependence. Traditionally explored in bioethics, relational autonomy must be integrated into public health policy to foster collective responsibility. Through

M. A. H. Zosa
Department of Philosophy, University of the Philippines Diliman, Quezon City, Philippines
e-mail: mhzosa1@up.edu.ph

M. Withers (✉)
Keck School of Medicine, University of Southern California, Los Angeles, CA, USA
e-mail: mwithers@usc.edu

L. P. Marquez et al. (eds.), *COVID-19 in the Pacific Rim and Beyond*,
Collaborative Bioethics 6, https://doi.org/10.1007/978-3-032-21218-4_13

these actions, we can build a more equitable and resilient society capable of responding effectively to future crises.

Keywords

COVID-19 · Relational justice · Health disparities · United States · Pandemic

Introduction

The COVID-19 pandemic is the deadliest disaster in the history of the USA. As of January 10, 2025, the United States has approximately 111.8 million confirmed COVID-19 cases and about 1.22 million confirmed COVID-19 deaths—the most of any country worldwide in terms of both cases and deaths [1]. Excess mortality refers to the additional number of deaths, from all causes, that are above the expected level in normal conditions [2]. When looking at excess deaths due to COVID-19 adjusted for population size, the USA had 140 excess deaths per 100,000 people, far above the global average of 96 per 100,000—making it one of the worst-performing countries among high-income nations [3].

Ron Klain, who led the efforts to fight Ebola in the USA in 2014, said, "The US response will be studied for generations as a textbook example of a disastrous, failed effort…what's happened in Washington has been a fiasco of incredible proportions" [4].

Factors contributing to the early spread in the USA include delays in widespread testing, failure to prepare the healthcare system for the outbreak, and lack of adherence to public health mitigation measures, such as limiting large gatherings and social distancing [4–7]. In this chapter, we describe the three main factors that explain why the USA was so severely impacted by the COVID-19 pandemic, including (1) societal inequities and social determinants of health; (2) dubious leadership and politicization of the national pandemic response; and (3) dominant social values of individualism versus collectivism.

First, we outline how the pandemic unfolded in the USA through these three factors. Second, we give an analysis of these three factors through a relational justice-based approach to public health ethics. The three aspects of relational health justice, based on (1) social justice, (2) relational solidarity, and (3) relational autonomy, were evaluated in relation to the three factors behind the severe impact of COVID-19 on the USA. We ultimately argue in this chapter that the relational justice approach in public health ethics is helpful to gather crucial post-pandemic lessons, directions, and policy recommendations for the country. Lastly, we conclude the chapter with some lessons learned and recommendations moving forward toward the new normal in the USA.

Societal Inequities and Health Disparities in the US COVID-19 Response

The first factor that contributed to the severe impact of COVID-19 on the USA is the prevalence of social inequities and health disparities in the country. As of 2023, the estimated population of the USA is approximately 340 million people, making it the third-most populous country in the world after China and India [8]. The median age of the population is 38.9 years, and about 18% of the population is aged 65 years or over. The USA is a multicultural society with a multitude of racial/ethnic groups. About 58% of the population is comprised of Whites,[1] with about 19% Hispanic (the largest racial/ethnic minority group and fastest-growing demographic), about 13% Black, about 6% Asian, and less than 2% Native American. The USA has one of the largest immigrant populations in the world; about 14% of the population is foreign-born [8]. About 83% of the US population lives in urban areas. The median household income is about US$78,000 per year, and the poverty rate is 11%.

In light of these demographics, the COVID-19 pandemic in the USA disproportionally affected racial/ethnic minority populations, specifically Black, Hispanic, and Native American populations. These historically marginalized groups experienced higher rates of infection, hospitalization, and death from COVID-19 than non-Hispanic Whites [9, 10]. Studies show that community transmission in 2020 was up to 28 times higher for Hispanics than non-Hispanic Whites, and, in hospital or healthcare settings, Hispanics had 1.3 to 7.7 times higher risk for COVID-19 than Whites [11]. The US Centers for Disease Control (CDC) estimates that some racial/ethnic minority populations experienced hospitalization due to COVID-19 at a rate more than three times that of Whites [12]. Age-adjusted hospitalization rates from COVID-19 during the first year of the pandemic for Black and Native American populations were five times that of Whites, and the rate for Hispanics was four times higher than that of Whites [9]. An analysis of over 1.6 million COVID-19 hospitalizations showed that Black, Hispanic, and Native American elderly populations had 1.5 to 2.3 times higher hospitalization rates than White populations [13].

Black, Hispanic, and Native American populations also suffered from a disproportionate number of the more than one million deaths in the USA due to COVID-19, in particular excess and premature deaths [11, 14]. Blacks account for 13% of the US population, yet 17% of the deaths, while Hispanics account for 17% of the US population, yet 22% of cases and 13.8% of deaths. Further, among non-Hispanic Whites, the largest number of deaths due to COVID-19 were among people aged 85 years or older. But among Hispanics, the largest number of deaths occurred in those aged 65–74 years. Among people under age 65, the percentage of Hispanic (35%) and non-White (30%) individuals who died from COVID-19 was much higher than that of Whites (13%) [15]. The risk of death from COVID-19 in people aged 25 to 54 years was approximately seven to nine times higher in Blacks and five to seven times higher among Hispanics than that in Whites of the same age group [16].

[1] By "Whites", we refer to the non-Hispanic white population.

These numbers are not surprising given that Blacks, Hispanics, and Native Americans have a disproportionately high prevalence of comorbidities that are risk factors for COVID-19 hospitalization and death, including diabetes, hypertension, obesity, and coronary artery disease. These populations are also more likely to die prematurely compared to Whites due to all causes [10]. Overall, COVID-19 deaths contributed to a 3.08-year decrease in life expectancy at birth among populations in the USA. However, life expectancy losses were highest in Hispanics (5.31 years at birth and 5.23 years at ages 35–45 years) and Blacks (3.79 years at birth and 3.75 years between the ages 35 and 45 years) [17].

In the next sections, we summarize how these minority groups have been severely affected in terms of poverty, employment, health insurance, living conditions, and education.

Poverty, Employment, Health Insurance, and Living Conditions

Black, Hispanic, and Native American populations in the USA have significantly lower income levels than Whites. Prior to the pandemic, poverty rates in the USA were about 24% for Native Americans, 21% for African Americans, and 17% for Hispanics, compared to only 9% for Whites [18]. Unsurprisingly, in the first year of the pandemic, US counties with a higher proportion of residents who were racial/ethnic minorities, non-English-speaking, or who had low household income reported more deaths from COVID-19 than other counties [19, 20].

In 2019, 53 million US workers, including 44% of all workers aged 18 to 64 years, were employed in low-wage jobs, earning a median hourly wage of $10.22 and translating into less than US$18,000 annually [21]. Nearly half of Black and Hispanic female healthcare workers earn less than $15 per hour [22].

Racial/ethnic minorities are more often employed in low-wage jobs without health insurance. Hispanic immigrants may also be ineligible for unemployment benefits or income relief initiatives if they have undocumented status [23]. It was estimated that 18.2 million people in the USA at the start of the pandemic were uninsured or underinsured [24]. Hispanics and Blacks were less likely to have health insurance compared to Whites [25–27]. In fact, one in five Hispanic and Native Americans and one in ten Black individuals under 65 years lacked health insurance in 2020 [25]. The lack of health insurance makes the costs of healthcare services prohibitive and is a key barrier to adequate healthcare access among racial/ethnic minority populations [26]. For example, by July 2020, Florida, Texas, Oklahoma, Mississippi, North Carolina, South Carolina, and Georgia had the greatest percentage of adults who were currently uninsured and the highest numbers of new COVID-19 cases [28]. Detecting and managing chronic conditions are more difficult among people who lack health insurance because they are less likely to have access to or utilize primary care [29].

Over 100 million Americans face barriers to accessing primary care, according to the National Association of Community Health Centers. Further, the estimated number of Americans who lack access to primary care due to an inadequate supply in their local community has nearly doubled since 2014 [30]. Lower-income communities, which tend to have a higher proportion of racial/ethnic minority populations, generally have fewer healthcare facilities, such as hospitals, clinics, and specialty care centers, as well as a shortage of healthcare professionals. These areas are often underserved because healthcare providers may be less likely to open practices or facilities in neighborhoods where they expect lower profitability or higher patient volumes, which are often characteristics of economically disadvantaged areas. While there are numerous barriers that limit utilization of healthcare services among disadvantaged populations, COVID-19 vaccines and treatment were generally available at little or no cost to people who accessed care. However, copays, deductibles, and co-insurance may have applied, depending upon the type of insurance coverage that the patient had.

Furthermore, working conditions undoubtedly contribute to the disproportionate impact of COVID-19 on racial/ethnic minority communities. Three-quarters of over 108 million workers have jobs that cannot be done from home; these jobs are disproportionately filled by racial/ethnic minority groups [31–33]. By April 2020, more than 26 million people in the USA had filed claims for unemployment benefits, representing over 16% of the total labor force—the highest since the Great Depression [28]. The hospitality, leisure, and healthcare industries were most impacted, affecting racial/ethnic minority workers disproportionally. Black and Hispanic populations were more likely to be essential/frontline workers with low-paying jobs and the inability to work remotely, leading to higher COVID-19 exposure due to frequent interaction with the public (i.e., cashiers, home health aides, food service workers, transit workers, etc.) [21]. For example, Blacks account for a larger percentage of the service sector as compared to Whites, including cashiers, bus drivers, taxi drivers, housekeeping, janitorial staff, and sanitation workers [34]. Hispanics are more likely than other racial/ethnic groups to be employed as essential workers in agriculture, construction, and food services [26]. For example, research shows that only 20% of Black workers could work from home, compared to 30% of Whites [34]. Essential workers often have less paid sick leave than workers in other professions [29, 35, 36]. For example, one study found that only 55% of workers in the retail and food service industries had paid sick leave [37].

Lastly, other structural and socioeconomic factors that increased risk of COVID-19 in racial/ethnic minority groups include neighborhood and household characteristics that predispose them to worse outcomes. Racial/ethnic minority populations generally have higher housing density and more multigenerational households, making social distancing and self-isolation difficult or impossible [10, 23, 26, 29, 35–40].

Education, Language, and Mistrust of the Healthcare System

Level of education is related to income and health insurance coverage [41]. Fewer racial/ethnic minority populations complete a four-year college degree as compared to Whites (26.1% vs. 40.1%), which leads to reduced income [34]. Low education is also related to low literacy. People with low health literacy may have difficulty navigating the complex healthcare system and/or understanding medical advice. Hispanic immigrants may also have limited English language proficiency [40, 42, 43]. In the USA, over 25 million people report limited English proficiency; limited English, low literacy, and limited internet access have also been shown to be barriers to online initiatives to vaccination rates [29].

Adding to these challenges, pervasive skepticism and mistrust of the medical field exist among racial/ethnic minorities, which also impedes accessing healthcare, testing, and vaccination rates. Compared with Whites, Blacks are less likely to use healthcare services because of a distrust in healthcare providers rooted in a history of systemic racism, even among those with health insurance and access to healthcare [41, 44]. The Tuskegee Syphilis Study, in which low-income Black men were not given treatment for syphilis for decades in order to study its progression, continues to negatively impact the trust of the medical community among Blacks [45]. In the USA, police kill nearly three per day, and they are disproportionately Black [21]. Wearing a mask is problematic for Blacks who may fear being mistaken for criminals, given the frequency of police shooting of Blacks and the long-standing association with race and criminality [41, 46]. Among Hispanic immigrant groups, a climate of fear exists, especially under the Trump administration. Because vaccination often requires documenting personal information, they may fear that their personal information might be shared with immigration authorities, leading to deportation [42].

The COVID-19 pandemic also saw the emergence of a large amount of misinformation and conspiracy theories [47]. These resulted in distrust of COVID-19 information and a lack of accurate, credible COVID-19 information reaching these communities, thereby leading to uncertainty or erroneous ideas about COVID-19 transmission, severity and treatment, and vaccination [26, 35, 36], as well as elevated levels of vaccine hesitancy.

US Government's Response to the COVID-19 Pandemic

The second factor that contributed to the severity of USA COVID-19 crisis is the suspect and politicized handling of the pandemic by the Trump Administration. The US government's response to the pandemic has been criticized, both in terms of the federal and state governments, for a variety of reasons. In the following sections, some of these reasons are discussed, including policies and practices related to travel restrictions, closures, and ignorance about the severity of the virus.

Travel Restrictions

Many thought that the US government's response to the pandemic was slow and misguided [4, 24]. The early US government's response focused on screening passengers returning from international travel and then travel restrictions/bans, which proved to be ineffective in stopping the transmission of the virus in the USA. The first case of COVID-19 in the United States was confirmed on January 20, 2020, in a 35-year-old male resident of the state of Washington who had recently returned from Wuhan, China. In fact, the first 14 COVID-19 cases detected in the United States were all people who had recently traveled to China. As a result, on January 17, 2020, passengers on direct and indirect flights from Wuhan, China, to major airports in the United States were screened for symptoms, and the CDC issued a Level 3 Travel Health Notice advising travelers to avoid all non-essential travel to China due to COVID-19. US citizens returning from China were screened at the airport and placed under a mandatory 14-day quarantine at an air force base. By January 31, 2020, travelers arriving from China who were not US citizens or residents were denied entry. Eventually, over the next few months, the ban extended to travelers from numerous countries in Europe, as well as people on cruise ships [48]. However, the travel restrictions did not apply to US citizens, permanent residents, or their immediate family members.

The travel restrictions and bans were ineffective for several reasons. By the time restrictions were put into place, China had also placed its own limitations on travel to and from Wuhan. However, by then, numerous cases had already occurred in the USA. Travel restrictions can buy governments time to enact other measures but, without other aggressive public health preventive actions, travel bans have little value [48]. The WHO also advised against travel restrictions because, in part, they may divert resources from other interventions [49]. Some believe that the travel limitations gave people a false sense of security that something major was being done to prevent COVID-19 from spreading to the USA. In addition, a large international analysis done by Think Global Health, a project of the Council on Foreign Relations, found no correlation between travel restrictions and nations' ability to successfully prevent the outbreak [48, 50].

Analyses of travel data of passengers arriving in the USA from China between December 2019 and February 2020 showed that a total of 759,493 people entered the USA; the majority arrived in January 2020 on more than 1300 direct flights to 17 cities in the USA from China, including thousands coming from Wuhan [48, 51]. More than 3.4 million people in total arrived during those dates from four countries which were hit hardest by the pandemic (China, Italy, Spain, and Britain) [51]. Despite travel restrictions, a *New York Times* report estimated that nearly 40,000 people arrived in the USA on direct flights from China within two months *after* the ban was announced [52].

In early March 2020, the US President, Donald Trump, said that restrictions he placed on travel to and from China "saved a lot of lives," a claim that grew to "hundreds of thousands" by early April [53]. However, studies show that the USA and other countries' travel restrictions regarding China had modest impacts, slowing the initial spread outside of China but not containing the pandemic.

Closures

On February 25, 2020, the CDC warned the public that COVID-19's "disruption to everyday life may be severe" and to expect efforts to contain COVID-19 including school closings, workplace shutdowns, and the canceling of large gatherings and public events. On March 13, 2020, the Trump Administration declared a nationwide emergency, and a few days later, the White House advised against any gatherings of more than ten people [54]. States begin to implement shutdowns in order to prevent the spread of COVID-19. The New York City public school system—the largest school system in the USA with more than one million students—was closed. The majority of states also began to enact some level of prevention measures, such as closure of schools and non-essential businesses, mask mandates, prohibition and cancellation of large gatherings such as festivals and sporting events, and/or mandatory shelter-in-place orders [55–57].

By the end of March 2020, the USA had the highest number of known cases of any country globally, with over 100,000 cases [58]. However, public health prevention and mitigation measures were often viewed as political, sparking protests and backlashes. Trump not only publicly dismissed the threat of COVID-19, but discouraged action and international cooperation, politicizing mask-wearing and re-openings of businesses [59]. For example, protests broke out in Southern California over closure orders. Some demonstrators waved American flags, Trump 2020 posters, and signs labelling California Governor Gavin Newsom a "terrorist" and "traitor" [60]. The local sheriff said that his department would not enforce mask mandates, stay-at-home orders, and business closures, citing constitutional rights [61, 62]. Many of those who attended such rallies did not wear masks as acts of defiance against what they perceived to be government intrusion [63].

Trump was eager to restart the paralyzed US economy, and by April 2020, calls to reopen non-essential businesses were being heard across many states. A handful of Republican-led states began reopening, despite warnings from health experts of potential surges in new infections [28]. In April, the Trump Administration released a plan outlining how states should gradually reopen [12]. Health professionals and leaders urged more stringent public health measures, including lockdowns. For example, in July 2020, over 150 concerned health professionals sent a letter asking the federal government to "take bold action to save lives." The letter called for closing all non-essential businesses and opening restaurants for takeout only, as well as the prohibition of interstate travel to keep the virus from spreading from state to state [64, 65]. For the most part, their pleas were ignored.

Studies show that state re-openings and lack of widespread enforcement of public health measures resulted in a sharp rise in cases across many states [66]. For example, in August 2020, close to 500,000 people attended a motorcycle rally that took place over ten days in South Dakota, which led to 300 people in more than 20 states becoming infected. In the neighboring state of Minnesota, the CDC investigated 51 confirmed primary event-associated cases, 21 secondary cases, and 5 tertiary cases, including one death [67]. More than two million people flew on airlines during the Thanksgiving period, despite the CDC pleas to avoid travel [68]. One study has found that between 2020 and 2022, excess death rates were higher in states with Republican governors and greater Republican representation in state legislatures [69]. Another analysis found that if all states had imposed restrictions similar to those used in the ten most restrictive states, excess deaths would have been an estimated 10% to 21% lower [70].

Downplaying Severity

In 2018, the pandemic unit in the US National Security Council tasked with preparing for health emergencies like the COVID-19 pandemic was disbanded, which set the stage for the botched response to the pandemic under President Trump [4]. Further, this left nearly 700 positions at the CDC open because of a hiring freeze [59].

On January 30, 2020, the day the US CDC confirmed the first person-to-person case in the USA, Trump held a national rally in Iowa with thousands of people in a crowded indoor space where he claimed the outbreak was "all under control" and "a very small problem in the US" [71]. The very next day, the leader of the Department of Health and Human Services (HHS) declared a public health emergency. However, mass testing for COVID-19 did not become widely available until more than six weeks after the first case was confirmed in the country, much later than in many other countries with successful pandemic response, such as South Korea [4].

Trump continued to downplay the severity of the threat of COVID-19, even holding political rallies with mass gatherings. On Feb 10, 2020, at another packed political rally, he declared, "By April, you know, in theory, when it gets a little warmer, it miraculously goes away." He assured the nation that the USA was "very, very ready" to respond to a COVID-19 outbreak [72]. This was at the same time that testing kits, personal protective equipment for frontline workers, hospital beds, and ventilators were in short supply [4].

Despite the ongoing pandemic, Trump proposed further budget cuts to public health agencies. Eleven days after the WHO declared COVID-19 a public health emergency, Trump proposed a 16% cut in CDC funding for 2021 [4]. By April 11, 2020, with over 20,000 confirmed deaths and more than 500,000 confirmed cases, the USA became the country with the most reported COVID-19 cases and deaths, surpassing Italy and Spain. Yet Trump announced that the USA would cease contributing funding to the WHO, later announcing the US's withdrawal from the WHO effective July 6, 2021 [73].

Individualism in US Culture During COVID-19

The last factor contributing to the severe impact of COVID-19 in the USA involves the individualistic culture in the country, as opposed to a more collective value system. The relationship between individualism or collectivism of a society and its corresponding rates of COVID-19 transmission and compliance to health protocols is well-documented. In fact, the USA is known for its "rugged individualism" as coined by President Herbert Hoover [74, 75]. In this section, we analyze two components of the "rugged individualism" of the USA: emphasis on social independence over interdependence, and an anti-statist and anti-science attitude.

Independence over Interdependence

The USA is considered a country that values independence over interdependence in society [76]. Geert Hofstede's model of cultural diversity is often cited in the determination of cross-cultural values around the world [77]. Out of the four cross-cultural values, "individualism versus collectivism" has been the most used in research on the global effect of COVID-19. Individualism is evident in societies that prioritize personal needs and goals over those of a group. Individualists harbor a loose connection with one another as they respectively value their own independence, freedom, and creative expression. Collectivism, on the other hand, is demonstrated in cultures that prioritize the interests and welfare of the group over the individual. Collectivists cultivate a strong sense of community bound by mutual interdependence, trust, and respect [78].

The case for individualism in societies during a pandemic at first glance may suggest a more efficient COVID-19 response due to heightened resourcefulness or innovation, stronger institutions, and relatively better economic growth [79]. However, the parasite stress theory of value implies that societies with higher degrees of pathogenic stress or prevalence tend to be more culturally collectivist. Collectivism serves as a social defense mechanism to impede the spread of infectious diseases. Conversely, as the more individualistic societies become today, the greater are their tendencies to be more vulnerable to infectious diseases. This is due to the non-negotiable need for individual protection of rights and freedoms over compliance with standard health protocols in individualistic cultures [79].

In the first year of the pandemic, the USA recorded the largest number of infected cases and suffered disproportionately higher fatalities relative to other nations [77]. This could be correlated with the fact that individualist societies like the USA were skeptical of health protocols such as wearing of face masks, social distancing, and vaccinations. Hence, they were negatively affected during the pandemic. Standard health guidelines, which were intended to stifle the spread of the virus, were viewed as affronts to personal freedom and rights [77]. This non-compliance with protocols led to higher infection, transmission, and fatality rates in individualistic societies than in collectivist societies. Higher disease prevalence and mortality rate were also observed in individualistic nations as opposed to collectivist ones [77]. Generally, non-adherence to systems of public health during the pandemic due to individualism is indicative of a lack of civil duty bound by pro-social values such as mutual cooperation, trust, and reciprocity [74].

Overall, the literature on individualism versus collectivism during the pandemic is telling of how the USA fared in battling the virus. Collectivist states such as

Thailand and South Korea showed high use of masks [80]. Community concern, trust in government, social norms of group conformity, and cultural tightness found in collectivist cultures also positively contributed to the compliance of other non-pharmaceutical interventions or NPIs (e.g., stay-at-home policies) and the lower COVID-19 positive rates and mortality in these areas [81]. Therefore, it is no surprise that COVID-19 spread faster in individualist cultures than in collectivist ones during the early transmission of the virus [79].

Anti-statism and Anti-science Attitude

Anti-statist sentiments that oppose government intervention help to define rugged individualism in the USA. Anti-mask protests, anti-vaccine movements, and neglect for social distancing were common in the USA even when masks, vaccines, and social distancing were recommended by the CDC to fight viral transmission [76, 80]. Americans, especially those with affiliations to the Republican Party, were more likely to refuse to wear masks and stage rallies in the name of safeguarding their right to comfort [80]. The USA has also had a long history of anti-vaccination sentiments, noted as early as the 1901 Smallpox outbreak in Boston, and which recurred during the COVID-19 pandemic [76]. Finally, location data retrieved through smartphones owned by affiliated Republicans showed that nonessential trips to barbershops, clothing shops, and restaurants were more frequent than in corresponding Democrat-affiliated Americans [76, 79]. Americans residing in counties with a history of frontier settlement wore masks, practiced social distancing, and observed NPIs, all to a lesser degree. This individualist mindset was not only common among Republican civilians but also among governors who issued fewer advisories to stay at home for social distancing [74]. It is no wonder then that any public health intervention would be met with strong resistance due to rabidly individualistic attitudes and behaviors.

Moreover, it is not only government institutions which American individualists mistrusted but medical and scientific institutions as well. Social institutions and the scientific communities were viewed as sources for violation of an American's "rights to freedom, control, and self-determination" [76]. This image and narrative of America as invincible amid "repressive" public governance and health establishments was exacerbated through the leadership of Trump and his relationship with science itself [76].

In January 2020, Trump created a special coronavirus taskforce and called on Vice President Mike Pence to lead it, despite the fact that Pence had no medical or science training [72]. Trump perpetuated mistruths and distorted facts, demonstrating a lack of understanding about scientific processes involved in clinical trials and vaccine development. In one meeting, he questioned why the flu vaccine was not being used to fight the COVID-19 pandemic. Trump personally recommended unproven treatment for COVID-19, in particular the anti-malarial drug hydrochloroquine [4]. Trump said hydroxychloroquine had "a real chance to be one of the biggest game changers in the history of medicine," even without evidence to back the claim [82]. He also held a briefing in May 2020 without wearing a mask, claiming that anyone who wanted a coronavirus test could get one, notwithstanding the CDC's masking recommendations and wide-scale shortages of testing equipment.

Such false claims sparked a wave anti-science sentiments and conspiracy theories across the country [4, 83]. Trump often questioned scientific experts within the US government and blatantly ignored their advice [83]. For example, despite the ongoing pandemic in early May 2020, Trump abandoned his daily briefings with the COVID-19 Taskforce. In particular, his feuds with the director of the National Institute of Allergy and Infectious Diseases, Dr. Anthony Fauci, were well-known and largely stemmed from Trump's anger at Fauci for correcting and contradicting him. Trump has gone so far as to call Fauci "a disaster" and other public health officials "idiots" [84, 85]. Fauci has since become a target of intense hatred and even death threats from people who blame him for mask mandates and other policies which they believe infringed on their rights [86]. An article published in *The Lancet* by Woolhandler et al. [59] said that Trump not only mismanaged the COVID-19 response but also called his response "appalling." Further, they found that his "disdain for science" and "health-damaging policies and pronouncements" caused tens of thousands of unnecessary deaths [59].

A Relational Justice Approach in Public Health Ethics to COVID-19 in the USA

The COVID-19 policies and practices in the USA posed various ethical issues due to the three factors mentioned: social inequities in the healthcare system, politicization of the national pandemic response, and the cultivation of an individualist over a collectivist mindset. We present the argument in this section that the relational justice approach or framework in public health ethics is most applicable to the US COVID-19 response. In forwarding this argument, we assess the nation's institutional and cultural limitations while identifying opportunities for future directions in public health.

Ethics in Public Health

Ethics can be broadly defined as the study of theories, systems, and approaches in search of guiding principles for right or wrong action [87]. Moreover, public health is "fundamentally concerned with promoting the health of populations through the prevention of disease and injury" [88]. Public health ethics, consequently, is concerned with establishing a framework that defines the ethical principles and values behind public health interventions, activities, policies, and programs [89]. Public health ethics also draws from social and political philosophy because the intersection between and among individuals in populations, the state, and health institutions and systems is also analyzed toward the attainment of public health goals.

There are three commonly applied approaches in terms of frameworks to public health ethics: consequentialism, principlism, and theories of health justice [87]. First, consequentialist approaches focus on the maximization of well-being in terms of health as an outcome of a public health policy, project, or program. An example of this would be optimization of healthy life years in a population through distributing resources that would yield the most "quality-adjusted life years" (QALYs) [88].

Second, principlist approaches hold that public health policies should follow guiding principles such as those in bioethics (e.g., beneficence, non-maleficence, autonomy, and justice). Other principles can also include respect for community, realizing solidarity, social justice, reciprocity, health maximization, efficiency, and proportionality. In the United Kingdom, listing ethical principles in fact aided the distribution of limited intensive care beds and ventilators as facilitated by various private and public organizations and individuals during the COVID-19 pandemic [87].

The third approach to public health ethics, which will be the basis of this chapter, is related to theories of health justice. In this framework, public health policies should address concerns of social justice related to the public health system, health inequalities, and social determinants of health. A public health system is tasked to serve the promotion, restoration, or maintenance of the health needs of populations, especially vulnerable ones [90]. There often arise differences in conditions and outcomes of health across various individuals and social groups within a health system. These differences are called health inequalities or disparities [91]. In order to uncover health inequalities, social determinants of health (SDOH) are necessary phenomena for analysis. Social SDOH are analyzed on a group level as these determinants look into "causes of distribution patterns of health across social groups" [91].

Many approaches to health justice have focused on fairly allocating health resources, maximizing health outcomes, providing equal opportunities for attaining health goals, and securing capabilities to be healthy [92]. We argue that a relational justice approach to public health best captures these aspects of health justice by prioritizing the worst off in society. A relational justice framework is based on three notions—(1) social justice, (2) relational solidarity, and (3) relational autonomy—as values for public health ethics [93, 94]. In the following section, each of the three relational values will be recommended as corrective measures to the three factors that aggravated the COVID-19 crisis in the USA. Relational social justice will be assessed in light of social inequities, relational solidarity will be evaluated in light of the problematic government leadership, and relational autonomy will be analyzed in light of the rugged individualism in the USA.

Relational Social Justice and Social and Health Inequities

The first component of relational justice is social justice, as opposed to the more common view of justice in public health, i.e., distributive justice. Distributive justice concerns the allocation of public funds or scarce resources like vaccines and hospital beds based on SDOH [93]. Social justice, however, aims to contextualize the emergence and perpetuation of health inequalities through the political structures, policy-making, and individual and collective social practices which underlie these inequities. In social justice, one is asked to view persons relationally as situated and constituted in society. Most importantly, social justice demands that patterns of systemic injustice must be rectified and not perpetuated. The main concern in social justice is the prioritization of the most disadvantaged and underprivileged in society [94].

In the case of social inequities during the pandemic in the USA, we take the most disadvantaged and vulnerable population to be racial/ethnic minorities and a major SDOH to be systemic racism [95]. The severe consequences of COVID-19 among minority groups detailed previously can be explained by systemic racism that created social inequities and health disparities, and increased the risk of infection and poor health outcomes due to COVID-19. According to Braveman et al., systemic racism is "deeply embedded in systems, laws, written or unwritten policies, and entrenched practices and beliefs that produce, condone, and perpetuate widespread unfair treatment and oppression of people of color, with adverse health consequences" [96]. These entrenched practices and beliefs disadvantage minority communities, limiting opportunities and resources in terms of education, employment, housing, and healthcare [29, 35, 36, 38, 97]. Social inequities also helped to fuel COVID-19 transmission and severe illness in these groups.

Systemic racism also impacts the normalization of discrimination, stereotyping, and prejudices within the healthcare setting [41, 98, 99]. For example, studies have shown that Blacks were less likely to be offered COVID-19 testing than Whites, even when presenting with similar symptoms such as fever and cough [100]. COVID-19 testing centers were predominantly located in wealthy and White neighborhoods, which further limited access to healthcare for people living in poor neighborhoods [99, 101].

Furthermore, a nursing model by Gray et al. best illustrates SDOH and COVID-19 through upstream, midstream, and downstream factors [102]. Upstream factors are "root causes and conditions that contribute to health disparities" [95]. Examples of upstream factors include racism, poverty, and discrimination. Midstream factors are a result of upstream factors going downward toward eventual health outcomes. Examples of midstream factors are access to healthcare and testing, education, employment, living conditions, food security, and medical mistrust. Finally, downstream factors are the health outcomes resulting from the upstream and midstream factors. These downstream factors include hospitalization or fatalities from COVID-19, obesity, heart and lung diseases, and diabetes.

A key takeaway from this model is the focus of the current US healthcare system on downstream health outcomes. There is less emphasis on addressing midstream and upstream SDOH [95]. Most healthcare providers are not fully equipped to evaluate factors of racism, discrimination, housing, and employment as resulting in poor health outcomes. Directly addressing poor health outcomes and not the upstream factors is already more commonly undertaken. Social justice requires various forms of structural competence, acknowledging that socio-economic and political factors influence and create health disparities to begin with. Structural competence should be incorporated in health-related contexts such as education, policy-making, research, and medical practice to combat systemic, institutional, and even interpersonal racism [95].

Relational Solidarity and Public Leadership

The second component of relational justice is relational solidarity. Solidarity is founded on mutual cooperation and reciprocity based on common interests and goals [93]. Relational solidarity is the recognition that interconnections are essential in public health towards a mutual interest of surviving and keeping safe. Awareness of shared vulnerability and interdependence is necessary for relational solidarity. However, concrete steps should be taken to address these vulnerabilities by pursuing public goods such as scientific knowledge and communicable disease control. The promotion of these public goods concretely actualizes relational solidarity [94]. Overall, solidarity represents an "us all" rather than a division between "us" and "them" [93, 94].

The US government's COVID-19 pandemic response was considered a mismanagement due to four factors: "lack of trust and utilization of expertise on pandemic response, slow decision-making, inconsistent communication on mitigation measures such as face covering, and an absence of collaboration and coordination" [103]. To contextualize, 22% of COVID-19 cases and 21% of fatalities in the world were attributed to the USA even if the country comprised only 4% of the global population [103]. For the first factor, presidential crisis leadership requires trust between the executive and the professional experts such as in the appointment of qualified emergency crisis personnel. As explained earlier, Vice President Mike Pence led the taskforce despite having no strong qualifications in the medical sciences.

Second, decision-making requires learning from the data from past disasters and anticipating future data to be collected. Crisis leadership should decipher the gravity of the threat in a timely manner. Trump's administration did not immediately acknowledge and even denied the severity of COVID-19. Critical decision-making also necessitates rational dependence on scientific evidence and health standards. As discussed, the CDC suffered budget cuts and resistance from the administration itself so as not to advise the public on social distancing and travel protocols for the sake of reopening the US economy.

Communication skills are a third vital factor in crisis leadership which Trump utilized to perpetuate conspiracy theories, divisiveness, avoidance of blame, and anti-government attitudes. Conspiracy theories contained disinformation about the virus such as hydroxychloroquine or even consuming bleach to remedy infection [103]. Furthermore, Trump's rhetoric to divide the public to divert issues of accountability was evident in his administration's branding of COVID-19 as a "Chinese virus," while criticizing the Chinese government for not having warned the world about the virus early on. Despite WHO guidelines that naming viruses after geographic locations or groups of people is inaccurate, inappropriate, and could foster racism and discrimination, Trump continued to refer to the virus as the "Chinese virus" and assert that the label was not racist because "it comes from China" [104]. In the same manner, such divisive statements also are intended to avoid or shift the blame toward certain government institutions, i.e., health institutions. Trump's leadership was in like manner heavily suspicious of institutions of government as evidenced by Trump's allegations of the Food and Drug Administration delaying vaccines to upset his chances of reelection [103].

Finally, coordination and collaboration refer to the processes of facilitating different action plans such as resource sharing and task alignment toward a shared outcome during a crisis. This can occur with both public and private organizations. Coordination is mandatory among institutions, whereas collaboration is more voluntarily cultivated. When Trump announced the plans to withdraw from the World Health Organization in July 2020 due to his accusations that the organization was too "pro-China", coordination and collaboration were hampered. Other delays in coordinating with distributing medical resources and vaccinations also transpired [103].

In terms of general leadership during the pandemic, some say that the urgent early warnings to prepare for the pandemic were ignored because Trump was focused on protecting the economy, his impeachment trial, re-election campaign, and protecting political interests like his relationship with the president of China, ironically enough. In the early days of the pandemic, he was also focused on protecting a robust economy which he "viewed as central to his reelection chances" [5]. At a meeting in January 2020, he "enthusiastically praised" China's President Xi's response to the pandemic; many said this was motivated by the desire to end a trade war to boost his reelection campaign chances [5]. By April 2020, the Senate had acquitted Trump on the impeachment charges but, instead of taking the pandemic seriously, he focused his efforts (and X postings) on complaints about his political rivals being out to get him [5]. In another example, Trump said at a July rally that he had asked officials to "slow the testing down" because too many positive cases were being found, which would reflect badly on his administration [105].

Given these, the Trump administration has been proven to be inadequate in terms of adhering to relational solidarity. Since relational solidarity must take into account promoting public health goods such as scientific knowledge and disease control mechanisms, it is difficult to defend this promotion due to Trump's inefficacy as a leader. Scientific knowledge could not have been disseminated truthfully due to the communication of conspiracy theories and distrust of institutions of government and health, with the goal of dividing attention and diverting accountability. Even disease control interventions such as face masks, social distancing, and vaccines were not encouraged by the administration. Instead, they were neglected due to lack of trust in the medical profession and poor decision-making.

Ultimately, constant collaboration and cooperation with the necessary organizations to mitigate the effects of the pandemic is needed to strengthen relational solidarity in a country. Relational solidarity is a crucial aspect in Trump's leadership that was strongly lacking due to the protection of his own self-interests for continuing his administration. In fact, the Trump administration's response to the pandemic was solitarily decided by the executive while disregarding the solidarity efforts by relevant and expert health institutions and state actors. Such prioritization of Trump's own political agenda and libertarian decision-making reinforces the divide between the individual with selfish ambition and others who look to government for health intervention. In the same way, divisive rhetorical tactics used by Trump further demarcate "us" from "them"—with the "us" directed toward liberty-loving, anti-establishment Americans and the "them" as civic-minded, pro-establishment Americans or even non-American carriers of the virus. The absence of the "us all" mentality undermines any semblance of relational solidarity.

Relational Autonomy and Individualism Versus Collectivism

The final component of relational justice is relational autonomy. Relational autonomy does not forego the idea of autonomy but instead recognizes that individuals and their autonomy are deeply situated in a social, political, and economic context. Any person's autonomous choices are dependent on their relations in society. A relational view on autonomy helps one to pay attention to the different social conditions that affect and are affected by the choices and opportunities of groups of individuals [93, 94].

The "relational turn" in autonomy has gained fairly recent traction in ethics and bioethics [106] and definitely requires more research especially in the field of public health. Nonetheless, insights into US rugged individualism during the pandemic can be gathered through an evaluation of relational autonomy. One aspect of US rugged individualism which emphasizes independence over interdependence is the traditional view of autonomy wherein the individual is the sole moral agent of their own lives [107]. In relational autonomy, the individual is still of moral importance but just not in isolation with others. An ethics of quarantine, isolation, and social distancing can surely be supported by relational autonomy [108]. One's individual choices to socially distance or not during the pandemic can impact and can be influenced by the individual choices of others. In comparison to individualistic states, collectivist ones may have reported less fatalities and transmission rates due to their interdependent and communal nature. However, the spirit of independence in the USA need not be necessarily compromised when autonomy is viewed relationally, especially in times of crisis and disaster.

The other aspect of rugged individualism is the distrust toward establishments of health and government. Similar to how relational solidarity requires collaboration and cooperation with various institutions, relational autonomy supports society's continuous dialogue and partnerships with the health profession and government. Public trust in these authorities is essential in the collective goal to prevent and further contain the spread of the virus [108]. As viral transmission is an interconnected and interdependent matter, relational autonomy requires that everyone is mutually cooperative and reciprocal of the efforts of every other individual and the expertise of health authorities. Anti-government and even anti-science sentiments should be given least priority during a global pandemic for the shared goal of survival.

Conclusion, Lessons Learned, and Recommendations

COVID-19 negatively impacted the USA in several ways, including the social inequities and health disparities brought about by upstream factors of systemic racism, the inefficient and divisive leadership under the Trump administration, and the predominance of rugged individualism in American society. We propose several recommendations to mitigate the lasting effects of these three factors post-COVID-19 and based on the three aspects of relational health justice.

In order to challenge systemic racism, we should begin to use a broader interpretive framework to understand why people are exposed to risk factors, and the social conditions under which individual risk factors are related to negative health outcomes [109]. Instead of focusing on biological explanations as causes of differential health outcomes or downstream factors of infections and diseases, we should focus on upstream and midstream factors to address SDOH. In terms of research, we need for more comprehensive data on the impact of the COVID-19 pandemic on racial/ethnic minorities, in particular for those living in disadvantaged neighborhoods [110]. Short-term policy reforms are recommended such as adequate monetary compensation for essential workers who serve marginalized sectors, expansion of the Affordable Care Act to cover COVID-19 treatment, adequate medical resources like protective equipment for racial/ethnic minorities, and provision of culturally sensitive care along with clearly understood language-appropriate health policies for non-English communities. Long-term solutions for vulnerable communities will have to include expanding educational access, sufficient employment generation, rectification of inequitable housing policies, funding for COVID-19 recovery programs, and continuous community engagement initiated by public healthcare systems [29].

Public health leadership should also serve as a valuable lesson to learn after the pandemic. Seven common themes for the next generation of public health leaders can be identified [111]. First, communication is key for leading effectively, but it should be based on building trust and accountability. Anti-science rhetoric and disinformation defeat the goals of trustworthy and accountable communication. Second, it is essential for public health leaders to forge, facilitate, and promote partnerships. Building and maintaining strategic partnerships are essential in public healthcare and practice to address crisis situations. Third, leaders should make an effort to connect public health with public healthcare systems rather than focusing on personal healthcare alone. Fourth, public health leaders should build information systems that provide accessible data at the individual, institutional, and systems levels. This is to ensure efficient and informed decision-making for the leadership as well. Fifth, it is vital to engage in systems and strategic thinking that involve the collaboration of agencies ranging from "clinical care delivery systems, community-based organizations, private nonprofit associations, educational institutions, private industry, and the media" [111]. Sixth, it is incumbent that public health leaders address structural racism and health inequities. This can be done through leadership trainings on equity, diversity, and inclusion alongside more collective efforts like trust-building with vulnerable communities and improving community-wide health infrastructures. Finally, public health leaders should achieve and maintain both emotional and psychological resilience in light of extremely stressful and distressing events during crises.

We would like to make a final recommendation in relation to relational autonomy. Cultural mindsets biased toward rugged individualism or excessive individualistic autonomy do not change overnight. Furthermore, there are many gaps and inadequacies to be filled and clarified in the research on relational autonomy [106]. The literature on relational autonomy alone has been more of a bioethical concern rather than a public health issue. Studies and reflections on relational autonomy gathered as a result of the COVID-19 crisis nonetheless are slowly emerging. Cultural anthropology, environmentalism, discourse ethics, philosophical anthropology, existential phenomenology, and hermeneutics have been identified as other sources of relational autonomy apart from bioethics and feminist ethics [106]. We encourage continued research on relational autonomy after the pandemic. Most importantly, we hope that the institutions of education, health, and government can instill and influence policies and programs which reflect the heart of relational autonomy—that we are all inextricably interdependent with one another as we go through, bounce back, and move forward from crises and disasters.

References

1. Mathieu E, Ritchie H, Rodés-Guirao L et al (2024) COVID-19 Pandemic. Our World in Data. https://ourworldindata.org/coronavirus
2. Checchi F, Roberts L (2005) Interpreting and using mortality data in humanitarian emergencies. Humanitarian Practice Network, p 52
3. Horton J (2022) Does US really have world's highest Covid death toll? BBC. https://www.bbc.com/news/61333847
4. Pilkington E, McCarthy T (2020) The missing six weeks: how Trump failed the biggest test of his life. The Guardian. https://www.theguardian.com/us-news/2020/mar/28/trump-coronavirus-politics-us-health-disaster?trk=public_post_comment-text
5. Lemire J, Miller Z, Colvin J et al (2020) Signs missed and steps slowed in Trump's pandemic response. Associated Press. https://apnews.com/article/virus-outbreak-donald-trump-us-news-ap-top-news-politics-6a8f85aad99607f313cca6ab1398e04d
6. Ollstein AM (2020) Trump halts funding to World Health Organization. Politico. https://www.politico.eu/article/trump-halting-fund-to-world-health-organization/
7. Whoriskey P, Satija N (2020) How U.S. coronavirus testing stalled: flawed tests, red tape and resistance to using the millions of tests produced by the WHO. The Washington Post. https://www.washingtonpost.com/business/2020/03/16/cdc-who-coronavirus-tests/
8. US Census Bureau (2024) Profiles-the United States. https://data.census.gov/profile
9. Acosta AM, Garg S, Pham H et al (2021) Racial and ethnic disparities in rates of COVID-19-associated hospitalization, intensive care unit admission, and in-hospital death in the United States from March 2020 to February 2021. JAMA Netw Open 4(10):e2130479. https://doi.org/10.1001/jamanetworkopen.2021.3047
10. Webb Hooper M, Nápoles AM, Pérez-Stable EJ (2020) COVID-19 and racial/ethnic disparities. JAMA 323(24):2466–2467. https://doi.org/10.1001/jama.2020.8598

11. Mackey K, Ayers CK, Kondo KK et al (2021) Racial and ethnic disparities in COVID-19-related infections, hospitalizations, and deaths: a systematic review. Ann Intern Med 174(3):362–373. https://doi.org/10.7326/M20-6306
12. CDC (2021) COVID-19 data tracker. https://covid.cdc.gov/covid-data-tracker
13. CMS (2021) Preliminary Medicare COVID-19 data snapshot. https://www.cms.gov/files/document/medicare-covid-19-data-snapshot-fact-sheet.pdf
14. Gross CP, Essien UR, Pasha S et al (2020) Racial and ethnic disparities in population-level Covid-19 mortality. J Gen Intern Med 35(10):3097–3099. https://doi.org/10.1007/s11606-020-06081-w
15. Wortham JM, Lee JT, Althomsons S et al (2020) Characteristics of persons who died with COVID-19 - United States, February 12-May 18, 2020. MMWR Morb Mortal Wkly Rep 69(28):923–929. https://doi.org/10.15585/mmwr.mm6928e1
16. Bassett MT, Krieger N, Chen JT (2020) The unequal toll of COVID-19 mortality by age in the United States: quantifying racial/ethnic disparities. Harvard. https://www.hsph.harvard.edu/social-and-behavioral-sciences/2020/06/23/the-unequal-toll-of-covid-19-mortality-by-age-in-the-united-states-quantifying-racial-ethnic-disparities/
17. Silva S, Goosby E, Reid MJA (2023) Assessing the impact of one million COVID-19 deaths in America: economic and life expectancy losses. Sci Reports 13(1):3065. https://doi.org/10.1038/s41598-023-30077-1
18. KFF (2019) Poverty rate by race/ethnicity. https://www.kff.org/other/state-indicator/poverty-rate-by-raceethnicity
19. Khazanchi R, Beiter ER, Gondi S et al (2020) County-level association of social vulnerability with COVID-19 cases and deaths in the USA. J Gen Intern Med. https://doi.org/10.1007/s11606-020-05882-3
20. Tan AX, Hinman JA, Abdel Magid HS et al (2021) Association between income inequality and county-level COVID-19 cases and deaths in the US. JAMA Netw Open 4(5):e218799. https://doi.org/10.1001/jamanetworkopen.2021.8799
21. Krieger N (2020) ENOUGH: COVID-19, structural racism, police brutality, plutocracy, climate change-and time for health justice, democratic governance, and an equitable, sustainable future. Am J Public Health 110(11):1620–1623. https://doi.org/10.2105/AJPH.2020.305886
22. Himmelstein KEW, Venkataramani AS (2019) Economic vulnerability among US female health care workers: potential impact of a $15-per-hour minimum wage. Am J Public Health 109(2):198–205. https://doi.org/10.2105/AJPH.2018.304801
23. Page KR, Flores-Miller A (2021) Lessons we've learned- Covid-19 and the undocumented Latinx community. New Engl J Med 384:5–7
24. Gaffney AW, Hawks L, Bor DH et al (2020) 18.2 million individuals at increased risk of severe COVID-19 illness are un- or underinsured. J Gen Intern Med 35(8):2487–2489. https://doi.org/10.1007/s11606-020-05899-8
25. Artiga S, Hill L, Orgera K et al. Health coverage by race and ethnicity, 2010-2019. https://www.kff.org/racial-equity-and-health-policy/issue-brief/health-coverage-by-race-and-ethnicity/
26. Andraska EA, Alabi O, Dorsey C et al (2021) Health care disparities during the COVID-19 pandemic. Sem Vas Surg 34(3):82–88. https://doi.org/10.1053/j.semvascsurg.2021.08.002
27. Moore JT, Pilkington W, Kumar D (2020) Diseases with health disparities as drivers of COVID-19 outcome. J Cell Mol Med 24(19):11038–11045. https://doi.org/10.1111/jcmm.15599
28. Mutikani L (2020) Millions of Americans join unemployment lines as coronavirus savages economy. Reuters. https://www.reuters.com/article/business/millions-of-americans-join-unemployment-line-as-coronavirus-savages-economy-idUSKCN2250CR/
29. Diaz AA, Thakur N, Celedón JC (2023) Lessons learned from health disparities in coronavirus disease-2019 in the United States. Clin Chest Med 44(2):425–434. https://doi.org/10.1016/j.ccm.2022.11.021
30. NACHC (2023) Closing the primary care gap: how community health centers can address the nation's primary care crisis. https://www.nachc.org/resource/closing-the-primary-care-gap-how-community-health-centers-can-address-the-nations-primary-care-crisis/

31. Baker MG (2020) Nonrelocatable occupations at increased risk during pandemics: United States, 2018. Am J Public Health 110(8):1126–1132. https://doi.org/10.2105/AJPH.2020.305738
32. Nana-Sinkam P, Kraschnewski J, Sacco R et al (2021) Health disparities and equity in the era of COVID-19. J Clin Transl Sci 5(1):e99. https://doi.org/10.1017/cts.2021.23
33. Pilkington E (2020) As 100,000 die, the virus lays bare America's brutal fault lines—race, gender, poverty, and broken politics. The Guardian. https://www.theguardian.com/us-news/2020/may/28/us-coronavirus-death-toll-racial-disparity-inequality
34. US Bureau of Labor Statistics (2020) Employed persons by detailed occupation, sex, race, and Hispanic or Latino ethnicity—2019. https://www.bls.gov/cps/cpsaat11.htm
35. Thakur N, Lovinsky-Desir S, Bime C et al (2020) The structural and social determinants of the racial/ethnic disparities in the U.S. COVID-19 pandemic. What's our role? Am J Respir Crit Care Med 202(7):943–949. https://doi.org/10.1164/rccm.202005-1523PP
36. Tai DBG, Shah A, Doubeni CA et al (2021) The disproportionate impact of COVID-19 on racial and ethnic minorities in the United States. Clin Infect Dis 72(4):703–706. https://doi.org/10.1093/cid/ciaa815
37. Schneider D, Harknett K (2020) Essential and vulnerable: service-sector workers and paid sick leave. Shift. https://shift.hks.harvard.edu/essential-and-vulnerable-service-sector-workers-and-paid-sick-leave/
38. Bailey ZD, Krieger N, Agénor M et al (2017) Structural racism and health inequities in the USA: evidence and interventions. Lancet 389(10077):1453–1463. https://doi.org/10.1016/S0140-6736(17)30569-X
39. Martin CA, Jenkins DR, Minhas JS et al (2020) Socio-demographic heterogeneity in the prevalence of COVID-19 during lockdown is associated with ethnicity and household size: results from an observational cohort study. EClinicalMedicine 25:100466
40. Singu S, Acharya A, Challagundla K et al (2020) Impact of social determinants of health on the emerging COVID-19 pandemic in the United States. Front Public Health 8:406. https://doi.org/10.3389/fpubh.2020.00406
41. Maness SB, Merrell L, Thompson EL et al (2021) Social determinants of health and health disparities: COVID-19 exposures and mortality among African American people in the United States. Public Health Rep 136(1):18–22. https://doi.org/10.1177/0033354920969169
42. Diaz AA, Celedón JC (2021) COVID-19 vaccination: helping the latinx community to come forward. EClinicalMedicine 35:100860. https://doi.org/10.1016/j.eclinm.2021.100860
43. Schillinger D (2021) Social determinants, health literacy, and disparities: intersections and controversies. Health Liter Res Pract 5(3):e234–e243. https://doi.org/10.3928/24748307-20210712-01
44. Kennedy BR, Mathis CC, Woods AK (2007) African Americans and their distrust of the health care system: healthcare for diverse populations. J Cult Div 14(2):56–60
45. Tobin MJ (2022) Fiftieth anniversary of uncovering the Tuskegee syphilis study: the story and timeless lessons. Am J Respir Crit Care Med 205(10):1145–1158. https://doi.org/10.1164/rccm.202201-0136SO
46. Bierria A (2020) Racial conflation: rethinking agency, black action, and criminal intent. J Soc Philos 53(4):575–594
47. Douglas KM (2021) The COVID-19 pandemic also saw the emergence of a large amount of misinformation and conspiracy theories. Group Process Intergroup Relat 24(2):270–275
48. Luthra S (2020) Are COVID-19 travel restrictions 'more critical in saving lives' than testing? Not really. PolitiFact. https://www.politifact.com/factchecks/2020/apr/03/jesse-watters/are-covid-19-travel-restrictions-more-critical-sav/
49. WHO (2020) Updated WHO recommendations for international traffic in relation to COVID-19 outbreak. https://www.who.int/news-room/articles-detail/updated-who-recommendations-for-international-traffic-in-relation-to-covid-19-outbreak/
50. Kiernan S, DeVita M, Bollyky TJ (2020) Tracking Coronavirus in countries with and without travel bans. https://www.thinkglobalhealth.org/article/tracking-coronavirus-countries-and-without-travel-bans

51. Thomas P, Date J, Salzman S et al (2020) Disaster in motion: 3.4 million travelers poured into the US as coronavirus pandemic erupted. ABC News. https://abcnews.go.com/Politics/disaster-motion-34-million-travelers-poured-us-coronavirus/story?id=69933625
52. Eder S, Fountain H, Keller MH et al (2020) 430,000 people have traveled from China to U.S. since Coronavirus surfaced. The New York Times. https://www.nytimes.com/2020/04/04/us/coronavirus-china-travel-restrictions.html
53. Robertson L (2020) Trump's snowballing China travel claim. FactCheck.Org. https://www.factcheck.org/2020/04/trumps-snowballing-china-travel-claim/
54. Liptak K (2020) White House advises public to avoid groups of more than 10, asks people to stay away from bars and restaurants. CNN. https://www.cnn.com/2020/03/16/politics/white-house-guidelines-coronavirus/index.html
55. Deb S, Cacciola S, Stein M (2020) Sports leagues bar fans and cancel games amid Coronavirus outbreak. The New York Times. https://www.nytimes.com/2020/03/11/sports/basketball/warriors-coronavirus-fans.html
56. Norwood C (2020) Most states have issued stay-at-home orders, but enforcement varies widely. PBS. https://www.pbs.org/newshour/politics/most-states-have-issued-stay-at-home-orders-but-enforcement-varies-widely
57. Shumaker L (2020) U.S. Coronavirus deaths top 20,000, highest in the world exceeding Italy. Reuters. https://www.reuters.com/article/us-health-coronavirus-usa-casualties-idUSKCN21T0NA/
58. McNeil DG Jr (2020) The U.S. now leads the world in confirmed coronavirus cases. The New York Times. https://www.nytimes.com/2020/03/26/health/usa-coronavirus-cases.html
59. Woolhandler S, Himmelstein DU, Ahmed S et al (2021) Public policy and health in the Trump era. Lancet 397(10275):705–753
60. Kallingal M, Moon S (2020) Large crowds in Huntington Beach protest beach closures by California governor. CNN. https://www.cnn.com/2020/05/01/us/california-orange-county-injunction-beaches/index.html
61. Money L (2020) 'We are not the mask police': Calif. Deputies won't enforce health order. Police 1. https://www.police1.com/coronavirus-covid-19/articles/we-are-not-the-mask-police-calif-deputies-wont-enforce-health-order-ZP2Iqh7WorWWuY81/
62. Nelson JQ (2020) California sheriff refuses to enforce stay-at-home orders: it 'doesn't pass constitutional test'. https://www.foxnews.com/politics/california-sheriff-refuses-home-orders-constitutional-test
63. Neergaard L, Alonso-Zaldivar R (2020) Fauci, Redfield, health officials testify on Trump administration's response to COVID-19. PBS. https://www.pbs.org/newshour/health/watch-live-fauci-health-officials-testify-on-trump-administrations-response-to-covid-19
64. Durkee A (2020) Medical experts tell government: 'Shut it down now, and start over.' Forbes. https://www.forbes.com/sites/alisondurkee/2020/07/24/medical-experts-tell-government-shut-it-down-now-and-start-over/
65. Board E (2020) America could control the pandemic by October. Let's get to It. The New York Times. https://www.nytimes.com/2020/08/08/opinion/sunday/coronavirus-response-testing-lockdown.html
66. Noori AF, Sheperd K, Hassan J et al (2020) U.S. surpasses 2 million coronavirus cases. The Washington Post. https://www.washingtonpost.com/nation/2020/06/11/coronavirus-update-us/
67. Firestone MJ, Wienkes H, Garfin J et al (2020) MMWR Morb Mortal Wkly Rep 69(47):1771–1776
68. O'Brien M (2020) Many Americans flying for Thanksgiving despite CDC pleas. Associated Press. https://apnews.com/general-news-ec9b3912f216708d04555e5171590732
69. Woolf SE, Lee JH, Chapman DA et al (2024) Excess death rates by state during the COVID-19 pandemic: United States, 2020–2023. Am J Public Health 114(9):882–891
70. Ruhm CJ (2024) US state restrictions and excess COVID-19 pandemic deaths. JAMA Health Forum 5(7):e242006. https://doi.org/10.1001/jamahealthforum.2024.2006.ne

71. Wayland M (2020) Trump says coronavirus outbreak is 'all under control' and a 'very small problem' in US. CNBC. https://www.cnbc.com/2020/01/30/trump-says-coronavirus-outbreak-is-all-under-control-and-is-a-very-small-problem-in-us.html
72. Santucci J (2020) What we know about the White House coronavirus task force now that Mike Pence is in charge. USA Today. https://www.usatoday.com/story/news/politics/2020/02/27/coronavirus-what-we-know-mike-pence-and-task-force/4891905002/
73. US Department of State (2020) Update on the US withdrawal from the World Health Organization. https://2017-2021.state.gov/update-on-u-s-withdrawal-from-the-world-health-organization/. Accessed 14 Jul 2025
74. Bazzi S, Fiszbein M, Gebresilasse M (2021) "Rugged individualism" and collective (in)action during the COVID-19 pandemic. J Public Econ 195:1–20. https://doi.org/10.1016/j.jpubeco.2020.104357
75. Webster GD, Howell JL, Losee JE et al (2021) Culture, COVID-19, and collectivism: a paradox of American exceptionalism? Per Ind Diff 178:1–10. https://doi.org/10.1016/j.paid.2021.110853
76. Kitayama S, Camp NP, Salvador CE (2022) Culture and the COVID-19 pandemic: multiple mechanisms and policy implications. Soc Issues Policy Rev 16(1):164–211. https://doi.org/10.1111/sipr.12080
77. Nair N, Selvaraj P, Nambudiri R (2022) Culture and COVID-19: impact of cross-cultural dimensions on behavioral responses. Encyclopedia 2(3):1210–1224. https://doi.org/10.3390/encyclopedia2030081
78. Güss CD, Tuason MT (2021) Individualism and egalitarianism can kill: how cultural values predict Coronavirus deaths across the globe. Front Psychol 12:1–13. https://doi.org/10.3389/fpsyg.2021.620490
79. Jiang S, Wei Q, Zhang L (2022) Individualism versus collectivism and the early stage transmission of COVID 19. Soc Ind Res 164:791–821. https://doi.org/10.1007/s11205-022-02972-z
80. Lu JG, Jin P, English AS (2021) Collectivism predicts mask use during COVID-19. Psych Cogn Sci 1(18):1–8. https://doi.org/10.1073/pnas.2021793118
81. Leong S, Eom K, Ishii K et al (2022) Individual costs and community benefits: collectivism and individuals' compliance with public health interventions. PLoS One 17(11):1–19. https://doi.org/10.1371/journal.pone.0275388
82. Erman M (2020) Malaria drug touted by Trump fails to prevent COVID-19 in a high profile study. Reuters. https://www.reuters.com/article/world/malaria-drug-touted-by-trump-fails-to-prevent-covid-19-in-high-profile-study-idUSKBN23B068/
83. Milman O (2020) How Trump's missteps undermined the US's recovery from pandemic. https://www.theguardian.com/us-news/2020/jun/17/trump-coronavirus-pandemic-recovery-missed-chance?utm_source=chatgpt.com
84. Bennett G, Cuevas K (2024) Fauci on his fraught relationship with Trump and the attacks he has faced. PBS. https://www.pbs.org/newshour/show/fauci-on-his-fraught-relationship-with-trump-and-the-attacks-he-has-faced
85. Behrmann S, Santucci J (2020) Here's a timeline of President Donald Trump's and Dr. Anthony Fauci's relationship. Trump and Fauci: a timeline of their relationship during COVID. USA Today. https://www.usatoday.com/story/news/politics/2020/10/28/president-donald-trump-anthony-fauci-timeline-relationship-coronavirus-pandemic/3718797001/
86. Long C, Miller Z (2020) Defending against possible Trump revenge Biden pardons Fauci, Milley and Jan. 6 committee members. PBS. https://www.pbs.org/newshour/politics/defending-against-possible-trump-revenge-biden-pardons-fauci-milley-and-jan-6-committee-members
87. Venkatapuram S (2023) Public health and ethics. In: Venkatapuram S, Broadbent A (eds) The Routledge handbook of philosophy of public health, 1st edn. Routledge, London, pp 70–84. https://doi.org/10.4324/9781315675411-7
88. Siegel AW, Merritt MW (2019) An overview of conceptual foundations, ethical tensions, and ethical frameworks in public health. In: Mastroianni AC, Kahn JP, Kass NE (eds) The Oxford

handbook of public health ethics, 1st edn. Oxford University Press, New York, pp 5–11. https://doi.org/10.1093/oxfordhb/9780190245191.013.1
89. Childress JF, Bernheim RG (2008) Public health ethics: public justification and public trust. Bundesgesundheitsblatt Gesundheitsforschung Gesundheitsschutz 51(2):158–163. https://doi.org/10.1007/s00103-008-0444-6
90. Merritt MW, Hyder AA (2019) An overview of the ethics of health systems. In: Mastroianni AC, Kahn JP, Kass NE (eds) The Oxford handbook of public health ethics, 1st edn. Oxford University Press, New York, pp 109–114. https://doi.org/10.1093/oxfordhb/9780190245191.013.10
91. Venkatapuram S (2019) Health disparities and the social determinants of health: ethical and social justice issues. In: Mastroianni AC, Kahn JP, Kass NE (eds) The Oxford handbook of public health ethics, 1st edn. Oxford University Press, New York, pp 266–276. https://doi.org/10.1093/oxfordhb/9780190245191.013.24
92. Smith MJ (2023) Social justice and public health. In: Venkatapuram S, Broadbent A (eds) The Routledge handbook of philosophy of public health, 1st edn. Routledge, London, pp 333–346. https://doi.org/10.4324/9781315675411-26
93. Baylis F, Kenny NP, Sherwin S (2008) A relational account of public health ethics. Public Health Ethics 1(3):196–209. https://doi.org/10.1093/phe/phn025
94. Kenny NP, Sherwin SB, Baylis FE (2010) Re-visioning public health ethics: a relational perspective. Can J Public Health 101(1):9–11. https://doi.org/10.1007/BF03405552
95. Robichaux C, Sauerland J (2021) The social determinants of health, COVID-19, and structural competence. Onl J Iss Nurs 26(2). https://doi.org/10.3912/OJIN.Vol26No02PPT67
96. Braveman PA, Arkin E, Proctor D et al (2022) Systemic and structural racism: definitions, examples, health damages, and approaches to dismantling. Health Aff (Proj Hope) 41(2):171–178. https://doi.org/10.1377/hlthaff.2021.01394
97. Egede LE, Walker RJ (2020) Structural racism, social risk factors, and Covid-19 - a dangerous convergence for Black Americans. N Engl J Med 383(12):e77. https://doi.org/10.1056/NEJMp2023616
98. Williams DR (2012) Miles to go before we sleep: racial inequities in health. J Health Soc Behav 53(3):279–295
99. Williams DR, Cooper LA (2019) Reducing racial inequities in health: using what we already know to take action. Int J Env Res Public Health 16(4):606
100. Farmer B (2020) The coronavirus doesn't discriminate, but US health care showing familiar biases. NPR. https://www.npr.org/sections/health-shots/2020/04/02/825730141/the-coronavirus-doesnt-discriminate-but-u-s-health-care-showing-familiar-biases
101. Kim SR, Vann M, Bronner L et al (2020) Which cities have the biggest racial gaps in COVID-19 testing access? FiveThirtyEight. https://fivethirtyeight.com/features/white-neighborhoods-have-more-access-to-covid-19-testing-sites/
102. Gray DM II, Anyane-Yeboa A, Balzora S et al (2020) COVID-19 and the other pandemic: populations made vulnerable by systemic inequity. Nat Rev Gastroenterol Hepatol 17:520–522. https://doi.org/10.1038/s41575-020-0330-8
103. Kapucu K, Moynihan D (2021) Trump's (mis)management of the COVID-19 pandemic in the US. Policy Stud 42(5):592–610. https://doi.org/10.1080/01442872.2021.1931671
104. Somvichian-Clausen A (2020) Trump's use of the term 'Chinese Virus' for coronavirus hurts Asian Americans, says expert. The Hill. https://thehill.com/changing-america/respect/diversity-inclusion/489464-trumps-use-of-the-term-chinese-virus-for/
105. Freking K (2020) Trump suggests US slow virus testing to avoid bad statistics. Associated Press. https://apnews.com/article/virus-outbreak-donald-trump-ap-top-news-joe-biden-tulsa-476068bd60e9048303b736e9d7fc6572
106. Gómez-Vírseda C, Usanos RA (2021) Relational autonomy: lessons from COVID 19 and twentieth century philosophy. Med Health Care Philos 24:493–505. https://doi.org/10.1007/s11019-021-10035-2
107. Lee J (2023) Relational approaches to personal autonomy. Philos Comp 18(5):1–14. https://doi.org/10.1111/phc3.12916

108. Jeffrey DI (2020) Relational ethical approaches to the COVID-19 pandemic. J Med Ethics 46:495–498. https://doi.org/10.1136/medethics-2020-106264
109. Wilder JM (2021) The disproportionate impact of COVID-19 on racial and ethnic minorities in the United States. Clin Infect Dis 72(4):707–709. https://doi.org/10.1093/cid/ciaa959
110. Geneviève LD, Martani A, Wangmo T et al (2022) Precision public health and structural racism in the United States: promoting health equity in the COVID-19 pandemic response. JMIR Public Health Surveill 8(3):e33277. https://doi.org/10.2196/33277
111. Helm-Murtagh SC, Erwin PC (2024) Building a new generation of public health leaders forged in a public health crisis. Am J Public Health 114(6):626–632. https://doi.org/10.2105/AJPH.2024.307633

Part VI

South America

14 Emergency and Manifestations of Problematic School Coexistence Among Students After Returning from Confinement Due to the COVID-19 Pandemic in Chile

Laura Rueda-Castro, Pumari Mena-Gutiérrez, Ana Francisca Pérez-Jara, Isidora Espinoza-Carrillo, and Arantza Kessi-Gutiérrez

Abstract

The COVID-19 pandemic caused unprecedented disruption to education systems worldwide. The prolonged school closures, lockdowns, and the abrupt transition to remote learning had profound effects on students' behavior, particularly their socioemotional development and school dynamics. The presented work captures the experiences of five public school teachers in the metropolitan region of Santiago, Chile, as a case study. They describe the disruptions in school coexistence observed upon returning to in-person activities.

Keywords

Pandemic · Quarantine · Social disruption · School violence · Mental health

Introduction

The COVID-19 pandemic in Chile triggered confinement, strict for the entire country for 8 months, with public outings only permitted with health authorization for a limited time, and in-person classes suspended entirely (in full between March and November 2020) measures in the population that had a significant impact on the teaching and learning processes for teachers and students throughout the country [1]. For 2 years, the educational community was forced to teach classes remotely,

L. Rueda-Castro (✉) · P. Mena-Gutiérrez · A. F. Pérez-Jara · I. Espinoza-Carrillo · A. Kessi-Gutiérrez
Universidad de Chile, Santiago, Chile
e-mail: lrueda@uchile.cl

L. P. Marquez et al. (eds.), *COVID-19 in the Pacific Rim and Beyond*,
Collaborative Bioethics 6, https://doi.org/10.1007/978-3-032-21218-4_14

and now (2025) that it has been possible to return to in-person classes, this has been marked by an increase in violence among students and its dissemination in the media [2].

So far this year (2025), reports of physical and psychological abuse in the school environment have increased by 60%, compared to the number of such accusations made in the same period, prior to the pandemic (2019). The breakdown by the Superintendency of Education revealed that in 2021, 1586 complaints were registered regarding problems in the area of school coexistence, of which 638 belong to schools in the Metropolitan Region [3].

In this sense, school coexistence is understood as a process of learning to live together that involves the interrelation of all educational actors [4, 5] and that is framed in—and at the same time generates—the daily life of the school [6]. School coexistence is a dynamic constructive process, in which people negotiate meanings, reach agreements, and recognize others, which causes a sense of belonging and group identity [6]. It occurs within the institutional policies and practices of each local school culture, that is, it considers both the disciplinary rules and the institutional history itself [7].

School coexistence, then, is built on the permanent interaction between the different members of the educational community (teachers, students, directors, families, and staff), in conversations of consensus and dissent, in sharing spaces and daily activities [4]. However, in research, there is a lack of analyses of school coexistence from the point of view of teachers, who are also active actors in the school community. This chapter aims to fill this gap.

The data analyzed are part of a study (degree thesis, School of Occupational Therapy, Faculty of Medicine, University of Chile, 2022) that aims to describe teachers' observations regarding problematic school coexistence upon returning to in-person classes, 2 years after the pandemic. Ethical safeguards were considered for the equitable selection of participants, the anonymization of personal information of reporting teachers, and the individualized application of informed consent when inviting them to join the study (the study was approved by the local ethics committee).

The study included the participation of five teachers (school teachers, northern area, metropolitan region), who were contacted through relatives who work in schools, through contact with teachers who taught classes to team members during their secondary education, and through an open call on social networks.

For the data collection of this research, a semi-structured interview (original and validated by an academic committee) was used, which was applied to the five chosen professors from different educational establishments in the Metropolitan region, after signing an informed consent which ensured the protection of their identity and confidentiality of their personal data. The questions asked were directed toward the themes of school coexistence, violence, and mental health of students. This data collection technique has the characteristics of being flexible and dynamic, being adaptable to each interviewee to thoroughly understand their experiences and make their participation more enjoyable [8]. The semi-structured interview provides the possibility of knowing in depth the perceptions of teachers in this scenario and thus reconstruct, in the set of individual interviews, a social sense [9].

The interviews were conducted and transcribed by the researchers themselves, and the recordings were stored on a personal computer. Participants' personal data were anonymized.

Results

To facilitate the management of the results found, they are presented based on three central categories emerging from the analysis of the transcripts of the interviews conducted with teachers: Characterization of the school context, manifestations of violence or school conflicts, and mental health and its dynamics upon returning to in-person classes. Then, the interpretation of each is presented, in addition to themes that emerged in the interviews.

Characterization of the School Context

Historically, the social construction of the western sector of the Metropolitan region is characterized by high levels of inequality in terms of health, living conditions, environmental conditions, and housing, among others. These inequalities also affect the educational field.

All participants report having at least one official in charge of school coexistence: 4 out of 5 participants report that this role is exercised by a psychologist and social worker, and of these 4, only 2 also have a counselor. In none of these establishments is there the presence of an occupational therapist as an official of the school community [10]. Regarding school coexistence protocols, this can be identified by 4 out of 5 teachers and only one of them has received formal training in this subject.

Manifestations of Violence or School Conflicts

Participants agree that the vast majority of problematic school coexistence expressions (verbal and physical aggression) occur in students between seventh grade and second year of high school, which is related to socialization among peers that was affected due to the absence of direct face-to-face contact and the creation of bonds during the confinement period.

Common among the testimonies of the participants as emerging categories from the analysis of the transcripts are the following: the occurrence of physical violence (4 out of 5 schools), verbal violence (5 out of 5 schools), sexual harassment (2 out of 5 schools), and cyberbullying (3 out of 5 schools) associated mainly with the creation of confession pages on social networks such as Instagram.

Regarding the occurrence of these acts of violence by students, there is a tendency toward aggression during breaks (4 out of 5 schools) and within the establishment (5 out of 5 schools), and manifestations of violence in the presence of adults responsible for care within the school (3 out of 5 schools).

Mental Health and Its Dynamics in the Return to In-Person Classes

Regarding this category, teachers highlight important changes in the mental health of students. In general, they recognize a greater deficit in terms of the identification and management of emotions (5 out of 5 participants), which is evidenced by an increase in the frequency of panic/anxiety crises (4 out of 5 schools) and depressive symptoms (3 out of 5 schools). This has triggered not only an increase in psychiatric medical leaves in the student body, but also a general deterioration in the mental health of the teaching staff (3 out of 5 participants). Along these same lines, a perception of academic and emotional overload has unleashed more situations of stress, tiredness, and exhaustion (4 out of 5 schools). For example,

> There were cases of penknives being used in fights inside, there were also many cases of people entering the school with toys, let's say… with toy guns, let's say blanks, and that was kind of the trend, but a lot of penknives, a lot of blades, that caught my attention (E2, 2022, p. 26 transcripts)

> Eh, on average, incidents of physical violence have occurred twice a week recently. Very intense fights inside the school, and what's also worrying has been that instead of separating them, their peers record them (E5, 2022, p. 70 transcripts)

On the other hand, with regard to the dynamics within the school context, difficulties in relating (3 out of 5 participants) and low performance (3 out of 5 participants) were identified.

Discussion

The purpose of this research was to describe the experiences of the educational community in relation to problematic school coexistence in the western sector of the Metropolitan Region. Based on the interviews conducted, it can be seen that the pandemic negatively impacted the school coexistence dynamics of some adolescents in this geographic sector (sector to which the participating teachers belong). Chile is a very socially segregated country, and the health crisis was able to demonstrate this—and accentuate it—with remote education: the most vulnerable communities of the population had less access and coverage to the Internet, thus losing the effectiveness of learning [11]. This is relevant to mention given that the western sector of the Metropolitan region manages less income than the eastern sector [12], and furthermore, the schools participating in the research are from municipal and private-subsidized entities, with two of these presenting a high rate of vulnerability. Furthermore, this inequality is not only evident in economic opportunities, but there are also differences in the interactions of family groups: as part of the educational process was delegated to support in homes, those students who do not have networks of this type and/or face situations of family vulnerability are more disadvantaged in terms of their learning [13].

Based on this, it can be seen that the pandemic brought about drastic changes in the teaching process. However, this was not the only area affected, as meeting and socializing spaces were also altered, and difficulties in relating and forming bonds, especially at the lower levels. Thus, from an academic and social perspective, it can be concluded that the health emergency caused a phenomenon of "stagnation" in these areas.

The adaptation process in telematics education, given the impossibility of maintaining in-person classes, was that online classes were implemented as the only viable alternative to ensure the continuity of the school educational process. Interestingly, that there were strengthening and closeness in the relationship between students and their families—referred to in the education superintendence—and reflected by school habits, as described by the teachers, are contrary to what was provided in the interviews, where in addition to family vulnerability, situations of loneliness in the students' homes are mentioned (2 of 5 interviews). On the other hand, the authors point out greater participation in schoolwork and in the completion of tasks, thus improving academic involvement. This, like the above, contradicts the learning difficulties described and a drop in academic performance (3 of 5 interviews).

In this sense, Garay et al. (2018) [14] warn that emotional intelligence—which considers components such as the ability to understand one's own emotions—facilitates the construction of healthy interpersonal relationships; thus, emotional intelligence constitutes a cognitive strength. This approach coincides with the ideas addressed by teachers, who warn of problems in the treatment of emotions and socio-emotional maturation by students, which they link to the lack of regulation of behavior and, therefore, to the manifestation of aggressive behavior. Emotional intelligence not only allows us to understand what we feel, but also how those feelings affect others. When this skill is cultivated, it develops the ability to resolve differences without violence and create safe spaces where relationships can be strengthened.

Faced with these problems, teachers say that they do not have the necessary tools to organize support groups for psychological assistance and, as a result, their mental health is compromised. Most of them assess an increase in psychiatric licenses, and even one of the interviewees refers to an increase in teacher resignations. Its relevance lies in the fact that, as part of the educational community, they also share responsibilities regarding school coexistence. According to Di Leo [15], when families do not fulfill their role of social support, this function is shifted to teachers. Therefore, the fact that the teaching team does not have knowledge on how to address emotional/behavioral problems constitutes a risk for the mental health of students and the social interactions they build.

Now, in relation to how to take action in the face of problematic school coexistence, the teachers in the interviews show interest in being agents of change and collaborating, but they recognize that they do not have the necessary tools or adequate training to address the situation. Thus, according to Di Leo [15], educational institutions must make efforts to incorporate specialized personnel in these emerging problems of their students and coordinate with other governmental and non-governmental networks, so that there is constant supervision and updating with respect to pedagogical practices.

Conclusion

As a final reflection, we can indicate that the results obtained and their subsequent analysis establish mental health as a fundamental factor in school coexistence, hand in hand with the family and socioeconomic context of young people. The lack of emotional regulation, socialization, and conflict resolution skills, exacerbated in part by the 2 years of confinement and the scarce interaction between peers, has generated an alarming increase in school violence that, although it has always existed, is much more evident today given the increase in its frequency and intensity.

Furthermore, confinement during the pandemic exacerbated violence—especially domestic and gender-based violence—due to a combination of social, psychological, and structural factors [16].

In the course of the research, in addition to what was mentioned above, other related phenomena were found, such as: differences in violent behavior according to the gender of the student, the coexistence of different cultural groups, the functioning of the Educational inclusion program (PIE) n schools, and especially the deterioration of the mental health of teachers after the pandemic. Although these topics were not explored in depth, they can serve as a starting point for future research. Another limitation that was found was the fact that none of the schools interviewed is private; therefore, there was no comparison between the type and frequency of school violence with an indicator at high and low socioeconomic levels.

Although in the case of Chile education has undergone various transformations throughout history in search of improving its quality for all young people, there are still pending challenges regarding daily relationships of coexistence, such as school violence, how teachers and families connect with the schools, and mental health, as exposed in this research. All this considering that the quality of education is not only measured from the academic point of view, but also from how socio-emotional skills that are significant for our lives are developed, as Humberto Maturana [17] states: "Today the problem of education is not intelligence, but emotion. If I do not meet the other and do not value it, (…) we cannot educate. Education is not about knowledge, it is about encounters."

References

1. Mineduc Studies Center (2020) Impact of COVID-19 on learning and schooling outcomes in Chile. Analysis based on simulation tool provided by the World Bank. https://www.mineduc.cl/wp-content/uploads/sites/19/2020/08/EstudioMineduc_bancomundial.pdf
2. UNICEF Chile (2022) We will not recover learning if the conditions in which we learn are not emotionally adequate. UNICEF. https://www.unicef.org/chile/historias/no-recuperaremos-aprendizajes-si-las-condiciones-en-que-se-aprende-no-son-adecuadas
3. Superintendency of Education (2022) Statistical Compendium of the Superintendency of Education, 1st, 2nd, 3rd and 4th quarter of 2021 [Information on the number of complaints

filed with the Superintendency of Education in the indicated period, by area, topic and subtopic during the corresponding quarter and year.]

4. Banz C (2008) School coexistence. You value UC. https://centroderecursos.educachile.cl/bitstream/handle/20.500.12246/55810/Clima%2005.pdf?sequence=1
5. Delors J, Amagi I, Carneiro R, Chung F, Geremek B, Gorham W, Kornhauser A, Manley M, Padrón M, Savané MA, Singh K, Stavenhagen R, Won M, Nanzhao Z (1996) There is a treasure in education: report to UNESCO by the international commission on education for the twenty-first century. UNESCO. https://redcampussustentable.cl/wp-content/uploads/2018/03/7-La-educaci%C3%B3n-encierra-un-tesoro.pdf
6. Mena M, Huneeus M (2017) School coexistence for learning and good treatment of all: towards a better understanding of the concept. Cult Educ Soc 8(2):9–20. https://doi.org/10.17981/cultedusoc.8.2.2017.01
7. Díaz S, Sime L (2016) School coexistence: a review of studies on basic education in Latin America. Revista Virtual Universidad Católica del Norte 49:125–145. https://www.redalyc.org/pdf/1942/194247574008.pdf
8. Diaz L, Torruco U, Martinez M, Varela M (2013) The interview, a flexible and dynamic resource. Res Med Educ 2(7):162–167. http://www.scielo.org.mx/scielo.php?script=sci_arttext&pid=S2007-50572013000300009&lng=es&tlng=es
9. Tonon G (2009) The semi -structured interview as a research technique. In: Tonon G, Alvarado S, Ospina H, Lucero P, Botero P, Luna M, Fabris F (eds) Latin American reflections on qualitative research, pp 47–68
10. Villagra AB (2024) Terapia ocupacional en escuelas: un análisis de las producciones argentinas escritas (1985–2015). Cadernos Brasileiros de Terapia Ocupacional 32(spe1):e3802
11. Quiroz C (2020) Covid-19 pandemic and territorial inequality: the worsening of educational inequalities in Chile. Int J Educ Soc Justice 9(3). https://www.researchgate.net/profile/Cristian-Quiroz-Reyes/publication/355424642_Pandemia_Covid-19_e_Inequidad_Territorial_El_Agravamiento_de_las_Desigualdades_Educativas_en_Chile/links/616f9512c891c4663aaa1d28/Pandemia-Covid-19-e-Inequidad-Territorial-El-Agravamiento-de-las-Desigualdades-Educativas-en-Chile.pdf
12. Seremi of Social Development and Metropolitan Family (2021) Metropolitan Region of Santiago index of social priority of communes 2020. https://www.desarrollosocialyfamilia.gob.cl/storage/docs/boletin_interno/INDICE_DE_PRIORIDAD_SOCIAL_2020(1).pdf
13. Murillo J, Duk C (2020) Covid-19 and educational gaps. Latin Am J Incl Educ 14(1):11–13. https://www.scielo.cl/pdf/rlei/v14n1/0718-7378-rlei-14-01-11.pdf
14. Garay O, De Sousa R, Fraile L, López N (2018) Relationship between emotional perception and the development of bullying behavior in adolescent bullies aged 11 to 16 years. In: Ruiz J, López E, Jiménez L, Lobo A, Támara M, Valencia O (eds) Legal psychology. Tools and challenges. The Modern Manual, pp 237–247. https://www.researchgate.net/profile/Ever-Lopez-Cantero-2/publication/354054777_PSICOLOGIA_JURIDICA_Herramientas_y_desafios/links/612170db0c2bfa282a603ecb/PSICOLOGIA-JURIDICA-Herramientas-y-desafios.pdf#page=257
15. Di Leo PF (2011) Violence and social climates in middle schools: experiences of teachers and principals. Educ Pesqui 37(3):599–612. https://doi.org/10.1590/s1517-97022011000300010
16. Medina-Gamero A, Regalado-Chamorro M (2021) Pandemia, confinamiento y violencia de género: un trinomio peligroso. Aten Primaria 53(10):102151
17. Maturana H (2017) To educate is to love [Main speech]. Conference by Humberto Maturana in the municipality of Chiguayante Inaugurating the School Year

Health Decision-Making During the Pandemic in Colombia: Implications for Public Health Ethics Research

15

Simon Turner

Abstract

This chapter addresses research audiences in the field of public health ethics by exploring the ethical implications associated with Colombia's response to the COVID-19 pandemic. The public health response to COVID-19 was led top-down, introduced abruptly, and had severe effects on social and economic activities. This chapter draws on interview data from a wider longitudinal study of health service and public health responses to the pandemic in Colombia. Qualitative findings were subjected to thematic analysis from a public health ethics perspective, resulting in four novel themes presented in this chapter: workforce effects, decision-making approaches, defining the "public" in health interventions, and inter-organisational activities. Following a summary of COVID-19's impact in Colombia, and associated health service and public health responses, the four themes are presented and analysed from a public health ethics perspective. This chapter contributes to the research literature on public health ethics by analysing the ethical implications of introducing interventions in crisis contexts, including the consequences for stakeholder involvement and effects upon different stakeholder groups within health systems, and society more broadly, and draws policy and practice lessons from Colombia's experiences for preparing for future pandemics.

Keywords

COVID-19 · Colombia · Qualitative · Pandemic · Public health ethics

S. Turner (✉)
School of Management, University of los Andes, Bogotá, Colombia
e-mail: s.turner@uniandes.edu.co

L. P. Marquez et al. (eds.), *COVID-19 in the Pacific Rim and Beyond*,
Collaborative Bioethics 6, https://doi.org/10.1007/978-3-032-21218-4_15

COVID-19 Arrives in Colombia

Colombia, with a population of 52 million, is an upper-middle-income country situated in the northwest of South America, bordering the Pacific Ocean and the Caribbean Sea. Like other countries along the Pacific Rim, Colombia was severely affected by a series of waves of COVID-19. Since March 2020, a variety of public health and health service initiatives were mobilised in response to COVID-19 at national and local levels. This chapter analyses some of Colombia's responses using an inductively constructed public health ethics framework. Next, Colombia's experiences of, and responses to, COVID-19 are described; then, the inductive research methods are outlined, followed by a policy and practice analysis from a public health perspective. The chapter concludes with lessons from Colombia's experiences for preparing for future pandemics.

The first case of COVID-19 in Colombia was reported on 6 March 2020; as well publicised at the time, it involved a Colombian national who had recently returned from a visit to Europe. By the end of April, there were 5949 confirmed cases. As of December 2024, 6.39 million cases of COVID-19 have been reported in Colombia, with 142,727 deaths confirmed [1].

Colombia's response to COVID-19 was characterised by a centralised approach, with legal decrees announced by the national government employed to influence social and economic behaviour top-down. The description of national and local public health measures is informed by a published timeline of events [2]. On 12 March 2020, a national "health emergency" was declared; this required the immediate cancellation of mass gatherings or events, with the closure of schools and universities on 15 March. A full state of national emergency was declared on 17 March, which paved the way for financial mechanisms, including an emergency tax measure, on 19 March, and an emergency mitigation fund, on 21 March, to support the economic response to COVID-19. Funds were used to support the health system, to aid vulnerable families, and to sustain businesses. A national lockdown was mandated initially for the period 24 March to 19 April 2020. Use of facemasks in public spaces became mandatory from 4 April. Restrictions to movement were introduced from 9 April, including measures known as *Pico y cedula*, that would only permit people to leave home on alternating days of the week (e.g. for essential grocery shopping) based on their identification card number.

By 12 April, measures were introduced to improve health system capacity. For example, a temporary hospital in the area Corferias, Bogotá, was opened on 20 April to care for non-COVID patients. On the same day, the national lockdown affecting the movement of the general population was extended until 11 May. From 27 April, the construction and manufacturing sectors were able to reopen, reflecting concerns about maintaining jobs, and economic output, in important sectors of the economy. Biosecurity protocols were announced for the technology, telecommunications, and education sectors in May and June allowing, in education, for classes to be conducted online. From May to August, various alerts concerning limited hospital capacity, including high ICU occupancy rates, were announced in Bogotá. In July, it was announced that the national lockdown would be extended until 31

August 2020. Lockdown ended on 1 September 2020 with a new phase of "selective isolation" to support economic reopening. From December 2020, Colombia was affected by further waves of the pandemic, triggering further lockdown and public health interventions from January 2021 to August 2021.

In summary, the public health response to COVID-19 was led top-down, introduced abruptly, and had severe effects upon social and economic activities, notably the initial extended lockdown from March to August 2020. In public statements, public health was prioritised in initial responses to the pandemic, with a particular emphasis on protecting vulnerable groups. Protecting our grandparents, or "nuestros abuelos", was part of the ethical justification of lockdown measures put to the population. This chapter analyses some of those responses from a public health ethics perspective and draws lessons from Colombia's experiences for preparing for future pandemics.

Research Methods

This perspective on public health ethics during the pandemic draws from a wider qualitative study of health system responses to COVID-19 in Colombia [3]. The study of early responses to COVID-19 involved 118 stakeholder interviews conducted between June 2020 and November 2020. These were later supplemented with focus group data, involving 18 participants drawn from the original sample, that were conducted from November 2021 to March 2022. Participants, drawn from national-level entities and from the major cities of Bogotá, Cali and Cartagena, included government entities, national health delivery organisations, health insurance providers, NGOs, scientific associations and guilds, local health planning authorities, integrated networks of providers, public, private and mixed healthcare providers, and professional associations involved in the response to COVID-19. All interviews/focus groups were conducted virtually using video conferencing software (e.g. Microsoft Teams). Interviewees provided informed consent, as required as part of the ethics approval, which was granted, with minimal risk judged, for both phases of the study by the Universidad de los Andes. All recordings, relating to interviews and to focus groups, were professionally transcribed.

The study resulted in a series of peer-reviewed publications; this chapter revisits the findings from relevant publications from a novel, public health ethics perspective. The analysis suggested that a series of ethical issues were involved in decision-making on COVID-19. To explore these issues, the original findings from each paper were summarised, novel connections to public health ethics were drawn, and relevant empirical cases were identified that suggested public health ethics implications. For example, one empirical case was organisational collaboration to improve surveillance of COVID-19. Through extracting and thematically analysing the findings presented in the published papers [4], four cross-cutting themes of relevance to public health ethics were identified: workforce effects, decision-making approaches, defining the "public" in health interventions, and inter-organisational activities. Extraction of potential themes was guided by attending to those that suggested public health ethics implications. Brief extracts from interview quotations used in papers already published from this dataset are used to illustrate the novel public health ethics argument. Where extracts from interview quotations are used, these are attributed to the original paper at the point in the text where they are cited.

Policy and Practice Analysis from a Public Health Ethics Perspective

Workforce

Healthcare professionals' work intensified during the pandemic. The demands of facing COVID-19 affected both front-line and managerial staff within the healthcare system, who bore the brunt of addressing its effects upon health systems. Front-line staff faced new or extended roles and responsibilities. For example, reallocating human resources to support COVID-19-related services, including home care and teleconsultations [5]. Managerial staff needed to work intensively to support the "pivoting" of the health system within and across organisations; this included attending and coordinating a higher intensity of meetings [6]. New responsibilities, which became normalised over an extended period of time, created potential for occupational burnout, as the following interviewee based in Bogotá's local government remarked: *"people were already totally tired and there was no recognition, motivation or compensation for their work"* [7].

What are the ethical implications of expecting more from health workers in emergency contexts? Our analysis suggested that both front-line and managerial level staff faced work-related stress and trauma from the heightened pace of work [5, 6]. However, it is critical through further studies to understand how exposure to risk was affected by role type and level of seniority. This analysis would help with decisions about allocating and supporting human resources to face future pandemics from a social equity perspective. Once that structure of effects across the workforce is mapped, this could inform the provision of monetary payments and non-monetary benefits to support workers based on the specific challenges they are likely to face during pandemics. For example, access to emergency government financing is likely to be required over a sustained period to scale up workforce capacity, and to continue to support health sector workers, so that existing staff are not overburdened by the pressures associated with pandemics with effects that accumulate over time. This is about maintaining organisational slack in allocation of roles and responsibilities to protect staff against workload pressures.

Existing and additional staff also require timely access to training associated with pandemic situations. A reassuring lesson from Colombia's experience was how different organisations across health ecosystems, including the medical training faculties of universities, came together to share resources associated with COVID-19 [5]. Those resources were made available online to develop an emerging educational curriculum that helped staff to manage the uncertainty around managing the disease.

Moral support among peers, at the team and organisational levels, was also an important factor in maintaining the wellbeing of staff, especially among front-line workers in clinical contexts. For example, one clinical team at a hospital in Bogotá reportedly placed additional focus on the health and wellbeing of the group, introducing the motto: *"if you protect yourself, you protect me"* [5]. The "informal" supportive role of clinical teams emerged in response to physical and emotional strains associated with dealing with COVID-19, as well as instances of stigmatisation in wider society, where health workers were seen as potential carriers of disease outside of work in their own communities.

At the organisational level, some healthcare leaders placed special emphasis on how they connected with workers on the front line, maintaining their visibility in relation to clinical spaces and seeking to understand the experiences and perceptions of workers in those spaces through more "open" or empathic forms of leadership, an approach that also characterised public health leadership [8].

Decision-Making Approaches

Processes were accelerated across a range of areas affecting public health. A quickened pace of decision-making was recognised as a critical feature to "get ahead" of the pandemic. Our study's analysis of early responses to the pandemic, both among national-level entities and those responsible for responding in Bogotá, suggested that planning and delivery decisions were characterised by radical uncertainty, as direct experiences and validated knowledge were lacking initially to inform responses. This meant that decisions needed to be made based on a trial-and-error approach, as this representative of a public health provider explained, "*we had to sit and acknowledge that we needed to do things adapting to what we have. It was very hard at the beginning until local guidelines started appearing*" [2]. Ethical implications can be drawn from this approach to decision-making, which are considered next.

On the one hand, this context of uncertainty forced recognition that decisions needed to be taken collectively, based on the perspectives of stakeholders that were ready to hand, rather than by following a fixed approach where relevant knowledge and experiences had already been set. The presence of uncertainty laid the foundations for democratic approaches to decision-making, which this representative of a local health planning organisation suggested was a lesson from the pandemic: *"socializing information and inviting people to have an opinion is something positive"* [9]. Navigating the uncertainty around decision-making extended to the use of peer-led social messaging platforms, such as WhatsApp groups. Peer groups shared COVID-19-related information; their constituency included members of latent international networks to support the response, as clinical peers located in Europe who were ahead temporally in accumulating experience of treating COVID-19 were consulted.

On the other hand, the pace of decision-making needed to respond to COVID-19 may have precluded the involvement of some types of stakeholders. For example, some concern was expressed about the relevance of telemedicine models, which were scaled up rapidly in Colombia as a delivery solution to COVID-19 in the context of physical distancing measures, for a significant proportion of Colombia's population, as this representative of a health professional association stated: *"more than 20 million Colombians do not have digital connectivity, they do not have broadband, how are you going to talk about teleconsultation?"* [6]. The rapidity of decision-making associated with COVID-19 raises ethical questions about (a) which types of stakeholders have an elevated role in decision-making in time-limited contexts and (b) how the criteria for making decisions in such contexts may vary from decision-making in non-urgent contexts.

With regard to service changes, it is critical that the quality of evidence and the level of stakeholder involvement that informed decision-making be audited, so that the validity of decisions can be judged. Given the novel threat posed by COVID-19, evidence and policy often needed to be constructed simultaneously. What are the pros and cons of novel approaches to decision-making from an ethical standpoint?

Defining the Public in Public Health

To address the spread and population health effects of COVID-19, population-wide intervention measures were introduced. These large-scale measures were distinctive in affecting the whole population, while having differential effects by population group. The simultaneous health, economic, and social effects of COVID-19 created challenges for constructing policy responses that addressed policy problems in one domain without having negative effects upon others. For example, prioritising public health through lockdown measures to control the circulation of COVID-19 can have further economic and social effects. How are the interests of different population groups acknowledged, ordered, and prioritised in large-scale interventions? Public health ethics should be informed by social equity criteria, e.g. a desire to protect the most vulnerable in society, as we see through universal or subsidised healthcare systems that seek to maximise access to care, unconstrained by the ability to pay. A policy challenge associated with COVID-19, as exemplified by the introduction of lockdown measures in Bogotá [5, 10], was that the whole population were affected by policy changes, in one domain of human activity or another.

The introduction of national and local lockdown measures to support parts of the population that were particularly vulnerable to COVID had a series of knock-on effects for other parts of the population. Enterprise owners and employees, especially in the retail sector, were affected by lockdown measures. Social effects included concerns about domestic violence associated with extended confinement at home as well as risks of physical and verbal abuse in public spaces (in Bogotá, there were some reports of aggression toward perceived immigrants as potential carriers of disease and concern about transphobia, when a policy to limit people in public space involved alternating permission to be outside by gender) [11–13].

Lockdown measures finally ended in response to public demonstrations in Spring 2021, when the social and economic effects of lockdown upon young people and women in particular were acknowledged, despite this coinciding with the highest ICU occupancy rate of 97% in Bogotá [8]. Mayor of Bogotá at the time, Claudia Lopez, who had prioritised evidence related to the spread of COVID-19 throughout the policy response in the city, openly acknowledged this irony when announcing the end of lockdown measures: "*It sounds absolutely contradictory, from an epidemiological point of view, to have 97% Intensive Care Unit (ICU) occupancy and announce a reopening; but from the point of view of the social, economic and political context, with deep institutional distrust, unacceptable poverty and unemployment that is especially affecting women and young people, it is necessary to do it*" [14].

With regard to public health ethics, this case highlights challenges associated with responding to the needs of different population groups in the context of a pandemic that has sudden, population-wide effects. In Colombia, the health, economic, and social consequences of the pandemic—each with differential effects across the population—were at least acknowledged, and attempts made at accommodating needs within each domain in policy responses. The need for collective public action to support the implementation policy responses was underpinned by "hard" and "soft" mechanisms; that is, a legal decree that enforced compliance with lockdown measures, on the one hand, and using espoused shared values to appeal to a "culture of citizenship", on the other, to garner support for complying with top-down action to protect the most vulnerable in society, including "nuestros abuelos", our grandparents. From an ethical standpoint, the pandemic highlighted the difficulties of ordering simultaneous health, economic, and social pressures, especially where placing policy emphasis on one domain appeared to have knock-on effects for the others. With the post-pandemic rise of authoritarian leadership in Europe and the United States, the approach to public health policymaking during the pandemic appeared to foreshadow more autocratic, decisive, and swift approaches to national policymaking. For example, the shift in international immigration policy and suspension of governmental diversity, equity, and integration programmes in the opening weeks of Trump's second administration.

Considering ongoing policymaking, the pandemic demonstrated that it was possible to change collective social behaviour through national-level intervention by the government. It is important to recall that different approaches were employed to influence collective behaviour: there was evidence of autocratic approaches to change enforced via legal decree, but alongside mandated change determined in the higher echelons of government, a desire to base decisions on evidence was also visible, such as monitoring of ICU occupancy rates to inform local lockdown decision-making. There was also an appeal to progressive values among the population, as illustrated by the suggested need to protect the most vulnerable in society through public participation in interventions.

It seems likely that population-wide measures were sustainable for so long in Bogotá, with the initial lockdown lasting from March to August 2020, due to a keen desire within government to communicate changes clearly, drawing on the best available evidence, and by appealing to progressive values. Continuing to employ autocratic approaches to policymaking in the absence of these other characteristics (trusted communication, validated evidence, and progressive values) is likely to (a) struggle to solve the policy problems described due to the neglect of validated evidence and (b) rapidly encounter resistance within the population due to concerns about communication and values. The evidence from Colombia suggests that top-down population-wide interventions are hard to sustain, and if traction is to be gained, they require complementary mechanisms for demonstrating success and maintaining population-level support.

Inter-organisational Collaborations

To support the response to COVID-19, collaborations were mobilised both within the health system and with other sectors of the economy. For example, improved community surveillance of COVID-19 was stimulated by informal coordination of different health insurers (known as "EPS" in Spanish), as various providers came to an agreement to coordinate testing of service users, such as members of the same family. Such informal agreements were then institutionalised. At the inter-sectoral level, various activities were pursued that involved contributions from the private sector (economic, strategic, and logistic roles), and academia (advisory, infrastructure, and networking roles) [9]. Inter-sectoral collaborations were the product of multiple forms of motivation, from intrinsic motivation in the context of crisis to make a contribution to society, through to economic and reputational interests that threatened particular organisations. One example of reputational interests was universities' desire to demonstrate to government entities their relevance in society. The crisis therefore prompted new forms of inter-sectoral collaboration that helped to build trust, communication, and accountability through the practice of interaction. In the process, elements of "fragmentation" across the health eco-system that beset collaboration in the pre-COVID system in Colombia were overcome [15, 16].

Ongoing health reform processes in Colombia, as well as financial challenges among providers within the current system, threaten to undermine new collaborations struck in response to COVID-19. What are the ethical implications of relying on "informal" linkages among economic units, especially where these depend on intrinsic motivation and shifting dispositions like collective goodwill? How should emergent collaborations be appropriately supported so they are sustained? These are not easy questions to answer. On the one hand, areas of progress with regard to collaboration should be formalised and institutionalised, but, on the other, such formalisation could outpace the intrinsic motivation to contribute in a context of crisis. Once crises pass, it is unclear how collaborations that demanded adapted or additional forms of work might be sustained. For example, coordinating insurance providers across the health system reportedly involved the holding of daily, rather than monthly, round table meetings [6]. While such intensity of interaction might have facilitated alignment across otherwise disconnected entities, the desire and wherewithal to maintain intensive interaction among normally competing entities is unlikely to be sustained without a motivating crisis.

Conclusion: Preparing for Future Pandemics

Being better prepared for the next pandemic means addressing known challenges with public health planning and health system organisation. Colombia's health system is currently undergoing restructuring that was stimulated by a health reform bill published in February 2023 by the incoming government [17]. It is critical that lessons from responding to COVID-19, including achievements and challenges, are captured to inform future responses. For example, while the reform bill spells out a desire to restructure the role of private insurance companies, in favour of an increased role for the state in governing service provision, those actors also gathered critical experiential knowledge in responding to the pandemic that should be acknowledged and captured in reform processes. For example, emergent forms of coordination to improve responses like surveillance of COVID-19. Internationally, where incumbent governments that managed COVID-19 have been succeeded by other parties, sometimes with differing ideologies concerning how to provide public health, drawing lessons from the pandemic should remain a key learning activity irrespective of party politics. Oversight of lessons past could leave countries avoidably underprepared for the next pandemic.

The themes presented suggest four key lessons from Colombia's experiences of the pandemic. First, relating to the workforce theme, it is clear that both front-line and managerial staff are the critical actors in responding to public health crises. The health workforce needs to be valued accordingly, in monetary terms, and with regard to their prosperity to aid their recovery from the last pandemic, which continues to have tracings that affect staff. This should include maintaining slack and giving access to wellbeing programmes so the workforce is better positioned to respond at the outset of the next public health challenge. Appropriate workforce measures are formalising durable workforce contracts, including health and social security benefits; encouraging leadership at the team, departmental, and organisation-wide level that emphasises improving morale, while staving off occupational burn-out; and ensuring clinical settings are adequately staffed, and replete with sufficient infrastructure, including biosecurity elements, to protect individual health workers as they go about their everyday activities.

Second, regarding decision-making approaches, decisions on service change were undertaken at a heightened pace during the pandemic. For future pandemics, it is critical that processes for facilitating rapid stakeholder involvement, and the criteria used for informing decisions in urgent contexts, are planned in advance. From an equity perspective, this should include an assessment of potential impact at different health system levels, from population health effects, through to operational level staff, and individual patients/carers.

Third, defining public need in public health interventions can become problematic where large-scale interventions are introduced that can have knock-on effects across health, economic, and social domains. The scale of impact of such interventions means that different stakeholder groups affected by interventions need to be defined and monitored. COVID-19 illustrated that, while sudden and drastic public health measures can be introduced population wide, the public's ongoing cooperation with such measures cannot be assumed. Concerning the role of the public in decision-making, methods for communicating clearly with, and involving the population in, the design and implementation of large-scale intervention measures such as "lockdowns" need to be a critical part of the approach. This extends to the use of digital stakeholder involvement methods to help order different public health needs, on the one hand, and collecting and sharing ongoing data about the effects of interventions across the health, economic, and social domains, on the other, so that the impact of interventions, including potential unintended consequences, can be tracked and addressed.

Fourth, relating to inter-organisational activities, crises can stimulate new forms of collaboration among otherwise "fragmented" entities, where trust, communication, and accountability had broken down, or at best was patchy, historically. New forms of collaboration, e.g. among insurance companies, became possible when a shared sense of purpose, coupled with collective fear about problem-solving in this case, was constructed among organisations. To prepare for future pandemics, a durable shared purpose across the health system needs to be constructed, which serves to bind professions and organisations together irrespective of the shifting pressures in the external environment. Naming that shared or binding purpose is not straightforward. It is unlikely to come from changes imposed upon stakeholders top down; instead, substantive deliberation and dialogue is needed among stakeholders to agree upon the values that should underpin the delivery of public health, including their own definition of "solidarity", that has been emphasised by the World Health Organization [18], and naming the barriers that need to be worked through so that solidarity can be prioritised. Areas of improved collaboration within the health system and across sectors, including private enterprise and academia, should be captured and formalised so that the learning from "learning together" is not lost. Such formalisation could be supported by durable infrastructure grants that support collaboration across health ecosystems around themes of common interest, at the national and regional levels.

In conclusion, the experience of COVID-19 in Colombia raised a series of ethical questions about what drives the provision of health services, from naming a collective purpose, through to how state-level actors approach population-level interventions, that is, how public priorities are ordered and how to approach communication with the public. Experiences of the pandemic also prompt reflection on what value is placed upon the health workforce, in the short and long term, and how lessons are carried over from one period to the next in the context of ongoing political turbulence.

References

1. Mathieu E, Ritchie H, Rodés-Guirao L et al (2020) COVID-19 pandemic. OurWorldInData.org. https://ourworldindata.org/coronavirus. Accessed 15 Nov 2022
2. Turner S, Ulloa AM, Valencia Godoy V et al (2021a) Learning from history or reacting to events? Colombia's navigation of major system change in response to COVID-19. In: Waring J et al (eds) Organising care in a time of Covid-19. Palgrave Macmillan, Cham, p 67

3. Turner S, Niño N (2020) Qualitative analysis of the coordination of major system change within the Colombian health system in response to COVID-19: study protocol. Implement Sci Commun 1:1–8
4. Bradley EH, Curry LA, Devers KJ (2007) Qualitative data analysis for health services research: developing taxonomy, themes, and theory. Health Serv Res 42(4):1758–1772. https://doi.org/10.1111/j.1475-6773.2006.00684.x
5. Turner S, Niño N, Segura C et al (2021b) Organisational responses to mitigate the effects of COVID-19 on healthcare workers: a qualitative case study from Bogotá, Colombia. BMC Health Serv Res 21:1–10
6. Turner S (2022) 'We are all vulnerable, we are all fragile': COVID-19 as opportunity for, or constraint on, health service resilience in Colombia? Public Manag Rev 25(10):1881–1902. https://doi.org/10.1080/14719037.2022.2052944
7. Turner S, Fernandez Gonzalez DP (2024) Strengthening health workforce well-being: lessons for research from the COVID-19 pandemic. In: Rezaei N (ed) Integrated science for sustainable development goal 3. Springer, Cham, p 313
8. Turner S, Schorch SL, Fernandez Gonzalez DP et al (2023) Public leadership of Covid-19: documentary analysis of the role of the Mayor's Office in shaping policy responses in Bogotá, Colombia. In: Brookes S (ed) Research handbook on public leadership. Edward Elgar, Cheltenham, p 178
9. Turner S, Ulloa AM, Niño N, Valencia Godoy V (2022) The role of intersectoral action in response to COVID-19: a qualitative study of the roles of academia and the private sector in Colombia. Int J Health Policy Manage 11(9):1913–1925. https://doi.org/10.34172/ijhpm.2021.100
10. Turner S, Wright J (2024) What role does social equity play in resilient public management? Implementing "lockdown" measures in response to COVID-19 in Australia and Colombia. In: Wiley K et al (eds) Social equity and public management theory: a global outlook. Routledge, New York, p 127
11. Sandoval T (2020a) New quarantine policy in the capital: Pico y género. The Bogotá Post. https://thebogotapost.com/new-quarantine-policy-in-the-capital-pico-y-genero/45796/ Accessed 15 Nov 2022
12. Sandoval T (2020b) What's been the impact of Bogotá's pico y género? The Bogotá Post, 24 April. https://thebogotapost.com/whats-the-impact-of-bogotas-pico-y-genero-been/46190/ Accessed 15 Nov 2022
13. Sandoval T (2020c) Domestic violence: the 'shadow pandemic'. The Bogotá Post, 10 May. https://thebogotapostcom/domestic-violence-the-shadow-pandemic/46472/ Accessed 15 Nov 2022
14. Ramírez LJ (2021) Alcaldesa presentó propuesta para reabrir Bogotá responsablemente desde junio 8. Alcaldía de Bogotá, 27 May. https://bogotagovco/mi-ciudad/desarrollo-economico/asi-propone-la-alcaldesa-reabrir-bogota-desde-el-8-de-junio Accessed 15 Nov 2022
15. Jaime S (2016) Participación social en sistemas de salud fragmentados: ¿una relación *virtuosa?* Revista Gerencia y Políticas de Salud 15:38–56
16. Hernández M (2022) Reforma sanitaria, equidad y derecho a la salud en Colombia. Cad Saude Publica 18:991–1001
17. Ministry of Health and Social Protection (2023) Health reform bill project. https://www.minsalud.gov.co/sites/rid/Lists/BibliotecaDigital/RIDE/DE/proyecto-ley-reforma-salud-msps.pdf Accessed 26 Mar 2025
18. World Health Organization (WHO) (2023) Zero draft of the WHO CA+. https://apps.who.int/gb/inb/pdf_files/inb4/A_INB4_3-en.pdf Accessed 19 Nov 2024

A Critical Overview of the COVID-19 Pandemic Response, Preparedness, and Management in Ecuador

16

Enrique Teran, Esteban Ortíz-Prado, Jonathan Guillemot, and Eduardo Vasconez

Abstract

COVID-19, caused by a novel SARS-CoV-2 virus, emerged in Wuhan, China, in December 2019 and spread globally at an alarming rate. Latin America experienced the most lethal outbreak, with Ecuador reporting the highest number of deaths per capita in a single day, making it one of the most severely affected countries worldwide. Despite a surge in infections, the country faced critical challenges, including limited access to diagnostic testing, overwhelmed hospital capacities, and a shortage of healthcare personnel, further exacerbated by insufficient protective equipment and inadequate infrastructure. Notably, excess mortality exceeded 88,000 deaths, including more than 100 healthcare workers. Ecuador's healthcare system also struggled with bureaucratic challenges, ineffective crisis management, and corruption, culminating in questionable handling of the national vaccination campaign. By mid-July 2022, the country had reported over 927,000 infections and 35,000 deaths. However, a subsequent decline in cases created a deceptive sense of safety, paving the way for another wave by December 2022. Despite these challenges, by September 2021, vaccination efforts led to nine million fully vaccinated individuals. The evolving pandemic

E. Teran (✉)
Escuela de Medicina, Colegio de Ciencias de la Salud, Universidad San Francisco de Quito USFQ, Quito, Ecuador
e-mail: eteran@usfq.edu.ec

E. Ortíz-Prado · E. Vasconez
One Health Group, Universidad de las Américas, Quito, Ecuador

J. Guillemot
Escuela de Medicina, Colegio de Ciencias de la Salud, Universidad San Francisco de Quito USFQ, Quito, Ecuador

Instituto de Medicina Social and Desafíos Globales, Universidad San Francisco de Quito USFQ, Quito, Ecuador

L. P. Marquez et al. (eds.), *COVID-19 in the Pacific Rim and Beyond*,
Collaborative Bioethics 6, https://doi.org/10.1007/978-3-032-21218-4_16

response in Ecuador underscores the complex interplay of health, policy, and sociopolitical factors in managing a public health crisis.

Keywords

COVID-19 · SARS-CoV-2 · Ecuador · Pandemic · Preparedness

Introduction

The first reports of a cluster of atypical pneumonia cases in Wuhan, China, caused by a new infectious agent responsible for causing COVID-19, were described in December 2019. The World Health Organization (WHO) declared the coronavirus disease 2019 a global pandemic on March 11, 2020 [1, 2]. This disease spread rapidly throughout the world, and by April 14, 2020, more than two million people had been infected, and 120,000 deaths had occurred in more than 210 countries [3]. According to WHO data, COVID-19 caused 15 million deaths worldwide. Recently, from July 17 to August 13, 2023, more than 1.4 million new cases of COVID-19 were reported, representing an increase of 63%; more than 2300 additional deaths occurred [4, 5].

Two months after the COVID-19 pandemic was declared, Latin America, along with the Caribbean, became the epicenter of this disease. While representing less than 10% of the global population, the region accounted for more than 27% of the deaths caused by COVID-19 worldwide [6]. By March 2022, 65.4 million confirmed cases and 1.65 million deaths had been reported for this region [7]. According to the WHO, numerous countries in Latin America and the Caribbean have been unable to meet adequate health service standards, falling below the global average of 2.9 hospital beds per 1000 inhabitants, leading to the system's collapse [8, 9].

Despite efforts to control its spread, SARS-CoV-2 proliferated exponentially, making it one of the hardest-hit countries in the world. The scarcity of diagnostic tests, with 7.46 PCR tests per 10,000 people, contributed to the difficulties encountered in the country, which finally led to the high mortality experienced in Ecuador [9, 10]. This led to a minimum of 80,000 excess deaths, culminating in 7 deaths per 100,000 people in a single day, a rate five times higher than the United States experienced during its worst wave [2, 10–12]. Lastly, the situation in Ecuador was worsened by the shortage of hospital beds in Ecuador (at that time had only 1.5 hospital beds for every 1000 inhabitants). The epicenter of COVID-19 at that time was located in the coastal city of Guayaquil, where patients with COVID-19 were left to die in their homes, and bodies appeared on the streets [13].

Timeline of COVID-19 in Ecuador: Initial Phase

In Ecuador, on January 26, 2020, the Ministry of Public Health of Ecuador (Ministerio de Salud Pública, MSP) announced a suspected case of COVID-19 in a 49-year-old Chinese citizen [14]. To confirm the case, a sample from the suspected case was sent for analysis to the Centers for Disease Control and Prevention (CDC) in Atlanta, United States [15]. However, it was not until February 4 that the MSP ruled out the presence of SARS-CoV-2 based on the CDC results. The Chinese citizen eventually died on February 7 from a respiratory disease, although the hospital blamed hepatitis B and severe pneumonia as the causes of death [16]. The medical community is convinced that this one was the first case of COVID-19 in Ecuador.

The MoH announced that the National Public Health Research Institute (INSPI), the WHO-accredited National Reference Center for Influenza and other Respiratory Viruses, had the necessary reagents for testing to diagnose possible SARS-CoV-2 cases [17]. Hence, the MOH should have generated a crisis management strategy. Instead, the ministry opted to take a back seat on the issue. The National Secretariat of Risk Management, a political entity void of health or epidemiology expertise, took the lead under the Emergency Operations Committee (Comité De Operaciones De Emergencia Nacional, COE) [18]. Inexplicably, and because of an apparent lack of national capacity, the COE sought assistance from the Pan American Health Organization (PAHO).

On February 14, 2020, an Ecuadorian citizen residing in Spain entered the country through the José Joaquín de Olmedo airport, Guayaquil, who later tested positive, and two weeks later, on February 29, the MOH officially reported the first case of COVID-19. Then, between his arrival and the confirmation, the patient attended several welcome parties, which led to an epidemiological encirclement of 177 people [19]. Doubts remain whether it was truly the first case, as posterior analyses suggested that a Chinese tourist attended a public hospital in Quito with an unusual respiratory infection had likely been infected several weeks before the sample was collected and later confirmed as positive by the CDC.

In retrospect, the initial response to the COVID-19 pandemic was inadequate. First, authorities were reluctant to declare the health emergency earlier and chose to wait until the WHO did so. Second, the ministry ignored expertise from academia, which, based on the evidence available at the time, suggested closing the airports or at least requiring travelers to quarantine upon arrival in the country. Additionally, the ministry also did not take advantage of the existing capability of universities to immediately provide support with their molecular biology laboratories to generate diagnostic tests for COVID-19 by RT-PCR. With full knowledge of the facts, such support was not recommended by the local representation of PAHO since PAHO expected PAHO-validated laboratories.

Timeline of COVID-19 in Ecuador: The Emergency

The Ministerial Agreement No. 00126–2020, issued on March 11, declared a State of Health Emergency in the National Health System [20]. On March 16, 2020, with 58 identified positive cases of COVID-19 and 2 deaths, the President of Ecuador, Lenín Moreno Garcés, declared a State of Emergency, through Executive Decree No. 1017 [21]. To contain the spread of the disease, the exercise of the rights to freedom of transit and freedom of association, and assembly was suspended. A curfew was declared, prohibiting the circulation on roads and public spaces nationwide. In subsequent days, Universidad San Francisco de Quito launched the campaign #Irestatmyhome (#yomequedoencasa, in Spanish) on social networks to raise awareness among the population about the importance of avoiding social contact to control the transmission of the virus.

On March 21, as 532 cases and 7 COVID-19-related deaths were reported and as confusion and public anxiety escalated, MOH Minister, Dr. Catalina Andramuño, resigned from her post, to be replaced by Dr. Juan Carlos Zevallos [22]. Concomitantly, a great number of epidemiology and public health "experts" appeared, each one defending a different point and approach. Such a cacophony undoubtedly blurred the objectives and desired outcomes. With daily mathematical trend projections, the lack of reliable and replicable scientific data became overwhelming [23]. To support the decision-making process, the Academy of Sciences of Ecuador (ACE) made its contingent of experts available to the COE. Once again, however, the MOH and COE ignored the resource [24, 25].

By April 18, 2020, 9468 positive cases of COVID-19 and at least 474 deaths had been officially reported in Ecuador, with men representing 55.4% of the reported cases. The overall infection rate was estimated to be 51.1 per 100,000 people, and the confirmed overall mortality rate was 2.7 per 100,000 people [1]. The maximum number of deaths in a single day in Ecuador occurred on April 4, 2020, with 1120 deaths. In Ecuador, the first wave hit the coastal provinces hardest during March, April, and May 2020. The second wave came concomitantly with accelerated infection rates in the central Sierra. By December 2020, infection reports began to decrease, which generated a false sense of security, which, in turn, led to declining respect for social distancing rules. This, in turn, fueled a surge in new cases and deaths, a prelude to the third and deadliest wave, which ultimately caused more fatalities than the previous waves combined [26].

To date, the issue of data management related to the epidemiological monitoring of the COVID-19 pandemic in Ecuador remains unresolved. Significant gaps persist in the availability and transparency of key indicators, including the number of diagnostic tests performed, confirmed cases, hospital bed occupancy rates, and mortality figures. Therefore, it remains impossible to correlate the authorities' claim of a "successful management" of the pandemic with data or hospital reality. If it were not for citizen contributions, such as ECUACOVID (https://github.com/andrab/ecuacovid), which compiles the various disaggregated data, Ecuador would likely not appear in world databases of the pandemic.

The availability of diagnostic tests for SARS-Cov2 infection by RT-PCR was insufficient for three reasons [27]: first, the decision to centralize the analysis of samples in the facilities of the Ministry of Public Health; second, the refusal to build a real and effective network of laboratories at the university level; and third, a lack of control by the Regulatory Agency (ARCSA) in the supply of the so-called "rapid tests" for antibodies. Thus, diagnoses were mainly linked to patients arriving at hospitals, with extremely high positivity rates and a backlog of unprocessed samples that made timely action impossible. The few universities that tried to support diagnosis encountered, in most cases, a bureaucratic barrier of requirements and a lack of access to low-cost supplies and reagents [28]. In parallel, the private sector responded with a supply of tests characterized by high cost, increasing inequity in the population [29]. The ARCSA irresponsibly granted sanitary registration to countless suppliers of "rapid tests" or antibody tests, without any type of validation or quality control. This opened doors to a lucrative and out-of-control business, yielding a series of false negative results and in turn increasing transmission rates [30].

Infections skyrocketed, without diagnostic tests, and hospital response capacity. The situation was made worse by recurrent issues of resources dedicated to health: lack of human resources (health personnel in general: doctors, nurses, support staff) and infrastructure. There were not enough professionals in health units, and those that existed were poorly paid, overworked, and reported low job satisfaction. The already limping healthcare system received countless patients requiring emergency and specialized care, which could not be provided. The investment in infrastructure was minimal, and supplies and medicines were scarce [31]. This abhorrent situation led to a high number of infected health workers: by mid-April 2021, it is estimated that more than 100 physicians, 100 nurses, 70 psychologists, 70 dentists, and many other health workers had died [32]. Naming them heroes should not mean that most of these deaths could have been prevented with an efficient and organized health policy. The collapse of the Ecuadorian healthcare system leaves a tremendous and long-lasting psychological impact on helpless staff.

By mid-July 2022, COVID-19 infections surpassed 927,000 cases and consequential deaths exceeded 35,000. However, as in other countries, reported cases have since decreasing, possibly due to underdiagnosis or because people not seeking hospital care. Then, by November 2022, in epidemiological week 45, only 661 new COVID-19 cases were registered. Subsequently, Ecuador experienced a new surge, and by mid-December 2022, cases had reached 6536; however, likely due to vaccination, hospitalizations and especially deaths remained low [33, 34]. By January 8, 2023, 2875 cases and 6 deaths were registered; and between February 19 and 25, 2023, only 284 cases were recorded. Finally, on May 11, 2023, the National Emergency Operations Center (COE) decided to endorse the decision of the World Health Organization that the public health emergency of international importance had ended, and therefore, the COE declared the end of the emergency in Ecuador [35–37].

Response and Preparedness Toward the Pandemic in Ecuador

On March 11, 2020, the MOH promulgated the State of Sanitary Emergency in the National Health System and issued Ministerial Agreement No. 00126–2020. On March 17, 2020, a national curfew was established, initiating a six-month confinement, which ended on September 11, except for social gatherings, schools, and access to beaches, among others. Starting on April 21, 2021, 16 provinces initiated a second curfew, which ended on May 20, 2021 [38, 39].

The country was not prepared for this disease, and even less so to face a pandemic. Medical personnel struggled due to the lack of personal protective equipment, hospitals lacked beds, and treatments failed repeatedly. To compensate for the demand for beds in hospitals and clinics, temporary tents were erected, and hospital areas such as endoscopy units or cafeterias were turned into clinics [40]. Occupancy of Intensive Care Units reached 95.7% in Pichincha, 95.6% in Guayas, and 93.3% in Azuay. In other provinces, such as Imbabura and El Oro, occupation reached full capacity [41]. The high mortality associated with COVID-19 caused city funeral homes to register a 50% demand increase, in addition to the appearance of so-called "garage" funeral homes. The latter unofficial services did not have municipal permits and often did not comply with legal and biosafety protocols, causing high risk of infection among funeral home workers [42].

Establishing an "effective" treatment for COVID-19 while physical isolation measures were still being inconsistently implemented was, for some people, a controversial and irresponsible decision. True, it was necessary to improvise and use drugs known for the management of other pathologies for which antiviral activity had been suggested at some point, and to test them "in vitro" against SARS-Cov-2. If it proved efficient, then we would make the treatment available! But we are now in the twenty-first century, and since 1992, we have been part of the world of evidence-based medicine (EBM) [43], hence "the proper use of the best evidence together with the best experience." A "titanic" battle began between clinicians who were desperately trying to save lives in hospitals and the methodologists who demanded double-blind and randomized clinical studies. This resulted in more than one treatment being questioned. In Ecuador, the development of a national "management protocol" took about a month after the declaration of the pandemic and was unfortunately used in a "discretionary" fashion [44, 45].

At such a critical moment, when Ecuador was named worldwide and pointed out as being the country with the worst response to COVID-19, it was imperative to answer questions about the behavior of the virus, transmission, genetics, prognosis, and treatment. However, since the Ministry of Public Health had decided to centralize any type of research related to COVID-19, previously approved bioethics committees were no longer allowed to evaluate such research. Worse, the MOH was to review and authorize biomedical research in three days, which became quickly practically impossible. Such a lack of self-awareness and responsibility led to a COVID-19 research standstill! [46].

The situation was made worse by corruption and speculation. While it did not appear with the pandemic, it certainly became publicly visible thanks to it.

Corruption in the health system, ranging from speculation and overpricing to the theft of medicines and business deals, became commonplace [47]. "Entrepreneurs" with doubtful moral standards took advantage of the anguish and need of the people and made unethical businesses out of the pandemic [48].

One of the most infamous cases of corruption was the January 2021 vaccination plan. It included people outside the priority population of phase zero; in addition, the Minister of Health at that time chose to first vaccinate their relatives in private centers using doses from the Pablo Arturo Suárez hospital, a public hospital specialized in the care of patients with COVID-19. Such a scandal forced the resignation of the health minister [49, 50].

During January and February 2021, despite a generally well-structured vaccination campaign later on, the initial rollout was marred by public health mismanagement, particularly in the prioritization of vaccine recipients. Although the supply of vaccines was limited at the time, relatively few doses were allocated to frontline healthcare workers, even though they were expected to be among the first protected. Ironically, subsequent analyses revealed that mortality among health workers was lower than in other high-risk groups, suggesting that early prioritization strategies may not have aligned with the actual burden of disease or vulnerability across the population [1]. Despite this suboptimal situation and thanks to the vaccination campaigns, cases began to recede from August 2021. By September 2021, nine million people had become fully vaccinated. The decline in cases and related deaths translated into a decrease in demand for hospital beds. By the end of 2021 and the beginning of 2022, new cases of COVID-19 started increasing again, reaching up to 5626 daily diagnoses in hospitals. This was most likely due to decreased motivation to receive booster doses [38, 39]. By December 2022, although 4.7 million citizens had received at least one dose of the COVID-19 vaccine, 2.6 million people had still not received a single dose and 500,000 had incomplete schemes. By 2023, the MOH planned to include the vaccine against COVID-19 in the regular vaccination scheme. On July 9, 2023, bivalent vaccination against COVID-19 began for health personnel working in management areas of respiratory diseases, and by July 24, 2023, vaccination began for adults over 65 years of age, patients with serious conditions, and persons with chronic pathologies [51–53].

The previous description also raises some bioethical concerns, like the inequitable distribution of resources, such as PPE, hospital beds, and vaccines, that violates the principle of justice. Prioritizing certain individuals outside the established criteria (e.g., vaccinating relatives of officials) also exemplifies unethical favoritism and a failure in distributive justice.

Also, improvising treatments without robust evidence, delaying the creation and application of standardized treatment protocols, and allowing unauthorized funeral services to operate posed risks to public safety, undermining the duty to act for the benefit of patients and avoid harm.

Finally, centralizing COVID-19 research under the Ministry of Health and restricting the roles of previously approved bioethics committees interfered with institutional autonomy and potentially obstructed ethically conducted research, delaying advancements that could benefit public health.

Research and Misinformation

It has been a difficult period for researchers. In particular, the situation was dire for those working in laboratories since their projects had been suspended or had mutated toward COVID-19-related topics, leading to a "covidization" of research, as Dr. Madhukar Pai called it [54]. It caused well-meaning scientists with real expertise in certain fields to meddle with others, making judgments when they lacked expertise, training, and knowledge. While in many countries the pandemic motivated the allocation of additional or special resources to encourage research on COVID-19-related topics, it was not the case in Ecuador: regulations and barriers unintentionally hindered research on the pandemic [55]. Yet, against all odds, Ecuadorian scientists found the opportunity to "dust off" their saved data, catch up with their pending projects, or simply maximize their collaborations, achieving a record scientific production, as reflected in PubMed: publications increased from 1136 records in all of 2019 to 1833 records until March 30, 2021 (+61%). According to the Ibero-American Observatory of Science, Technology, and Society, there were only 117 publications in Ecuador related to COVID-19, despite being one of the countries most severely affected by this pandemic [56]. However, when searching for similar information in the "Scopus" database for 2020 and up to June 2021, there were approximately 442 publications on COVID-19 with an author from Ecuador, of which 32 belong to Universidad San Francisco de Quito USFQ [57]. The pandemic was also an unfortunate opportunity for abuse of certain spaces for scientific dissemination, such as the so-called "preprints." Such are nothing more than portals where researchers can place their manuscripts before formally submitting them to a medical journal for peer review. As it was shown, only 6% of them finally became formal publications [58].

With isolation in place, the economic crisis gradually settled, and the eighth problem appeared: planning challenges. It reflected the structural weaknesses that hindered the transition from emergency response to strategic recovery. Physical distancing aimed to famously "flatten the curve," meaning in other words to slow down the speed of contagion and ensure the availability of health services. Unfortunately, gains from physical distancing were mostly lost to the misuse of the "traffic light" tool. Cautionary behavior (yellow ligth) quickly became misunderstood and omitted as free rains (green ligth), and, consequently, contagions became out of control [59]. During the isolation, there was not enough information, and there was no systematic search for or use of strategic alliances with academia, sports, the church, politics, and even show business to prepare for the "post-isolation" and the "new normality" subsequent phase. There were only isolated initiatives and, in some cases, controversial messages that generated more confusion.

Due to a lack of ARCSA management, misleading advertising of supposed "immunomodulators" or "magic cures" started to take advantage of people's ignorance and anguish [60, 61]. In particular, the control of false and misleading information in social networks, or rather the lack thereof, left the circulation of "treatments" to control COVID-19. This would have had little consequences if drugs sold only with prescription had been respected and not sold over the counter.

This is but another myth to which the authorities were unable to respond. The risks of self-medication, particularly with products that do not even have the status of pharmaceuticals, were immense, and the requirements regarding education and information were not met outside of individual initiatives [62, 63].

Learned Lessons

- Ecuador lacks a structured prevention system, which led to most COVID-19 cases reaching hospitals when the appropriate strategy should have been to intervene earlier through primary health care. This situation was further aggravated by the fact that not all individuals had access to health services, and the country did not have universal health coverage nor an integrated digital medical records system that allows public and private providers to access and share patient information virtually. These structural deficiencies severely limited the ability to implement timely and coordinated responses during the pandemic [10].
- Local multisectoral structures, such as universities, could function as health and surveillance groups that record epidemiological data, do contact tracing, and support close surveillance of mild and asymptomatic symptoms in people with confirmed or suspected infection [10].
- The Ecuadorian government seems to neglect the value of data for decision-making. The research produced by the Ministry of Health is limited, and there is no system or agenda for research collaborations with academic institutions [64].
- Having a structure that defines the basic contents of primary prevention before and during a global public health emergency, and investing in developing a National Health System resilient to crises, is paramount [65].
- The infodemic damaged trust in science. Social networks appeared as one of the most damaging conductors whereby false news, fueled by fear, was easily and widely shared. They also highlighted false news about new medicines or practices that claimed to prevent infection or cure coronavirus, like using hydroxychloroquine or diluted bleach [66, 67].

Conclusions

The COVID-19 pandemic left a profound mark on Ecuador's health system, society, and institutions. It exposed long-standing structural weaknesses, including insufficient investment in public health, fragmented governance, and limited preparedness for large-scale health emergencies. While much criticism has rightfully focused on the national government's actions—such as delays in declaring the emergency, centralized and opaque data management, and ethical lapses in the early phases of the vaccination campaign—the crisis unfolded within a broader context of institutional fragility, inequality, and a lack of coordinated multisectoral engagement.

Yet, this experience also offers critical lessons. First, it underscores the urgent need to strengthen primary health care as the frontline for prevention, community surveillance, and early intervention. Second, the pandemic highlighted the importance of developing interoperable and integrated health information systems that bridge public and private sectors, enabling real-time data sharing and evidence-based decision-making. Third, institutionalizing science-policy dialogue through independent, multidisciplinary advisory bodies is essential to ensure that health policy is guided by technical and ethical standards rather than improvisation. Additionally, transparency, accountability, and efforts to rebuild public trust must be at the core of any public health strategy.

Importantly, the pandemic also revealed the vital, yet underutilized, contributions of other stakeholders. Universities, research institutions, and civil society played a key role in filling gaps in diagnostics, surveillance, and public communication. Citizen-led initiatives, such as independent data platforms, helped compensate for official information shortcomings and ensured that Ecuador remained visible in global pandemic tracking.

References

1. Ortiz-Prado E, Simbaña-Rivera K, Barreno LG, Diaz AM, Barreto A, Moyano C et al (2021) Epidemiological, socio-demographic and clinical features of the early phase of the COVID-19 epidemic in Ecuador. PLoS Negl Trop Dis 15(1):e0008958
2. Garzon-Chavez D, Romero-Alvarez D, Bonifaz M, Gaviria J, Mero D, Gunsha N et al (2021) Adapting for the COVID-19 pandemic in Ecuador, a characterization of hospital strategies and patients. PLoS One 16(5):e0251295
3. Ortiz-Prado E, Simbaña-Rivera K, Gómez- Barreno L, Rubio-Neira M, Guaman LP, Kyriakidis NC et al (2020) Clinical, molecular, and epidemiological characterization of the SARS-CoV-2 virus and the coronavirus disease 2019 (COVID-19), a comprehensive literature review. Diagn Microbiol Infect Dis 98(1):115094
4. BBC News Mundo (2022) Covid: el número real de muertes por la pandemia en todo el mundo es de casi 15 millones (y qué países de América Latina tienen mayor exceso de mortalidad). BBC News Mundo [Internet].. [cited 2023 Aug 27]; Available from: https://www.bbc.com/mundo/noticias-61333739
5. OPS. Actualización sobre la variante de interés EG.5 y la variante bajo vigilancia BA.2.86—OPS/OMS | Organización Panamericana de la Salud [Internet]. 2023 [cited 2023 Aug 27]. Available from: https://www.paho.org/es/noticias/23-8-2023-actualizacion-sobre-variante-interes-eg5-variante-bajo-vigilancia-ba286
6. PAHO. PAHO Weekly COVID-19 Epidemiological Update—27 September 2022—PAHO/WHO | Pan American Health Organization [Internet]. 2022 [cited 2022 Oct 6]. Available from: https://www.paho.org/en/documents/paho-weekly-covid-19-epidemiological-update-27-september-2022
7. JHU. COVID-19 Map—Johns Hopkins Coronavirus Resource Center [Internet]. 2020 [cited 2020 Apr 14]. Available from: https://coronavirus.jhu.edu/map.html
8. Anaya-Covarrubias JY, Pizuorno A, Mirazo S, Torres-Flores J, Du Pont G, Lamoyi E et al (2022) COVID-19 in Latin America and the Caribbean region: symptoms and morbidities in the epidemiology of infection. Curr Opin Pharmacol 63:102203
9. Schwalb A, Armyra E, Méndez-Aranda M, Ugarte-Gil C (2022) COVID-19 in Latin America and the Caribbean: two years of the pandemic. J Intern Med 292(3):409-427. https://doi.org/10.1111/joim.13499
10. Torres I, Sacoto F (2020) Localising an asset-based COVID-19 response in Ecuador. Lancet 395(10233):1339

11. Ortiz-Prado E, Fernandez Naranjo RP, Vasconez E, Simbaña-Rivera K, Correa-Sancho T, Lister A et al (2021) Analysis of excess mortality data at different altitudes during the COVID-19 outbreak in Ecuador. High Alt Med Biol 22(4):406–416
12. Cronin CJ, Evans WN (2021) Excess mortality from COVID and non-COVID causes in minority populations. Proc Natl Acad Sci 118(39):e2101386118
13. A Pandemic Tragedy in Guayaquil | The New Yorker [Internet]. 2022 [cited 2023 Oct 11]. Available from: https://www.newyorker.com/magazine/2022/03/14/a-pandemic-tragedy-in-guayaquil
14. Ciudadano chino que fue sospechoso de coronavirus falleció en Quito [Internet]. Primicias. [cited 2023 Oct 11]. Available from: https://www.primicias.ec/noticias/sociedad/ciudadano-chino-sospechoso-coronavirus-ecuador-fallecio-quito/
15. Forbes Colombia. 2020 [cited 2023 Aug 24]. Ecuador estudia un caso sospechoso de coronavirus. Available from: https://forbes.co/2020/01/27/actualidad/ecuador-estudia-un-caso-sospechoso-de-coronavirus/
16. Ministerio de Salud Pública EC [@Salud_Ec]. Comunicado #Coronavirus 08-02-2020 | El Ministerio de Salud Pública en Ecuador, informa que falleció el ciudadano extranjero cuyo diagnóstico para coronavirus dio negativo. Expresamos nuestras condolencias y reiteramos a la ciudadanía informarse por fuentes oficiales. https://t.co/Mbf1b9kgir [Internet]. Twitter. 2020 [cited 2023 Aug 24]. Available from: https://twitter.com/Salud_Ec/status/1226168623854694400
17. Andramuño C. Interim guidelines for persons with risk exposure to the new 2019 coronavirus (Covid-19) in community or travel-related settings [Internet]. 2020 [cited 2023 Aug 24]. Available from: https://www.salud.gob.ec/wp-content/uploads/2020/02/Lineamientos_Coronavirus_2020MSP.pdf
18. Ministerio de Salud Publica. Emergencies Operations Committee, Release 4.—Dirección General de Aviación Civil [Internet]. 2020 [cited 2023 Aug 24]. Available from: https://www.aviacioncivil.gob.ec/emergencies-operations-committee-release-4/
19. Espinosa C. El Comercio. 1AD [cited 2023 Aug 24]. Ministerio de Salud confirma primer caso de coronavirus en Ecuador. Available from: https://www.elcomercio.com/tendencias/sociedad/salud-confirma-primer-caso-coronavirus.html
20. Ministerio de Salud Publica del Ecuador. Registro Oficial del Ecuador [Internet]. 2020 [cited 2023 Aug 24]. Available from: https://www.registroficial.gob.ec/
21. Moreno Garcés L. Decreto_presidencial_No_1017 [Internet]. 2020 [cited 2023 Aug 24]. Available from: https://www.defensa.gob.ec/wp-content/uploads/downloads/2020/03/Decreto_presidencial_No_1017_17-Marzo-2020.pdf
22. Moreno Garcés L. DECRETO_No._1018 [Internet]. 2020 [cited 2023 Aug 24]. Available from: https://www.salud.gob.ec/wp-content/uploads/2020/03/DECRETO_No._1018_20200221174230_20200221174413.pdf
23. Espinosa P, Quirola-Amores P, Teran E (2020) Application of a susceptible, infectious, and/or recovered (SIR) model to the COVID-19 Pandemic in Ecuador. Front Appl Math Stat 6:571544
24. Coello C. Edición. 2020 [cited 2023 Aug 24]. Academia de Ciencias se pone a disposición del Gobierno para asesorarlo en la emergencia. Available from: https://www.edicionmedica.ec/secciones/profesionales/academia-de-ciencias-recomienda-crear-un-comite-cientifico-nacional-que-asesore-a-las-autoridades-95538
25. Academy of Sciences of Ecuador. Google Docs. 2020 [cited 2023 Aug 24]. Final Comunicado ACE coronavirus. Available from: https://drive.google.com/file/d/10b6ZiK487bGV9QlaJ9xqdOpyeSPJhdVT/view?usp=embed_facebook
26. Ortiz-Prado E, Aguilar AC, Gómez Jaramillo AM (2021) La enfermedad del coronavirus-2019 (COVID-19) y el virus del SARS-CoV-2: una visión multidisciplinaria de la pandemia, su origen, sus implicaciones médicas, sociales y sanitarias. Primera edición. In: Landázuri Navas S, Navarrete Santacruz D, Noboa Lasso M d L, Salvador Crespe S, Vascones JE (eds). Facultad de Ciencias de la Salud, UDLA, Quito, p 728. p. (Núcleo / UDLA Ediciones)

27. Diario La Hora. 2021 [cited 2023 Aug 24]. Ecuador debería recoger 20.000 PCR diarias, pero apenas toma 4.000. Available from: https://www.lahora.com.ec/pais/ecuador-deberia-recoger-20-000-pcr-diarias-pero-apenas-toma-4-000/
28. El Comercio. 1AD [cited 2023 Aug 24]. Entre la ciencia, producción de insumos y servicios, universidades se suman a la pelea contra el coronavirus en Ecuador. Available from: https://www.elcomercio.com/actualidad/universidades-ecuador-invetigacion-coronavirus-covid19.html
29. Ministerio de Salud Publica. Comunicado oficial: el costo de las pruebas RT-PCR será de USD 45.08—Ministerio de Salud Pública [Internet]. 2021 [cited 2023 Aug 24]. Available from: https://www.salud.gob.ec/comunicado-oficial-el-costo-de-las-pruebas-rt-pcr-sera-de-usd-45-08/
30. Arroyo MB, Torres A, Pérez A, Cheng K, Torres C, Machado J. Tras las huellas del dragón en la pandemia. [cited 2023 Aug 24]. Tras las huellas del dragón en la pandemia. Available from: https://www.connectas.org/especiales/tras-las-huellas-de-china-en-la-pandemia/
31. El Universo. 2021 [cited 2023 Aug 24]. Agotados, con estrés y con otras dolencias viven médicos que 'batallan' a diario contra el COVID-19 en Ecuador. Available from: https://www.eluniverso.com/noticias/ecuador/agotados-con-estres-y-con-otras-dolencias-viven-medicos-que-batallan-a-diario-contra-el-covid-19-en-ecuador-nota/
32. Chejín SR. En la pandemia han muerto 407 médicos en el país, según las Federaciones Unidas de Profesionales de la Salud [Internet]. GK 2021 [cited 2023 Aug 24]. Available from: https://gk.city/2021/04/18/medicos-400-muertos-ecuador/
33. Statista. 2022 [cited 2023 Aug 25]. COVID-19: casos y muertes semanales Ecuador 2020–2022. Available from: https://es.statista.com/estadisticas/1110063/numero-casos-muertes-covid-19-ecuador/
34. Pinchao J. El Comercio. 2022 [cited 2023 Aug 25]. 21 muertes se registraron en el último brote del 2022 de covid-19 en Ecuador. Available from: https://www.elcomercio.com/tendencias/sociedad/muertes-covid19-brote-noviembre-ecuador.html
35. Oñate S. El Comercio. 2023 [cited 2023 Aug 25]. Seis fallecidos por covid-19 se registraron en la primera semana de 2023 en Ecuador. Available from: https://www.elcomercio.com/tendencias/sociedad/fallecidos-covid19-primera-semana-ecuador-2023.html
36. Prensa Latina. Ecuador registra cifra más baja de contagios por Covid-19—Prensa Latina [Internet]. 2023 [cited 2023 Aug 25]. Available from: https://www.prensa-latina.cu/2023/03/03/ecuador-registra-cifra-mas-baja-de-contagios-por-covid-19
37. Ministerio de Salud Publica. Ecuador se suma a la decisión de la OMS de poner fin a la emergencia en salud pública por COVID-19—Ministerio de Salud Pública [Internet]. 2023 [cited 2023 Aug 25]. Available from: https://www.salud.gob.ec/ecuador-se-suma-a-la-decision-de-la-oms-de-poner-fin-a-la-emergencia-en-salud-publica-por-covid-19/
38. Alava JJ, Guevara A (2021) A critical narrative of Ecuador's preparedness and response to the COVID-19 pandemic. Public Health Pract 2:100127
39. Coral-Almeida M, Carrington SJ, Carrión-Yaguana VD, Mascialino G (2022) How not to manage a pandemic, and how to recover from it: lessons from Ecuador. FACETS 7:1367–1384
40. Molina GA, Ayala AV, Endara SA, Aguayo WG, Rojas CL, Jiménez GE et al (2021) Ecuador and Covid-19: a pandemic we won't be able to run away from. Int J Infect Dis 109:33–35
41. Sandoval C. El Comercio. 1AD [cited 2023 Aug 25]. Ministro de Salud habla de la ocupación de las UCI en Ecuador y dice que 86 000 dosis de vacunas se aplicarán en 43 000 personas de grupos prioritarios. Available from: https://www.elcomercio.com/tendencias/sociedad/ministro-salud-vacunacion-ecuador-coronavirus.html
42. Ortiz-Prado E, Vásconez JE, Vallejo-Janeta AP, Morales-Jadán D, Henriquez-Trujillo AR, Rivera-Olivero IA, et al (2022) High rates of SARS-CoV-2 infection in funeral home workers in Ecuador: is it an occupational risk for COVID-19? Front Public Health 10:1012434
43. Nicolalde B, Añazco D, Mushtaq M, Aguilar A, Terán E (2020) Current Pharmacological Therapy against COVID-19: A Latin American perspective. Latin Am J Clin Sci Med Technol 2(2):59–68
44. Watson J. An open letter to Mehra et al and The Lancet [Internet]. Zenodo; 2020 May [cited 2023 Aug 24]. Available from: https://zenodo.org/record/3864691

45. Coello C. Edición Médica. 2020 [cited 2023 Aug 24]. Salud actuó sin fundamento al suspender la hidroxicloroquina en el tratamiento de la CoVID19. Available from: https://www.edicionmedica.ec/secciones/profesionales/salud-actuo-sin-fundamento-al-suspender-la-hidroxicloroquina-en-el-tratamiento-de-la-covid19-95904
46. Zevallos López J. Reglamento de investigaciones en salud durante emergencia sanitaria [Internet]. 2020 [cited 2023 Aug 24]. Available from: https://www.salud.gob.ec/wp-content/uploads/2020/04/Reglamento-Investigaci%C3%B3n-en-salud-RO-504.pdf
47. Yépez J. Hospitales públicos, el foco de la corrupción en la emergencia sanitaria [Internet]. Primicias. 2021 [cited 2023 Aug 24]. Available from: https://www.primicias.ec/noticias/politica/hospitales-publicos-corrupcion-pandemia-covid/
48. Ortiz-Prado E, Fernandez-Naranjo R, Torres-Berru Y, Lowe R, Torres I (2021) Exceptional prices of medical and other supplies during the COVID-19 pandemic in Ecuador. Am J Trop Med Hyg 105(1):81–87
49. BBC News Mundo. El escándalo por la decisión del ministro de Salud ecuatoriano de vacunar primero a sus familiares contra covid-19. BBC News Mundo [Internet]. 2021 [cited 2023 Aug 25]; Available from: https://www.bbc.com/mundo/noticias-america-latina-55830752
50. BBC News Mundo. Renuncia el ministro de Salud de Ecuador tras el escándalo de vacunación. BBC News Mundo [Internet] 2021 [cited 2023 Aug 25]; Available from: https://www.bbc.com/mundo/noticias-america-latina-56210659
51. Machado J, Castillo D. El 15% de ecuatorianos aún no está vacunado contra el Covid-19 [Internet]. Primicias. 2022 [cited 2023 Aug 25]. Available from: https://www.primicias.ec/noticias/sociedad/covid-influeza-vacuna-enfermedades/
52. Machado J. Ecuador cierra emergencia por Covid-19, pero mantiene la vacunación [Internet]. Primicias. [cited 2023 Aug 25]. Available from: https://www.primicias.ec/noticias/sociedad/covid-emergencia-coe-vacunas/
53. Ministerio de Salud Publica. Vacunación bivalente contra la Covid-19 inició con personal sanitario en riesgo—Ministerio de Salud Pública [Internet]. 2023 [cited 2023 Aug 25]. Available from: https://www.salud.gob.ec/vacunacion-bivalente-contra-la-covid-19-inicio-con-personal-sanitario-en-riesgo/
54. Pai M (2020) Covidization of research: what are the risks? Nat Med 26(8):1159–1159
55. Sisa I, Mena B, Teran E (2021) The negative impact of ad hoc committees for ethical evaluation: the case of COVID-19-related research in Ecuador. Dev World Bioeth 21(1):3–6
56. Explorador de la investigación latinoamericana en COVID-19. Explorador de la investigación latinoamericana en COVID-19 [Internet]. 2020 [cited 2023 Aug 24]. Available from: http://octs.ricyt.org/reportes/explorador.html
57. Terán E (2021) Ciencia y producción científica en el Ecuador en tiempos de la COVID-19. Rev Ecuat Med Cienc Biol Aug 17 [cited 2023 Aug 24];42(1). Available from: https://remcb-puce.edu.ec/remcb/article/view/907
58. Añazco D, Nicolalde B, Espinosa I, Camacho J, Mushtaq M, Gimenez J et al (2021) Publication rate and citation counts for preprints released during the COVID-19 pandemic: the good, the bad and the ugly. PeerJ 9:e10927
59. Veletanga J. Edición Médica. 2020 [cited 2023 Aug 24]. Nuevas reglas para contener la COVID19 en Ecuador entrarán en vigencia desde junio. Available from: https://www.edicionmedica.ec/secciones/salud-publica/nuevas-reglas-para-contener-la-covid19-en-ecuador-entraran-en-vigencia-desde-junio%2D%2D95896
60. Agencia Nacional de Regulación, Control y Vigilancia Sanitaria » ARCSA informa a la ciudadanía sobre el Dióxido de Cloro. [Internet]. 2020 [cited 2023 Aug 24]. Available from: https://www.controlsanitario.gob.ec/arcsa-informa-a-la-ciudadania-sobre-el-dioxido-de-cloro/
61. Machado J. Científicos ecuatorianos alertan sobre el peligro del uso de dióxido de cloro contra el Covid-19 [Internet]. Primicias. 2020 [cited 2023 Aug 24]. Available from: https://www.primicias.ec/noticias/sociedad/cientificos-alertan-uso-dioxido-cloro/
62. Oñate S. El Comercio. 1AD [cited 2023 Aug 24]. Tratamientos para el covid-19 todavía son experimentales. Available from: https://www.elcomercio.com/actualidad/tratamientos-experimentales-coronavirus-medicamentos-vacuna.html

63. El Universo. 2020 [cited 2023 Aug 24]. Riesgos de automedicación enfrentan indígenas en Ecuador. Available from: https://www.eluniverso.com/noticias/2020/08/10/nota/7936551/riesgos-automedicacion-enfrentan-indigenas-ecuador
64. Torres I, López-Cevallos DF (2018) Institutional challenges to achieving health equity in Ecuador. Lancet Glob Health 6(8):e832–e833
65. Edición Medica. 2021 [cited 2023 Aug 27]. ¿Qué lecciones deja la pandemia de la COVID-19 al sistema de salud ecuatoriano? Available from: https://www.edicionmedica.ec/secciones/profesionales/-que-lecciones-deja-la-pandemia-de-la-covid-19-al-sistema-de-salud-ecuatoriano%2D%2D98358
66. Mena MB, Sisa I, Teran E (2022) Misleading Advertising of Health-Related Products in Ecuador during the COVID-19 Pandemic. Diseases 10(4):91.
67. Instituto Salud Global Barcelona. ISGlobal. 2020 [cited 2023 Aug 27]. Infodemia: ¿Cómo ha afectado la epidemia de desinformación a la respuesta frente a la COVID-19? Available from: https://www.isglobal.org/-/infodemia-como-ha-contribuido-la-epidemia-de-desinformacion-a-la-respuesta-frente-a-la-covid-19-

Part VII

Conclusion

Is the Asia-Pacific Region Ready for the Next Pandemic? 17

King L. Chow, Alexander Atrio Lim Lopez,
María de Jesús Medina-Arellano, Leander Penaso Marquez,
and Mellissa Withers

Abstract

The COVID-19 pandemic stands as one of the most profound global health crises in modern history. The pandemic unfolded in three distinct phases. The initial outbreak in 2020 led to widespread border controls and regional lockdowns, causing a global economic shock. In 2021, vaccine rollouts coincided with the emergence of new variants, and vaccines were approved through emergency procedures to accelerate their availability. However, vaccine hoarding by wealthier nations and the political relations among nations created significant global inequities in distribution. The situation evolved with the appearance of the highly transmissible but less severe Omicron variants, which ultimately led to the gradual lifting of restrictions worldwide. The WHO officially declared the end of the global health emergency in May 2023. Focusing on the Asia-Pacific region,

K. L. Chow (✉)
Division of Public Policy, Division of Life Science, Department of Chemical and Biological Engineering, Hong Kong University of Science and Technology, Clear Water Bay, Hong Kong
e-mail: bokchow@ust.hk

A. A. L. Lopez
School of Law, Ateneo de Manila University, Makati City, Philippines

M. Medina-Arellano
Instituto de Investigaciones Jurídicas, Universidad Nacional Autónoma de México, Mexico City, Mexico

L. P. Marquez
College of Education, University of the Philippines Diliman, Quezon City, Philippines

M. Withers
Keck School of Medicine, University of Southern California, Los Angeles, CA, USA

L. P. Marquez et al. (eds.), *COVID-19 in the Pacific Rim and Beyond*,
Collaborative Bioethics 6, https://doi.org/10.1007/978-3-032-21218-4_17

which accounts for 60% of the world's population, we take note of how these global impacts are manifested locally. We ask, further, are we ready for the next pandemic?

Keywords

Ethical governance · Transparency · Trust · Multilateral cooperation · Political leadership

The COVID Pandemic and Its Impact

The COVID-19 pandemic stands as one of the most profound global health crises in modern history. Since its emergence in December 2019, COVID-19 has claimed more than seven million lives worldwide as of early 2024 [1]. COVID-19 was declared a global pandemic by the World Health Organization (WHO) on March 11, 2020 [2]. The virus spread rapidly through international travel, triggering successive waves of infections worldwide.

The pandemic unfolded in three distinct phases. The initial outbreak in 2020 led to widespread border controls and regional lockdowns, causing a global economic shock. In 2021, vaccine rollouts coincided with the emergence of new variants, and vaccines were approved through emergency procedures to accelerate their availability. However, vaccine hoarding by wealthier nations and the political relations among nations created significant global inequities in distribution. The situation evolved with the appearance of the highly transmissible but less severe Omicron variants, which ultimately led to the gradual lifting of restrictions worldwide. The WHO officially declared the end of the global health emergency in May 2023 [3]. As detailed in previous chapters, the pandemic has disrupted economies, strained healthcare systems, and reshaped the fabric of societies across every continent. During this period, national responses varied significantly due to differing objectives, priorities, constraints, and capabilities, resulting in varying degrees of success.

While each nation faced unique challenges shaped by its local context, the pandemic's toll was, nonetheless, staggering at approximately 800 million confirmed cases and over seven million deaths globally [1]. WHO estimates suggest an additional 15–20 million excess deaths occurred, while over 60 million people continue to suffer from Long COVID symptoms. The International Monetary Fund (IMF) calculated global economic losses exceeding $12 trillion [4], with severe impacts across multiple sectors. Education systems were disrupted, affecting 1.6 billion students [5], and the WHO reported a 25% global increase in mental health cases [6]. As the pandemic subsided, it left indelible marks on public health systems, economies, and daily life worldwide. This ordeal offers crucial lessons for future public health crisis management.

Focusing on the Asia-Pacific region, which accounts for 60% of the world's population, we take note of how these global impacts are manifested locally. Official figures report over 1.3 million COVID-19 deaths in the region within the 3 years of the pandemic, representing about 15% of global fatalities [7]. However, these numbers likely represent significant underreporting, particularly in areas with limited healthcare infrastructure and surveillance systems.

Different Strategies Taken by Different Countries

Strict containment and surveillance are two dominating strategies adopted by most of the regions studied here, as outlined in previous chapters. China implemented a "Zero-COVID" policy, enforcing lockdowns of cities like Shanghai and Wuhan, mass testing, and digital surveillance via health codes. While effective in curbing early outbreaks, the policy led to economic strain, testing the patience of citizens, before its abrupt relaxation in late 2022, which was followed by a massive surge in cases with weaker symptoms. The adopted approach under this People-Centered Integrated Management and Control Model prioritizes the health of individuals and the broader community. A similar approach was taken in the Philippines, with more stringent control measures, but it struggled with weak social support, inadequate healthcare infrastructure, low testing rates, and vaccine hesitancy. Singapore initially pursued a Zero-COVID approach but later pivoted to "living with COVID" after achieving a high vaccination rate. Its strong healthcare system managed outbreaks well, though migrant worker dormitories were early hotspots and created another ethical concern. These differences highlight that similar measures would not necessarily serve different communities to achieve the same outcome, and the local social and governing context must be carefully considered.

During these years, alternative policy adoption still faced frequent challenges in privacy, equity, and democratic governance issues. They triggered similar resistance in many regions in Asia, such as Malaysia. In contrast to China, Singapore, or the Philippines, South Korea solely relied on mass testing, contact tracing, and quarantine enforcement, without resorting to full lockdowns. Yet, its early success was later challenged by outbreaks in 2022. Fortunately, high vaccination rates in Korea helped mitigate severe outcomes. Japan did not put in place a stringent lockdown policy. It encouraged behavioral changes through "state of emergency" declarations which resulted in good compliance among citizens. Its healthcare system, however, faced strain during the Omicron wave, challenging the prior preparation of medical capacity and amplifying stigmatization across the community. Vietnam experienced early success with lockdowns implemented in a strict top-down approach, but faced economic disruption and vaccine shortages before gradually recovering in 2022. The lack of transparency and candid communication of government strategy with the public also hindered the effective adoption of public health measures. In turn, it created more distrust between the government and the public. Thailand relied on tourism-dependent policies, imposing quarantine rules that hurt its economy until its border reopening in late 2021. Several South Asian countries, e.g., Indonesia and the Philippines, saw high death rates in 2021 due to slow vaccine rollouts. They all faced unprecedented pressure due to the demand for medical support when protective facilities were not readily available.

Balancing economic costs with public health has been challenging. In regions where the Zero-COVID policy was adopted, strict implementation of different measures became unsustainable due to the drastic reduction in economic activities resulting from the absence of trade and travel across countries. Hong Kong was certainly struggling between these two competing forces, and this struggle was also compounded by its positioning, which requires alignment with China and a connection to the international community through trade. In regions where their economy was strained, such a dilemma was further heightened when healthcare inequality is severe and economic disparity is high. In such cases, ethical issues associated with healthcare inequality often arose and were frequently raised by critics within the community and the media. When such situations were not managed well, further widening of social and economic inequality resulted, as seen in Costa Rica (see Chap. 11).

The interesting contrast is that Australia and New Zealand, like China and the Philippines, initially pursued "COVID-zero" through strict border closures and lockdowns. New Zealand successfully eliminated early outbreaks with hard lockdowns (adopting an Alert Level system). However, it faced criticism for prolonged isolation, which led to labor shortages and indirect economic pressure until it reopened in late 2021 after a high vaccination rate was achieved. Australia used state-level lockdowns and internal border controls. The delayed vaccine rollout caused frustration, but high uptake later allowed reopening. Again, public fatigue from prolonged restrictions and economic impacts from border closures consistently skewed the public's sentiment, raising questions about whether the government had made the appropriate decision to balance population health protection with the health of the economy.

Small island nations, such as Fiji and the Cook Islands, faced unique challenges. They managed to control outbreaks through lockdowns and later vaccinations. They had low testing and vaccine uptake for a much smaller but isolated population, which also led to underreported outbreak cases. Weak healthcare systems increase mortality risks. Limited access to healthcare and weak financial support led to heavy reliance on international aid for vaccines. Their vulnerability is mainly due to their remote geography and poor national resources.

On the other side of the Pacific, Mexico, Chile, Colombia, and Ecuador faced a vastly different challenge. The fragmented healthcare system, with the social, economic, and political structures, contributes to the ineffectiveness in managing the pandemic and amplifying the inequities in public health management and support, possibly also affecting other civic activities, such as education, workforce deployment, and institutional/organizational collaboration. All these shortcomings highlight the need for a structured and coordinated effort across multiple agencies, enabling governments to develop a clear strategy and priority for delivering basic and minimal support in a coordinated manner.

On the other hand, while resources and efficient deployment are key, cultural differences would make a stark difference in effective coping with the pandemic. The USA, in a much more affluent condition, still faced mounting challenges during this period. While coping strategies, such as those adopted by other countries, were implemented, effective compliance was unsatisfactory. Without a strong public leader or authority to manage the outbreak with a well-articulated framework, individualism trumped the collective benefit. It created distrust of authority and scientists. Political motivation also interfered with health policy. Since social and health inequities were long-entrenched problems in the USA, society became more polarized. The healthcare system could not cope with the pressing demands of its people.

While East Asia and Oceania managed early waves well, countries there faced economic and social costs from prolonged restrictions. Southeast Asia struggled with healthcare capacity and vaccine distribution, while the Pacific Islands dealt with isolation and resource shortages. In contrast, the American system had ample resources and connections, but its pandemic response generated different issues. It highlighted constraints that lead to a different level of failure in managing this health crisis.

The pandemic reveals both resilience and systemic weaknesses across the Asia-Pacific region, proving that a one-size-fits-all approach is ineffective for a region of such vast systems and cultural diversity. However, common themes emerge in each of these regions, and some common ethical concerns are brought to the forefront.

Practical and Ethical Challenges Encountered

Privacy and Surveillance: The Trade-off Between Safety and Freedom

Several East Asian countries employed digital surveillance to enforce lockdowns and contact tracing, raising concerns about government overreach and privacy violations. For example, China's Health Code System tracked citizens' movements and restricted access based on infection risk. While effective, it may lack transparency, with reports of arbitrary restrictions. South Korea's contact tracing utilized credit card transactions, GPS data, and CCTV, which critics argued infringed on personal privacy. Singapore's TraceTogether app initially claimed that data would only be used for COVID-19 purposes, but later admitted that police could access it for criminal investigations, thereby breaking public trust. Hence, one has to wonder: is mass surveillance justified if it saves lives, or does it create a dangerous precedent for post-pandemic authoritarian control? On the issue of individual liberty versus public health enforcement, how much personal freedom should be sacrificed for the greater good?

Lockdowns and Inequality: Who Bears the Burden?

Strict lockdowns helped control outbreaks but disproportionately harmed vulnerable groups, including low-income workers, migrants, and informal laborers. For example, India's 2020 lockdown left millions of migrant workers stranded without transport or income, leading to a humanitarian crisis with deaths from hunger and exhaustion. Singapore's handling of foreign labor, with the containment restricted to their own quarters, faces the challenge of discrimination. Australia's strict border closures kept cases low but separated families for years, raising ethical concerns about emotional suffering versus public health. The Philippines' militarized lockdowns saw poor communities harshly penalized for violations while elites flouted rules, highlighting class-based enforcement disparities and injustice. These issues exposed the fact that lockdowns may have protected the majority at the expense of marginalized groups, and whether appropriate measures to mitigate these harms had been put in place. The controversy becomes: should lockdowns be enforced even if they devastate the livelihoods of the minority?

On the other hand, Indonesia's reluctance to lockdowns avoided the necessity of using strict measures to protect the economy. However, this strategy led to high death tolls and overwhelmed hospitals in providing sufficient medical support for a large population. In the case of tourism-dependent nations, such as Thailand, border closures saved lives but caused mass unemployment and economic crises. In such circumstances, who should bear the economic burden of pandemic control: workers, businesses, or the State?

Personal Needs or the Common Good?

Corollary to travel restrictions and lockdowns is the basic issue of survival. Without the freedom to go out to buy food, water, medicine, and other daily necessities, not only was quality of life severely diminished, but personal safety itself was threatened. Even those with financial means could not do much but struggle, because they had no access to stores or marketplaces. Initiatives like China's and Malaysia's volunteer food delivery network, the Philippines' Ayuda (help) Programs, or Vietnam's deployment of the military-led grocery distribution, provided some relief. Yet, these measures were far from sufficient, leaving people in a dilemma, either risking personal safety to venture out to secure necessities for their families while contracting the virus and spreading it to family members or staying home to avoid exposure to the virus. Either choice carried dire consequences: starvation and malnutrition on one hand, or the threat of contracting and spreading COVID-19 on the other.

Vaccine Nationalism and Equity: Who Gets Priority?

Vaccine distribution also revealed glaring inequalities. While wealthier nations hoarded doses, poorer countries struggled. Australia and New Zealand secured timely access to vaccines but were slow to distribute them to Pacific neighbors, such as the Pacific Islands, where outbreaks raged unchecked. China's vaccine diplomacy provided doses to Southeast Asia and the Pacific, but the vaccines might not have reached many of the intended recipients. One may wonder whether countries should prioritize their citizens or if there is a moral duty to ensure global vaccine equity. Is it acceptable for wealthy nations to stockpile vaccines while poorer countries suffer? Should countries prioritize their citizens or contribute to global vaccine equity?

Freedom of Movement Versus Public Health: Border Closures and Discrimination

Travel restrictions were necessary, but they were sometimes unfairly targeted at specific groups. Australia and New Zealand's border closures left thousands of citizens stranded abroad, raising questions about the right to return home. Hong Kong's flight bans were applied to nations like the UK and India. Japan and South Korea imposed stricter rules and more extended quarantines on foreign travelers as compared to locals, leading to accusations of xenophobia in pandemic policies. In Singapore, migrant workers in cramped dormitories during the COVID outbreaks were exposed to less desirable living conditions of much higher risk. Such measures would have protected the public health of the majority, but at the same time, they might have unfairly discriminated against certain groups. Does science justify selective restrictions, or does it reinforce discrimination?

Censorship and Misinformation: Controlling the Narrative

Some governments suppressed criticism of their COVID-19 responses, limiting free speech. China censored discussions of Wuhan's early outbreak, which delayed the global response to the pandemic. Thailand and Vietnam arrested critics of their pandemic policies under "fake news" laws, stifling dissent. On the other hand, the strong circulation of anti-vax movement information in some countries also heavily undermined the trust in the government when the implementation of a holistic strategy was in progress, dampening the effectiveness of vaccination and different preventive measures. One may argue about where to draw the line regarding governments' control of pandemic information to prevent panic and to promote the efficient implementation of credible professional advice and government policy versus the impact on undermining civic trust and democratic rights. Does controlling pandemic narratives protect public order, or does it hide government failures?

Ethical Governance and Public Health Management

The pandemic responses in the Asia-Pacific region revealed deep-seated ethical tensions within each country, balancing public health against personal freedom, efficiency against equity, national interest against global solidarity, and transparency against control. Moving forward, policymakers must strike a balance between these dilemmas, prioritizing fairness, accountability, and respect for human rights. The pandemic was not just a test of medical systems but of moral leadership, and the ethical failures and successes will shape future crisis responses.

A few immediate questions can be raised: when the lockdown has hurt many who were affected, should governments compensate? Would digital surveillance be more widely adopted in future pandemics? What needs to be managed to ensure the protection of privacy and freedom? What obligations do wealthy nations have to address the needs of poorer ones amid global health crises? After all, a global pandemic took its toll, with over ten million dead (including excess mortality) and huge economic cost: \$4–5 trillion+ in lost GDP, long-term damage, and social cost represented in education gaps, mental health decline, and deepened inequality. There remain debates about which policy serves the community best: (1) Zero-COVID, a practical early step, but unsustainable under economic and social pressure, (2) Herd immunity, a controversial approach with high number of early deaths, or (3) Vaccine-based approach mitigation, which highlights the inequitable distribution of medical support and vaccines across countries and regions, which also led to vaccine nationalism at the early stage. These policies are often compounded by the infrastructure of each region and systemic support. Meanwhile, misinformation circulated in the media had a severe negative impact on the success of their implementation, as trust in authority was damaged.

The COVID-19 pandemic exposed critical weaknesses in the Asia-Pacific region's preparedness, including fragmented surveillance systems and inequitable access to diagnostics, vaccines, therapeutics, healthcare capacity, critical situation management, and policy coordination. However, it also highlighted successful strategies that can be refined for future outbreaks. To prepare for future pandemics or health crises, some key areas need immediate attention for improvement and upgrade:

(a) Strengthen regional cooperation and data sharing. Countries acted in isolation, leading to uneven responses, travel bans without coordination, and vaccine hoarding. We cannot afford to have individual countries acting alone, primarily with their own interest in mind. An Asia-Pacific Pandemic/Health Crisis Task Force (under ASEAN, APEC, or WHO) would be in high demand to provide real-time outbreak alerts, avoid censorship delays, facilitate joint stockpiling of PPE, vaccines, and antivirals, and establish standardized travel protocols (rather than abrupt border closures). Transparency agreements would be necessary during peacetime when and where nations share genomic data (e.g., viral variants) without political interference. Under such auspices, a response fund could be established with a strong, binding commitment from all partners to share capabilities and resources.

(b) The entire Asia-Pacific region (and elsewhere) needs to invest in equitable and resilient healthcare systems that cover both urban and remote regions, including those less affluent communities, so that they can receive proper healthcare and medical support. Countries may consider setting up a universal healthcare expansion initiative. Medical production should be expanded and decentralized to avoid local hoarding of material, while production sites can help share the burden when the outbreak is more localized. For remote areas, mobile health units should become a routine practice (e.g., in the Pacific Islands and other remote locations) so that they are not left out in times of emergency.

(c) Ethical and adaptive lockdown policies should be discussed in advance as a crisis management protocol to avoid a sudden lockdown when an overly rigid policy is applied. As a manageable practice, staged lockdowns with safety nets, such as cash/food aid for daily wage workers, e.g., financial support programs, and designated transportation and accommodation for migrants, short- or long-term containment facilities with sufficient support, with special assistance (e.g., Fangcang in China) should be in place. To better prepare citizens, a staged and gradual adjustment of alert levels would enhance community preparedness, as seen in New Zealand's implementation of its alert-level system under the COVID-19 Protection Framework.

(d) In times of pandemics, effective dissemination of credible information is important. Measures should be in place to combat misinformation without suppressing free speech. As observed in the practice of some countries, strict censorship may breed distrust. However, non-authoritative, non-expert information circulation can worsen any outbreak when the general public cannot understand the rationale behind each measure implemented. The government's practice of transparently sharing information, candidly articulating the rationale behind its actions, supported by independent fact-checking coalitions, social media collaborations, and professional and community leaders serving as trusted messengers, would all be essential for a successful crisis management campaign.

(e) To prepare for future zoonotic threats and pandemics, we should take note of the spread of viruses or infectious agents from animals. Some cultural practices may need to be reconsidered for better control; for example, wet markets that sell live wild animals (China, Southeast Asia) and deforestation that leads to the relocation of wildlife species (Indonesia) could increase spillover risks. To better manage such situations, stricter wildlife trade regulations should be enforced in contained and well-managed facilities. Active coordination of different government branches, e.g., health programs involving agencies attending to environmental issues, custom control of cross-border transportation of food, plants, and animals, units handling wildlife monitoring, and healthcare/medical establishments with units investigating emerging risks of zoonotic spill, could be expanded into a research and surveillance hub to cover the Asia-Pacific region. As a result, early warning systems can be established to inform the community about various risks and hazards. Such proactive health preservation and preventive measures outweigh reactive disease management.

(f) It is also important that leaders of the region put people before politics. When COVID-19 struck with devastating and unforeseen force, governments worldwide faced an unprecedented challenge. Leaders were put at a difficult position: grappling with the urgent suffering of their citizens with thousands falling ill and hospitals overwhelmed, while they were confronting the weight of their own leadership legacies and their reputation. In the midst of this chaos, some resorted to misinformation or hollow reassurances in an effort to project control, or to show that actions were taken to address the crisis, even as communities lived a starkly different reality. During such a crisis, priorities are revealed. Transparency is what builds trust. While preventing panic is crucial, so is honesty—acknowledging uncertainties, listening to experts, and putting people's lives above political calculations. Governments that prioritized collective well-being over election results fostered something far more powerful than compliance: public confidence and unity. After all, leadership is not about having all the answers; it is about empowering all stakeholders with the truth and the tools so that communities can persevere together.

Conclusion

Scenarios during COVID-19, as described in previous chapters, prove that no single country is safe until all regions are safe. A more united, equitable, and transparent approach is the only way for countries to adapt to ensure a safe future. A mix of science, solidarity, and smart governance will determine whether the Asia-Pacific region leads or stumbles in the next pandemic. Hence, the future enhancement of health systems, global cooperation on shared information, updated data, and equitable access to medical resource supply are all critical. Globally, a pandemic treaty under the stronger leadership of the WHO is necessary. The Pandemic Accord was proposed by country leaders, UN Secretary-General, and Nobel Laureates aiming to strengthen global coordination [8]. In parallel, the WHO's International Health Regulations (IHR) was amended in 2024 precisely to address this need of solidarity to fight future health crises [9]. This Pandemic Agreement was finally adopted on May 20, 2025, at the 78th World Health Assembly by the 194 member states [10], excluding the USA, which opted out from the WHO in early 2025. When the WHO lacks authoritative power over individual countries, challenges to implementation persist. It is a general principle of international law, that once this legal instrument comes into force, it is binding on the signatory parties, and WHO member states can decide on compliance mechanisms. While the adoption of this accord is a significant step forward, it is ultimately the political will and moral stance of the leaders of member states that can safeguard us from the next devastating pandemic. After all, it is the building up of this mutual trust between leaders, governing bodies, the media, and the public that makes it possible for the smooth execution of pandemic prevention and eradication measures when the next public health crisis arrives.

References

1. WHO (2024) World Health Statistics
2. WHO News (2020) Director-General's opening remarks at the media briefing on COVID-19. March 2020 WHO Director-General's opening remarks at the media briefing on COVID-19 - March 20, 2020
3. WHO chief declares end to COVID-19 as a global health emergency. UN News. https://news.un.org/en/story/2023/05/1136367

4. IMF sees cost of COVID pandemic rising beyond $12.5 trillion estimate. Reuters News. https://www.reuters.com/business/imf-sees-cost-covid-pandemic-rising-beyond-125-trillion-estimate-2022-01-20/
5. Pearson H (2022) COVID derailed learning for 1.6 billion students. Here's how schools can help them catch up. Nature 605:608–611. https://doi.org/10.1038/d41586-022-01387-7
6. Kupcova I, Danisovic L, Klein M, Harsanyi S (2023) Effects of the COVID-19 pandemic on mental health, anxiety, and depression. BMC Psychol 11(1):108. https://doi.org/10.1186/s40359-023-01130-5
7. WHO (2024) COVID-19 Dashboard
8. WHO News (2024) Call for urgent agreement on international deal to prepare for and prevent future pandemics. March 20, 2024. https://www.who.int/news/item/20-03-2024-call-for-urgent-agreement-on-international-deal-to-prepare-for-and-prevent-future-pandemics
9. WHO News (2024) World Health Assembly agreement reached on wide-ranging, decisive package of amendments to improve the International Health Regulations. June 1, 2024. https://www.who.int/news/item/01-06-2024-world-health-assembly-agreement-reached-on-wide-ranging%2D%2Ddecisive-package-of-amendments-to-improve-the-international-health-regulations%2D%2Dand-sets-date-for-finalizing-negotiations-on-a-proposed-pandemic-agreement
10. WHO News (2025) World Health Assembly adopts historic Pandemic Agreement to make the world more equitable and safer from future pandemics. May 20, 2025. https://www.globalsecurity.org/security/library/news/2025/05/sec-250520-who01.htm

The manufacturer's authorised representative in the EU is Springer Nature Customer Service Centre GmbH, Europaplatz 3, 69115 Heidelberg, Germany. If you have any concerns regarding our products, please contact ProductSafety@springernature.com

Printed and bound by CPI Group (UK) Ltd, Croydon, CR0 4YY
05/10/2026
02228057-0007